Advances in Experimental and Computational Rheology

Advances in Experimental and Computational Rheology

Special Issue Editors

Maria Teresa Cidade
João Miguel Nóbrega

MDPI • Basel • Beijing • Wuhan • Barcelona • Belgrade

MDPI

Special Issue Editors

Maria Teresa Cidade
Universidade Nova de Lisboa
Portugal

João Miguel Nóbrega
University of Minho
Portugal

Editorial Office
MDPI
St. Alban-Anlage 66
4052 Basel, Switzerland

This is a reprint of articles from the Special Issue published online in the open access journal *Fluids* (ISSN 2311-5521) from 2018 to 2019 (available at: https://www.mdpi.com/journal/fluids/special_issues/advances_rheology)

For citation purposes, cite each article independently as indicated on the article page online and as indicated below:

LastName, A.A.; LastName, B.B.; LastName, C.C. Article Title. *Journal Name* **Year**, *Article Number*, Page Range.

ISBN 978-3-03921-333-7 (Pbk)
ISBN 978-3-03921-334-4 (PDF)

Contents

About the Special Issue Editors vii

Maria Teresa Cidade and João Miguel Nóbrega
Editorial for Special Issue "Advances in Experimental and Computational Rheology"
Reprinted from: *Fluids* 2019, 4, 131, doi:10.3390/fluids4030131 1

Miguel A. Delgado, Sebastien Secouard, Concepción Valencia and José M. Franco
On the Steady-State Flow and Yielding Behaviour of Lubricating Greases
Reprinted from: *Fluids* 2019, 4, 6, doi:10.3390/fluids4010006 3

Priscilla R. Varges , Camila M. Costa, Bruno S. Fonseca, Mônica F. Naccache and Paulo R. de Souza Mendes
Rheological Characterization of Carbopol® Dispersions in Water and in Water/Glycerol Solutions
Reprinted from: *Fluids* 2019, 4, 3, doi:10.3390/fluids4010003 18

Luis G. Baltazar, Fernando M.A. Henriques and Maria Teresa Cidade
Rheology of Natural Hydraulic Lime Grouts for Conservation of Stone Masonry—Influence of Compositional and Processing Parameters
Reprinted from: *Fluids* 2019, 4, 13, doi:10.3390/fluids4010013 38

Francisco-José Rubio-Hernández
Rheological Behavior of Fresh Cement Pastes
Reprinted from: *Fluids* 2018, 3, 106, doi:10.3390/fluids3040106 57

Manuel Félix, Alberto Romero, Cecilio Carrera-Sanchez and Antonio Guerrero
A Comprehensive Approach from Interfacial to Bulk Properties of Legume Protein-Stabilized Emulsions
Reprinted from: *Fluids* 2019, 4, 65, doi:10.3390/fluids4020065 73

Alexander Kurz, Jörg Bauer and Manfred Wagner
Piezo-Plunger Jetting Technology: An Experimental Study on Jetting Characteristics of Filled Epoxy Polymers
Reprinted from: *Fluids* 2019, 4, 23, doi:10.3390/fluids4010023 84

Mercedes Fernandez, Arrate Huegun and Antxon Santamaria
Relevance of Rheology on the Properties of PP/MWCNT Nanocomposites Elaborated with Different Irradiation/Mixing Protocols
Reprinted from: *Fluids* 2019, 4, 7, doi:10.3390/fluids4010007 99

Mª Carmen García González, María del Socorro Cely García, José Muñoz García and Maria-Carmen Alfaro-Rodriguez
A Comparison of the Effect of Temperature on the Rheological Properties of Diutan and Rhamsan Gum Aqueous Solutions
Reprinted from: *Fluids* 2019, 4, 22, doi:10.3390/fluids4010022 111

Luis A. Trujillo-Cayado, Jenifer Santos, Nuria Calero, Maria del Carmen Alfaro and José Muñoz
Influence of the Homogenization Pressure on the Rheology of Biopolymer-Stabilized Emulsions Formulated with Thyme Oil
Reprinted from: *Fluids* 2019, 4, 29, doi:10.3390/fluids4010029 119

Sónia Costa, Paulo F. Teixeira, José A. Covas and Loic Hilliou
Assessment of Piezoelectric Sensors for the Acquisition of Steady Melt Pressures in
Polymer Extrusion
Reprinted from: *Fluids* **2019**, *4*, 66, doi:10.3390/fluids4020066 . . . . **130**

Salvatore Costanzo, Rossana Pasquino, Jörg Läuger and Nino Grizzuti
Milligram Size Rheology of Molten Polymers
Reprinted from: *Fluids* **2019**, *4*, 28, doi:10.3390/fluids4010028 . . . . **142**

Oumar Abdoulaye Fadoul and Philippe Coussot
Saffman–Taylor Instability in Yield Stress Fluids: Theory–Experiment Comparison
Reprinted from: *Fluids* **2019**, *4*, 53, doi:10.3390/fluids4010053 . . . . **153**

J. Paulo García-Sandoval, Fernando Bautista, Jorge E. Puig and Octavio Manero
Inhomogeneous Flow of Wormlike Micelles: Predictions of the Generalized BMP Model with
Normal Stresses
Reprinted from: *Fluids* **2019**, *4*, 45, doi:10.3390/fluids4010045 . . . . **165**

Eden Furtak-Cole and Aleksey Telyakovskiy
A 3D Numerical Study of Interface Effects Influencing Viscous Gravity Currents in a Parabolic
Fissure, with Implications for Modeling with 1D Nonlinear Diffusion Equations
Reprinted from: *Fluids* **2019**, *4*, 97, doi:10.3390/fluids4020097 . . . . **180**

Joseph A. Green, Daniel J. Ryckman and Michael Cromer
A Continuum Model for Complex Flows of Shear Thickening Colloidal Solutions
Reprinted from: *Fluids* **2019**, *4*, 21, doi:10.3390/fluids4010021 . . . . **198**

About the Special Issue Editors

Maria Teresa Cidade belongs to the Polymeric and Mesomorphic Materials Group of the Faculty of Sciences and Technology of the New University of Lisbon (FCT NOVA), Portugal. She graduated in Chemical Engineering (IST/Technical University of Lisbon,1983) and obtained a Ph.D. degree from FCT NOVA (1994). In 2006, she was appointed with the Habilitation in Polymer Engineering. Currently, she is Assistant Professor with Habilitation at the Materials Science Department (DCM) of FCT NOVA. She is the Coordinator of the Polymeric and Mesomorphic Materials Group of DCM, Coordinator of the Rheology Sub-Group of the Soft and Bifunctional Materials Group of the Materials Research Centre (Cenimat) of DCM, Coordinator of the Doctoral Program in Materials Science and Engineering, and Coordinator of FCT/UNL in the Doctoral Program in Advanced Materials and Processing (a Doctoral Program in Association with six other Portuguese Universities: Lisbon, Coimbra, Beira Interior, Aveiro, Porto, and Minho, supported by the Portuguese Foundation for Science and Technology). She is the President of the Portuguese Society of Rheology, Associate Editor of *Physica Scripta* (IOP), and a member of the Editorial Board of *Fluids* (MDPI). Her main scientific interests are the rheology (including electrorheology and rheo-optics) of complex systems (polymers and polymeric base systems, liquid crystals, nanocomposites, biomaterials, building materials, etc.), the mechanical characterization of polymers and polymer composites, and polymer processing. During her career, she was involved in the supervision of more than 30 researchers, coauthored four book chapters and 79 ISI papers, lodged 1 patent, and presented more than 100 communications in conferences.

João Miguel Nóbrega is Associate Professor at the Polymer Engineering Department of the University of Minho, and member of the Institute for Polymers and Composites. In 2004, he received his Ph.D. degree from the University of Minho in Polymer Science and Engineering. He is Vice-President of the Portuguese Society of Rheology, Editor of OpenFOAM® Wiki, and a founder member of the Iberian OpenFOAM® Technology Users. His research activity lies in three overlapping areas: product development, polymer processing, and material rheology. For this purpose, he has been developing computational rheology tools to model the flow of complex fluids in various polymer processing techniques. Regarding the product development area, he has been involved in the design and manufacture of polymeric products across several fields, comprising applications for health, textiles, sensing/monitoring, construction, and mobility. In 2014, he joined the OpenFOAM® Extend community, and has focused, since then, on the main numerical developments in this open source computational library. In 2016, he was the chair of the 11th Workshop OpenFOAM, which took place in Guimarães, Portugal. During his career, he was involved in the supervision of more than 50 researchers, working both in fundamental and applied research projects; he coedited 2 books and 22 book chapters, published 83 papers in international refereed journals, lodged 9 patents (3 international), and presented approximately 200 communications in conferences.

𝖆𝖆𝖆 *fluids*

MDPI

Editorial

Editorial for Special Issue "Advances in Experimental and Computational Rheology"

Maria Teresa Cidade [1,*] **and João Miguel Nóbrega** [2,*]

[1] Departamento de Ciência dos Materiais and Cenimat/I3N, Faculdade de Ciências e Tecnologia, Universidade Nova de Lisboa, 2829-516 Caparica, Portugal
[2] Institute for Polymers and Composites/I3N, University of Minho, Campus de Azurém 4800-058 Guimarães, Portugal
*   Correspondence: mtc@fct.unl.pt (M.T.C.); mnobrega@dep.uminho.pt (J.M.N.)

Received: 9 July 2019; Accepted: 11 July 2019; Published: 12 July 2019

Rheology, defined as the science of deformation and flow of matter, is a multidisciplinary scientific field, covering both fundamental and applied approaches. The study of rheology includes both experimental and computational methods, which are not mutually exclusive. Its practical importance embraces many processes, from daily life, like preparing mayonnaise, spreading an ointment, or shampooing, to industrial processes like polymer processing and oil extraction, among several others. Practical applications include also formulation and product development.

The special issue "Advances in Experimental and Computational Rheology" joins fifteen works covering some of the latest advances in the fields of experimental and computational rheology applied to a diverse class of materials and processes, which can be grouped into four main topics: rheology [1–5], effect of process variables [6–9], rheometry and processing [10,11], and theoretical modeling [12–15]

The characterization of rheological behavior is the main topic of five contributions, covering the following material systems: lubricating greases (Delgado et al. [1]), Carbopol® dispersion in water and in water/glycerol solutions (Varges et al. [2]), natural hydraulic lime grouts (Baltazar et al. [3]), fresh cement pastes (Rubio-Hernández [4]), and legume-protein-stabilized emulsions (Félix et al. [5]).

The effect of process variables is covered in four papers. Kurz et al. [6] studied the droplet formation of Newtonian fluids and suspensions modified by spherical, non-colloidal particles. Fernandez et al. [7] investigated the effect of different irradiations/mixing on the rheology and electrical conductivity of PP/MWCNT nanocomposites. García et al. [8] evaluated the effect of temperature on the rheology of diutan and rhamsan gum aqueous solutions. Trujillo-Cayado et al. [9] described the effect of homogenization pressure on the rheological behavior of biopolymer-stabilized emulsions formulated with thyme oil.

Two of the special issue works are dedicated to rheometry and processing. Costa et al. [10] assessed the employment of piezoelectric sensors on the acquisition of steady melt pressures in polymer extrusion, and Costanzo et al. [11] evaluated the possibility of performing the linear and non-linear rheological characterization of samples with just a few milligrams.

Theoretical modeling is the main topic of the four remaining works. Fadoul and Coussot [12] proposed and performed a set of experiments to assess a theoretical model developed to predict the flow Saffman–Taylor instability in yield stress fluids. García-Sandoval et al. [13] studied the capability of the Bautista–Manero–Puig model to predict shear banding in polymer-like micellar solutions. Furtak-Cole and Telyakovskiy [14] resorted to 3D modelling techniques to assess the applicability of a simple 1D model for the flow in aquifers and fissures. Green et al. [15] proposed modifications of a previously developed constitutive model for shear thickening colloidal solutions, which explicitly accounts for the evolution of its microstructure during flow, and assessed its accuracy.

Finally, it is very important to recognize and acknowledge the effort put forth by the large number of anonymous reviewers, which has been essential to assuring the high quality of all the contributions of this special issue.

Conflicts of Interest: The authors declare no conflict of interest.

References

1. Delgado, M.A.; Secouard, S.; Valencia, C.; Franco, J.M. On the Steady-State Flow and Yielding Behaviour of Lubricating Greases. *Fluids* **2019**, *4*, 6. [CrossRef]
2. Varges, P.R.; Costa, C.M.; Fonseca, B.S.; Naccache, M.F.; De Souza Mendes, P.R. Rheological Characterization of Carbopol® Dispersions in Water and in Water/Glycerol Solutions. *Fluids* **2019**, *4*, 3. [CrossRef]
3. Baltazar, L.G.; Henriques, F.M.; Cidade, M.T. Rheology of Natural Hydraulic Lime Grouts for Conservation of Stone Masonry—Influence of Compositional and Processing Parameters. *Fluids* **2019**, *4*, 13. [CrossRef]
4. Rubio-Hernández, F.J. Rheological Behavior of Fresh Cement Pastes. *Fluids* **2018**, *3*, 106. [CrossRef]
5. Félix, M.; Romero, A.; Carrera-Sanchez, C.; Guerrero, A. A Comprehensive Approach from Interfacial to Bulk Properties of Legume Protein-Stabilized Emulsions. *Fluids* **2019**, *4*, 65. [CrossRef]
6. Kurz, A.; Bauer, J.; Wagner, M. Piezo-Plunger Jetting Technology: An Experimental Study on Jetting Characteristics of Filled Epoxy Polymers. *Fluids* **2019**, *4*, 23. [CrossRef]
7. Fernandez, M.; Huegun, A.; Santamaria, A. Relevance of Rheology on the Properties of PP/MWCNT Nanocomposites Elaborated with Different Irradiation/Mixing Protocols. *Fluids* **2019**, *4*, 7. [CrossRef]
8. González, M.G.; García, M.C.; García, J.M.; Alfaro-Rodriguez, M.-C. A Comparison of the Effect of Temperature on the Rheological Properties of Diutan and Rhamsan Gum Aqueous Solutions. *Fluids* **2019**, *4*, 22. [CrossRef]
9. Trujillo-Cayado, L.A.; Santos, J.; Calero, N.; Alfaro, M.D.C.; Muñoz, J. Influence of the Homogenization Pressure on the Rheology of Biopolymer-Stabilized Emulsions Formulated with Thyme Oil. *Fluids* **2019**, *4*, 29. [CrossRef]
10. Costa, S.; Teixeira, P.F.; Covas, J.A.; Hilliou, L. Assessment of Piezoelectric Sensors for the Acquisition of Steady Melt Pressures in Polymer Extrusion. *Fluids* **2019**, *4*, 66. [CrossRef]
11. Costanzo, S.; Pasquino, R.; Läuger, J.; Grizzuti, N. Milligram Size Rheology of Molten Polymers. *Fluids* **2019**, *4*, 28. [CrossRef]
12. Fadoul, O.A.; Coussot, P. Saffman–Taylor Instability in Yield Stress Fluids: Theory–Experiment Comparison. *Fluids* **2019**, *4*, 53. [CrossRef]
13. García-Sandoval, J.P.; Bautista, F.; Puig, J.E.; Manero, O. Inhomogeneous Flow of Wormlike Micelles: Predictions of the Generalized BMP Model with Normal Stresses. *Fluids* **2019**, *4*, 45. [CrossRef]
14. Furtak-Cole, E.; Telyakovskiy, A.S. A 3D Numerical Study of Interface Effects Influencing Viscous Gravity Currents in a Parabolic Fissure, with Implications for Modeling with 1D Nonlinear Diffusion Equations. *Fluids* **2019**, *4*, 97. [CrossRef]
15. Green, J.A.; Ryckman, D.J.; Cromer, M. A Continuum Model for Complex Flows of Shear Thickening Colloidal Solutions. *Fluids* **2019**, *4*, 21. [CrossRef]

fluids

MDPI

Article

On the Steady-State Flow and Yielding Behaviour of Lubricating Greases

Miguel A. Delgado *, Sebastien Secouard, Concepción Valencia and José M. Franco *

Pro2TecS-Chemical Process and Product Technology Research Center, Universidad de Huelva, Campus El Carmen, 21071 Huelva, Spain; sebastien.secouard@fresenius-kabi.com (S.S.); barragan@uhu.es (C.V.)
* Correspondence: miguel.delgado@diq.uhu.es (M.A.D.); franco@uhu.es (J.M.F.); Tel.: +34-959-219-865 (M.A.D.); +34-959-219-995 (J.M.F.)

Received: 6 December 2018; Accepted: 4 January 2019; Published: 9 January 2019

Abstract: Practical steady-state flow curves were obtained from different rheological tests and protocols for five lubricating greases, containing thickeners of a rather different nature, i.e., aluminum complex, lithium, lithium complex, and calcium complex soaps and polyurea. The experimental results demonstrated the difficulty to reach "real" steady-state flow conditions for these colloidal suspensions as a consequence of the strong time dependence and marked yielding behavior in a wide range of shear rates, resulting in flow instabilities such as shear banding and fracture. In order to better understand these phenomena, transient flow experiments, at constant shear rates, and creep tests, at constant shear stresses, were also carried out using controlled-strain and controlled-stress rheometers, respectively. The main objective of this work was to study the steady-state flow behaviour of lubricating greases, analyzing how the microstructural characteristics may affect the yielding flow behaviour.

Keywords: lubricating grease; rheology; steady-state and transient flow; microstructure

1. Introduction

Lubricating greases are generally highly structured suspensions consisting of a thickener, usually a metal soap such as lithium, calcium, sodium, barium, or aluminum, dispersed in mineral or synthetic oils. In addition, lubricating greases usually contain some performance additives [1–3]. Thickener molecules combine to form tridimensional networks consisting of fibers, small spheres, rods, or platelets in which the oil is trapped, conferring the appropriate rheological and tribological behaviour to the grease. The main purpose of the thickener is to prevent the loss of lubricant under operating conditions, providing gel-like characteristics to the grease, despite this evidently implying a considerable resistance to the flow of these materials [2,4,5].

Special attention has been paid to the steady-state flow behaviour of lubricating greases because of the complex strain response to stress and the time-dependent behaviour exhibited [3–9]. It is accepted that lubricating greases are yielding materials, characterized by a discontinuity in the flow curve, particularly in the yielding stress range, where the deformation process is characterized by a sudden drop in viscosity when the shear stress slightly increases. The existence of three different regions in the viscous flow curve of lubricating greases has been previously discussed [6–10]: firstly, a tendency to reach a limiting viscosity, at very low stress values; secondly, a dramatic drop in viscosity; and finally, a tendency to reach a constant high-shear stress–limiting viscosity. The appropriate characterization of this flow behaviour has an evident practical importance, above all in relation to average life-time, resistance under operating conditions, and pumpability [2]. Indeed, this characteristic flow behaviour allows the "practical" yield stress value to be determined, which plays an important role in the design of automatic grease centralized pumping and distribution systems [5].

Very recently, different studies have pointed out the importance of the flow behavior in the correct evaluation of the fluid dynamics of greases in practical situations, for instance, by performing numerical simulations. In this sense, numerical simulations applying computational fluid dynamics (CFD) procedures or Runge-Kutta methods have been performed to predict the particle motion/migration in pockets between two-rotating cylinders with different restriction geometries [11,12], the grease flow in journal bearings as a function of surface texture [13], the dynamics of greases in tapered roller bearings [14], or the laminar flow in rectangular cross section channels [15], among other practical situations. Additionally, the dynamics of greases was experimentally studied in both a labyrinth seal geometry [16] and concentric cylinder configurations with a rotating shaft to simulate the grease flow in a double restriction seal geometry [17] using the microparticle image velocimetry technique. In most of these studies, the relevance of the yielding flow behavior, mainly approached by the Herschel-Bulkley model; the plastic viscosity; or the shear-thinning character of greases was emphasized.

In this regard, great efforts have been made to correctly determine the yield stress value in greases [18,19]. Thus, in order to get reproducible and accurate results, it is necessary to conveniently define all test conditions since the flow curve data and the yield stress values may be widely dependent on them. Although it is worth mentioning that some controversy about the concept of yielding flow and associated experimental difficulties has been traditionally found in the literature [20–22], it is obvious that the yield stress is a parameter commonly used in the industry because of the importance of knowing the stress-time relationship to predict the effective flow conditions [23].

Several investigations on the rheological properties of lubricating greases have dealt with the application of pre-shearing conditions to simplify and accelerate the consecution of the steady-state flow [4,5]. The application of pre-shearing results in both the absence of stress overshoot and a relatively rapid achievement of the steady-state in transient experiments. However, this procedure leads to significant microstructural changes that affect the original rigidity of the network formed by the thickener. Thus, the ionic and van der Waals forces that contribute to the viscoelasticity of this network are highly affected by pre-shearing, also dramatically influencing the entanglements among the structural units (fibers, rods, platelets, . . .) that make up the structural skeleton.

This paper attempts to describe and understand the "real" steady-state flow behaviour of greases without applying pre-shearing conditions. The main objective is to elucidate the effect of the time scale and shear protocol on the consecution of the steady-state flow conditions for a variety of the most commonly used types of lubricating greases.

2. Materials and Methods

2.1. Materials

Lubricating greases were kindly supplied by Total France (Lyon, France). They consist of mineral oils and different metal soaps or polyurea thickeners. Some relevant physical properties provided by the manufacturer are presented in Table 1. All studied greases have an NLGI 2 grade, except the calcium complex soap-based formulation, which is an NLGI 1 grade grease.

Table 1. Lubricating grease properties.

Soap	Oil Viscosity at 40 °C (cP)	NLGI Grade	Drop Point (°C)
Aluminum Complex	130	2	>250
Lithium	150	2	>150
Lithium Complex	160	2	>250
Calcium Complex	150	1	>250
Polyurea	110	2	>240

2.2. Rheological Characterization

The rheological characterization of this group of greases was carried out with both a RS-150 (Thermohaake, Karlsruhe, Germany) controlled-stress rheometer, and an ARES (Rheometrics Scientific, Leatherhead, UK) controlled-strain rheometer. All tests were performed using serrated plate–plate geometries (35 mm in controlled-stress, 25 mm in controlled-strain; 1mm gap, 0.4 relative roughness) in order to prevent the wall slip effects typically found in these materials [10]. No gap influence was detected when performing tests with different plate-plate gaps (0.5–3 mm), thus confirming the absence of wall slip. All measurements were conducted at 25 °C, following the same recent thermal and mechanical history, i.e., 30 min resting time at the selected temperature, and replicated at least three times using new un-sheared samples.

2.2.1. Stepped Shear Rate and Shear Stress Ramps

Upward stepped ramps of shear rates and shear stresses traditionally employed to obtain steady-state flow curves were applied using both controlled-stress, in a 10–3000 Pa range of shear stress, and controlled-strain, in a 0.0125–100 s^{-1} range of shear rate, rheometers. Each point within the flow curve was acquired after shearing for 3 min in each step. This curve will be called the "practical" steady-state flow curve in the text.

2.2.2. Transient Experiments at Constant Shear Rate

Stress-growth experiments were performed, at 25 °C, in the controlled-strain rheometer at different constant shear rates (0.0125, 0.1, and 1 s^{-1}). The evolution of shear stress with time was monitored until a steady state was reached. However, some experiments, especially at 1 s^{-1}, were eventually stopped before a steady state was reached when fracture and/or partial expelling of the sample was detected.

2.2.3. Creep Tests

Different constant shear stresses were applied to grease samples, using a controlled-stress rheometer, and the corresponding shear rate was measured for a period of time. These creep experiments were carried out between 100 and 1000 Pa, depending on the consistency of the lubricating grease.

2.2.4. SAOS Experiments

In addition, small-amplitude oscillatory shear (SAOS) tests, inside the linear viscoelasticity range, were carried out in a frequency range between 0.01 and 100 rad/s, using a plate-plate geometry (35 mm diameter, 1 mm gap) in the controlled stress rheometer. Stress sweep tests, at 1 Hz, were previously performed on each sample to determine the linear viscoelasticity region.

2.3. Atomic Force Microscopy

The microstructural characterization of greases was carried out by means of atomic force microscopy (AFM) using a multimode apparatus connected to a Nanoscope-IV scanning probe microscope controller (Digital Instruments, Veeco Metrology Group Inc., Santa Barbara, CA, USA). All images were acquired in the tapping mode using Veeco NanoprobeTM tips.

3. Results

3.1. Linear Viscoelastic Response and Microstructure

Figure 1 shows the mechanical spectra, in the linear viscoelasticity range, for the different lubricating greases studied. These mechanical spectra are qualitatively similar to those shown by other particle gels [24], supporting the idea that lubricating greases are highly structured systems, as has also been detected by AFM microscopy (Figure 2). Some useful information about the microstructural network of these greases may be extracted from small-amplitude oscillatory shear measurements.

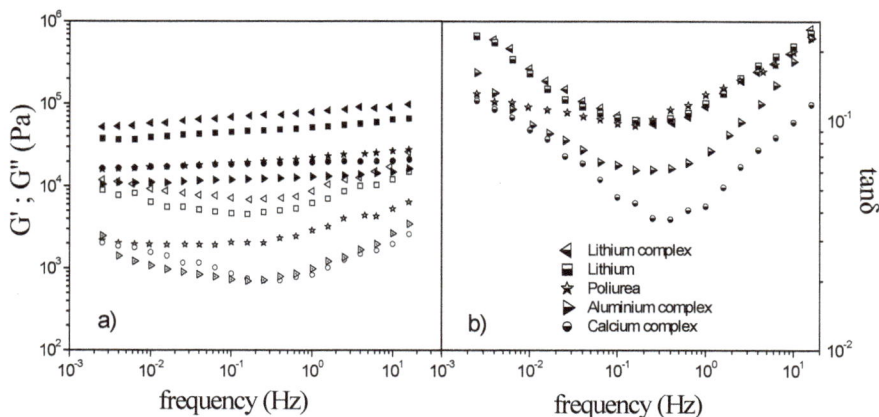

Figure 1. Frequency dependence of the storage and loss moduli (**a**) and tan δ (**b**), in the linear viscoelasticity region, for the five lubricating greases studied (filled symbols, G′; empty symbol, G″).

As typically found in commercial lubricating greases [25], all samples studied show values of the storage modulus (G′) around one decade higher than those measured for the loss modulus (G″), as well as a minimum in G″ at intermediate frequencies. However, a much softer evolution was found for the polyurea-based grease, with almost constant values of G″ within the frequency range of around 10^{-3}–10^{-1} Hz, and a gradual increase afterwards. On the other hand, very similar values of the loss tangent (tan δ) were obtained in most cases, with slightly lower values for the lithium and especially the complex calcium greases. In general, this viscoelastic behaviour is the result of both interparticle and thickener-oil interactions [7,26].

Although most of the samples studied are NLGI grade 2 greases (except the calcium complex soap-based grease), significant differences among them have been found from both viscoelastic and microstructural points of view. As can be observed in Figure 2, grease microstructures are very different, depending on the type of thickener, which determines the rheological behaviour of these lubricating greases. While most of the greases show a high density of soap particles in the form of small or long interconnected fibers, the calcium complex grease (Figure 2d) discloses a number of polydisperse large aggregates. Consequently, it may be assumed that this microstructure can exhibit a certain brittle character under shear, which can justify the relatively low values of tan δ compared with the rest of the samples.

3.2. Stepped-Shear Rate and Stress Ramps: The "Practical" Steady-State Flow Curve

Figure 3 shows the viscous flow curves obtained from stepped-shear rate and stress ramps for the five grease samples studied. The shear stress versus shear rate curves were acquired by applying both the control stress (CS) and the controlled shear rate (CR) modes and plotted together, in a wide range of shear rates, to obtain the complete flow curves. In all cases, a strong shear rate dependence is appreciated. Initially, shear stress increases with shear rate up to around 10^{-3} s^{-1}, and shear stresses ranged from 500 Pa, for the lithium grease, to 1000 Pa, for the polyurea-based grease, which was only measurable in the CS mode. Then, the flow curves for the aluminum complex, lithium complex, and simple lithium greases exhibit several decades of almost constant values of shear stress between 10^{-3} and 10 s^{-1}, representative of the characteristic yielding flow behaviour of structured materials [6,10,27]. In addition, a slight non-monotonic evolution of the stress can be observed in this shear rate range (see, for instance, the lithium grease in Figure 3). However, calcium complex soap- and polyurea-based greases clearly show a minimum in the shear stress vs. shear rate plots inside this range. These trends were only observed in the CR curves, whereas a sudden jump in the shear rate was observed in the CS curves in this shear rate range. Afterwards, above $10 \, s^{-1}$, shear stress starts to increase again with shear rate, leading to coinciding CR and CS curves.

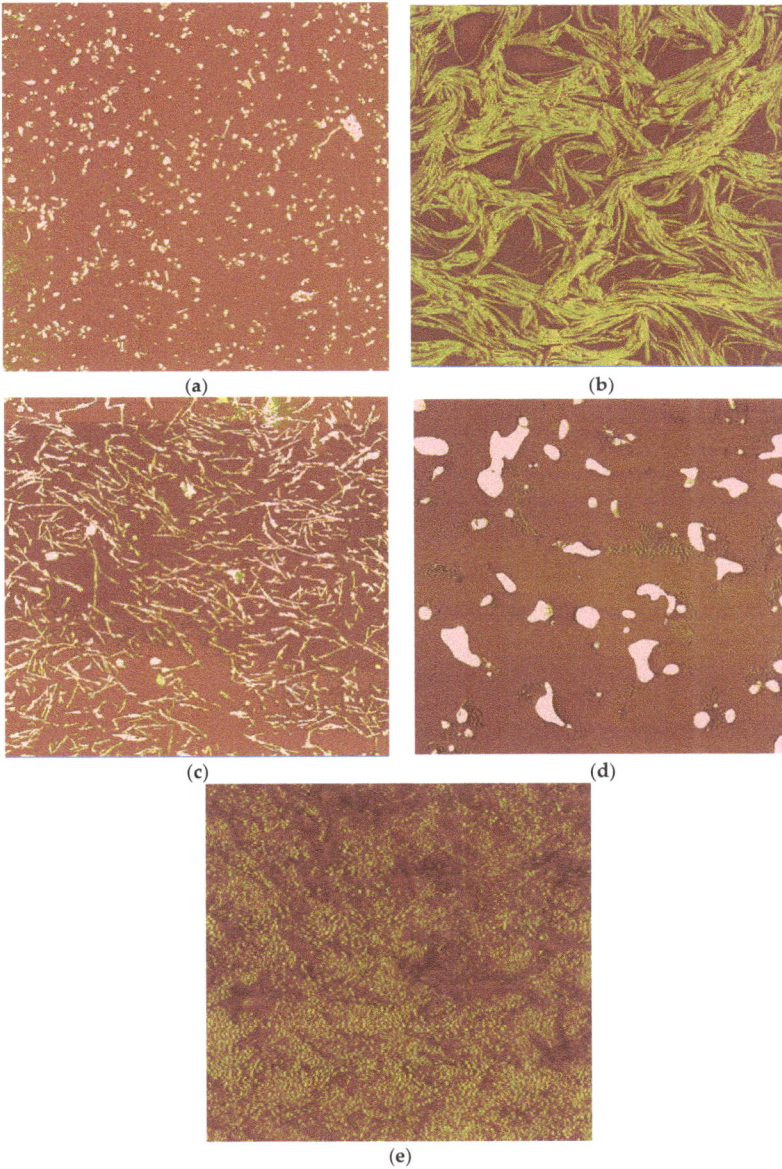

Figure 2. AFM photomicrographs for the five different lubricating greases studied (window size corresponds to 20 μm): (**a**) aluminum complex, (**b**) lithium, (**c**) lithium complex, (**d**) calcium complex, and (**e**) polyurea * greases (* window size corresponds to 5 μm).

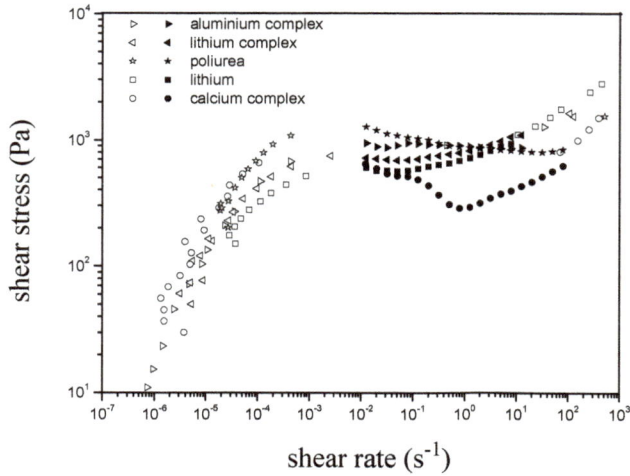

Figure 3. Stepped-stress and shear rate flow curves for the lubricating greases studied (filled symbols, CR curves; empty symbol, CS curves).

Traditionally, this behaviour has been attributed to a dynamically non-stable region and may be related to a non-homogeneous field of velocities during the viscometric flow of yielding materials, as, for instance, derived from wall depletion and shear banding phenomena, which can finally induce the fracture of the sample [28–31]. Thus, these experimental results, i.e., the jump in the stress in CS measurements and/or the minimum in stress eventually appearing in CR tests, have been explained by attending to the coexistence of three different shear rates at a given constant shear stress. In fact, only two possible steady flow regimes are possible since the other one corresponds to an unstable regime, which coincides with the part of the model in which the stress may decrease with shear rate [10]. Considering this model, a sudden transition from the low-shear rate to the high-shear rate regime was found in controlled-stress experiments and the fluid behaves like a typical yield stress fluid (Figure 3). This experimental flow curve has also been recently explained with simple mathematical arguments in the case of highly shear-thinning fluids, where the CR and CS shearing protocol modes play a decisive role in obtaining a minimum in the stress vs. shear rate plot [32]. This explanation does not need to be supported with the occurrence of flow instabilities as previously discussed, although the condition of a very pronounced shear-thinning behavior can easily promote such instabilities, since the apparent viscosity of the fluid will show huge variations along the gradient direction because of its enormous dependence on small shear stress variations. Nevertheless, as is discussed below, the non-consecution of steady conditions in stepped shear rate or stress ramps and the appearance of instabilities need to be additionally considered to explain the yielding behaviour of lubricating greases.

3.3. Transient Flow

In order to check the achievement of steady-state conditions in the stepped-shear rate and stress ramps, transient experiments at a constant shear rate or shear stress were performed for all the greases studied. Figure 4 shows the stress-growth curves obtained in transient experiments at constant shear rates of 0.0125, 0.1, and 1 s^{-1}. The evolution of the transient stress with time is similar to that observed for other thixotropic colloidal systems [20,33,34]. In all cases, a non-linear viscoelastic response was observed with two distinct regions: the first one between the onset of the transient test and the maximum shear stress, the so-called stress overshoot (τ_{max}); and the second one ranging between this maximum and the equilibrium or steady-state shear stress (τ_{eq}). The first part of these curves is mainly the result of the well-known viscoelastic response of greases, with elastic deformation of the prevailing component. Thus, the stress overshoot has been related to the energy level that has to be applied on the grease to produce a significant structural breakdown, when the shear flow process begins to be predominant [35].

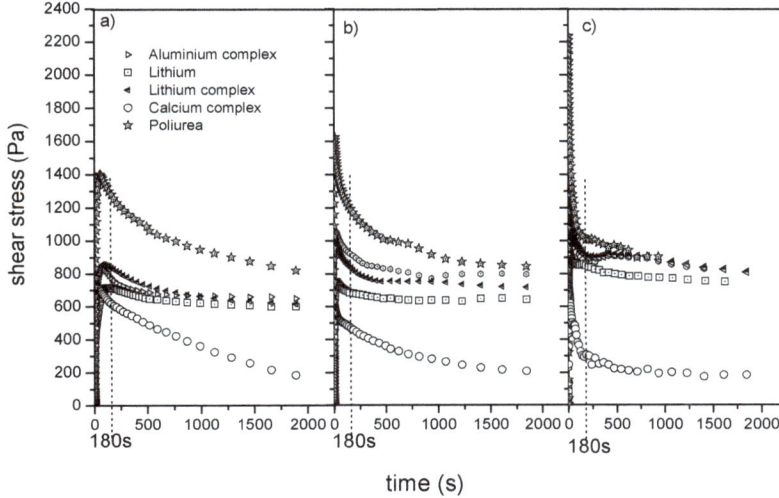

Figure 4. Stress-growth experiments for the greases studied at $0.0125\ s^{-1}$ (**a**), $0.1\ s^{-1}$ (**b**), and $1\ s^{-1}$ (**c**).

As can be appreciated in Figure 4, polyurea-based grease displays the highest values of stress overshoot, as a result of a larger mechanical resistance to flow. This fact is in agreement with the almost parallel trend observed between the storage and loss moduli curves at low frequencies, characteristic of stronger gel-like properties [7,22]. Moreover, both calcium complex and polyurea greases exhibit higher differences between the stress overshoot and the steady-state shear stress, for all the shear rates applied. This fact evidences the important shear-induced structural breakdown occurring in these greases, which can be evaluated by means of the amount of overshoot, defined as follows:

$$S^+ = \frac{\tau_{max} - \tau_{eq}}{\tau_{eq}} \tag{1}$$

Table 2 shows the values of the amount of overshoot and the elapsed time necessary to reach this overshoot peak (t_{max}) for the lubricating greases studied, as a function of shear rate. As already mentioned, this time is related to the beginning of the structural breakdown process [35], which decreases with shear rate for all greases.

Table 2. Values of the amount of overshoot (S^+) and time to reach the stress overshoot at different shear rates for all the lubricating greases studied.

Lubricating Greases	$0.0125\ s^{-1}$		$0.1\ s^{-1}$		$1\ s^{-1}$	
	S^+	t_{max} (s)	S^+	t_{max} (s)	S^+	t_{max} (s)
Aluminum Complex	0.282	74	0.321	16.8	0.439	1.91
Lithium	0.172	128	0.157	22.1	0.238	3.08
Lithium complex	0.392	85	0.397	15.7	0.351	1.67
Calcium complex	2.946	26	1.635	2.6	3.043	0.24
Polyurea	0.708	52	0.890	6.8	1.477	0.65

The stress growth curves of lubricating greases have been previously related to shear-induced microstructural changes and associated frictional energy dissipation [36,37]. Thus, the aluminum complex grease, which shows low values of the amount of overshoot and a microstructure based on small dispersed particles (Figure 2a), seems to be relatively stable under low shear rates, similarly to the lithium complex grease, which shows a high density of small fiber-like particles (Figure 2c). The lithium grease shows the highest structural stability against shear, providing the lowest values of the amount of overshoot for all the shear rates applied. These structural features bring about a high level of interactions among soap particles, conferring high elasticity, associated with high t_{max} values, and consistency, backed up by higher values of the stress overshoot and storage modulus in the entire frequency range studied (Figure 1). On the other hand, although the calcium complex grease shows relatively low values of stress overshoot in comparison with other greases (Figure 4), which must be attributed to its lower NLGI grade, it also shows the most pronounced time-dependent behaviour, with the highest value of the amount of overshoot (Table 2); exhibiting stress overshoot values at least three times higher than the corresponding steady-state value, for each shear rate evaluated (Figure 4). Furthermore, the lowest elapsed time at the overshoot peak (t_{max}) was detected for this grease. These transient flow properties, followed by those exhibited by the polyurea grease, are a consequence of the lower particle dispersion degree and not so entangled structures like those observed in the case of fibers.

As it has been mentioned above, these transient experiments at a constant shear rate allow us to evaluate the achievement of the steady-state in the previously discussed stepped-ramp flow tests. As can be appreciated in Figure 4, the equilibrium state is reached at 0.0125 and 0.1 s^{-1} much longer than the 3 min applied in the stepped ramps, for all greases. Consequently, the "practical" flow curves displayed in Figure 3 must be significantly corrected in this shear rate range in order to get a more realistic steady-state flow behaviour. On the other hand, fracture, followed by the expulsion of the sample to some extent, was observed when some of the greases were submitted to moderate constant shear rates. This phenomenon hinders the consecution of the final equilibrium stress value. As a consequence, in those cases, the equilibrium time was considered to be reached when a variation of less than 5% of shear stress was maintained for 400 s. Thus, equilibrium times of around 1800, 1400, and 800 s were taken at 0.0125, 0.1, and 1 s^{-1}, respectively, for most of the samples. However, for calcium complex and polyurea greases, variations of up to 19% and 13%, respectively, have to be considered at 0.0125 s^{-1} after 2000 s.

Additionally, creep experiments were also performed. The main benefit of these tests was the relatively low stress values able to be applied, which do not significantly perturb the grease structural skeleton. When the applied shear stress was inferior to the yield stress, an equilibrium time of around eight hours was fixed to reach and effectively verify the steady-state, whereas 4 h allowed us to obtain the steady-state when the applied stress was slightly higher than the yield stress. These equilibrium times were imposed to obtain a variation of the shear rate lower than 2% for at least 300 s.

As can be observed in Figure 5, creep experiments evidence different transient responses, depending on the type of grease. All greases reveal a noticeable jump of the shear rate at a given value of shear stress, which again is indicative of the yielding behaviour of these materials. In addition, it is worth pointing out that for aluminum complex, lithium, and polyurea greases, the shear rate may increase suddenly and sharply from a zero-near value to a high-shear rate value in the same test after a certain shearing time. This fact is appreciated at 700, 490, and 1100 Pa, respectively, which roughly correspond to the apparent yield stress values, i.e., the stress values for which a sudden jump in shear rate was observed in the stepped-stress ramps obtained in the CS mode. Coussot et al. [38] observed a similar phenomenon with a bentonite-water mixture, which was described as "heavy creep instability". This fact could be associated with the well-known shear banding effect, for which different values of shear rate can coexist for the same shear stress value applied [10,28,31]. Thus, at very low shear stress, the material behaves as a stiff material; however, when grease starts to flow, it seems to do so heterogeneously, and the shear banding phenomenon finally promotes the fracture of the sample.

Figure 6 shows the viscous flow behaviour for the calcium complex grease obtained from the creep test. As can be observed, a quite pronounced drop in viscosity, from 5×10^8 to 3 Pa·s, was noticed when the shear stress was slightly increased from 590 Pa to 595 Pa, i.e., the yield stress. This particularly dramatic drop in viscosity is favored by the relatively weak structural skeleton that, as already discussed, is greatly affected by this critical shear stress value.

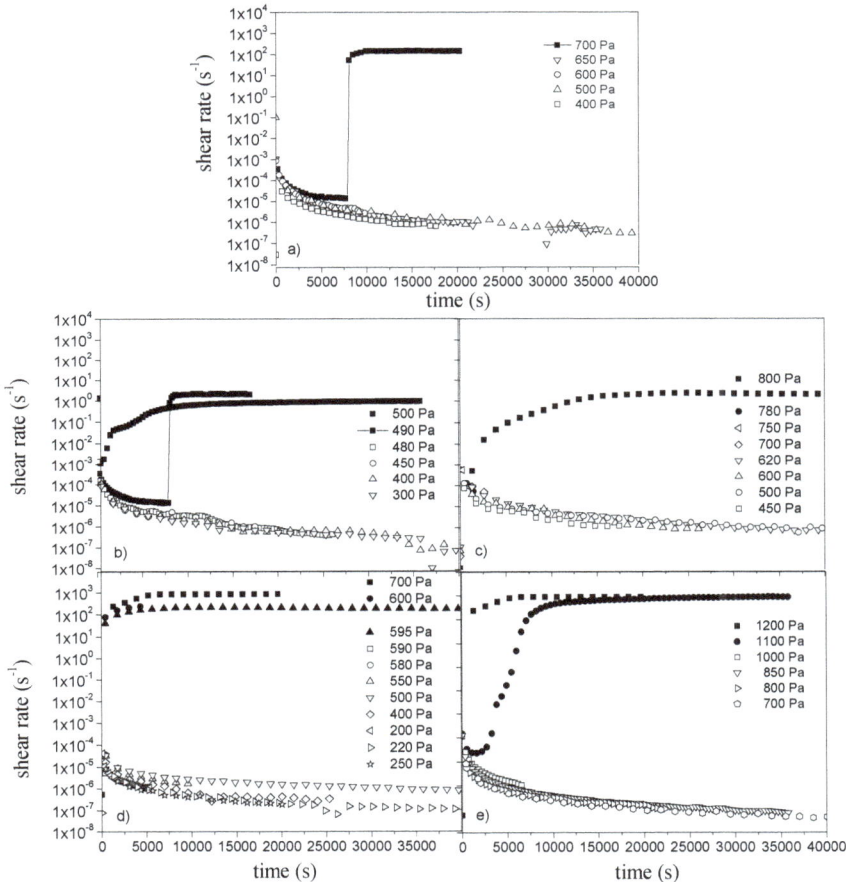

Figure 5. Creep tests for the lubricating greases studied: aluminum complex (**a**), lithium (**b**), lithium complex (**c**), calcium complex (**d**), and polyurea-based grease (**e**).

3.4. Correction of the "Practical" Steady-State Flow Curve

Figures 7 and 8 display the correction of the "practical" steady-state flow curves for the five lubricating greases studied, by inserting the equilibrium shear rate/stress values obtained from both creep and stress-growth tests, respectively. The transient stress values at a constant shear rate obtained in stress-growth experiments are not included in Figure 8 because of the difficulty to reach the equilibrium values for both the calcium complex and polyurea greases, as a consequence of the fracture and subsequent partial sample expelling from the measuring gap.

As can be observed in Figures 7 and 8, the non-monotonic evolution observed in stepped-shear rate ramp flow curves was not clearly observed, or significantly dampened, when the equilibrium (or pseudo-equilibrium) stress or shear rate values obtained from stress-growth and creep experiments, respectively, were plotted. Once again, it must be noticed that the pseudo-equilibrium stress value is that obtained previously for the detection of a significant fracture on samples submitted to a specific constant high shear rate.

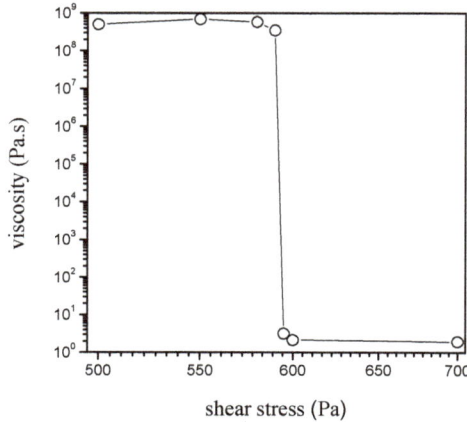

Figure 6. Viscosity vs. shear stress plot obtained from creep tests for the calcium complex grease.

On the other hand and more remarkably, the particularly relevant time-dependent flow behaviour of lubricating greases needs to be considered, especially at low shear rates. Thus, when the shear stress vs. shear rate plots obtained from the stepped-stress or shear rate ramp tests are compared with the equilibrium values obtained from transient experiments, minor correction is necessary for shear rate values higher than 0.1 s^{-1}. However, significant deviations were found below 10^{-3} s^{-1}, as a consequence of the extremely long time required to achieve the steady-state regime at low shear rates or shear stress values. This implies that stress values are noticeably underestimated in stepped-shear rate/stress ramp tests before the yielding flow behavior is apparent (see Figure 3).

Therefore, although stepped-shear rate/stress ramps provide reasonably accurate steady-state data at high shear rates, i.e., above 0.1 s^{-1} for lithium and aluminum greases, long-term experiments are necessary to obtain the steady-state values for lower shear rates and particularly to satisfactorily determine the extension of the yielding region. Overall, the discrepancies of data found in the medium shear rate range, i.e., 10^{-3}–0.1 s^{-1}, are due to both the time effects already discussed and the non-monotonicity found in the stress evolution, caused by flow instabilities which are favored at shorter times in the stepped shear rate tests, especially in polyurea and calcium greases, but also in the aluminum grease.

In all cases, the Herschel-Bulkley model (Equation (2)) has been used to fit these corrected steady-state flow curves:

$$\tau = \tau_0 + k_H \cdot \dot{\gamma}^n \tag{2}$$

where τ_0 is the yield stress (Pa), k_H is the plastic viscosity (Pa·s^n), and n is the flow index. These fitting parameters are shown in Table 3.

Table 3. Herschel-Bulkley fitting parameters.

Lubricating Greases	τ_0 (Pa)	k_H (Pa·s^n)	n
Aluminum Complex	654	232	0.28
Lithium	448	301	0.35
Lithium Complex	626	105	0.72
Calcium Complex	543	40	0.15
Polyurea	859	47	0.20

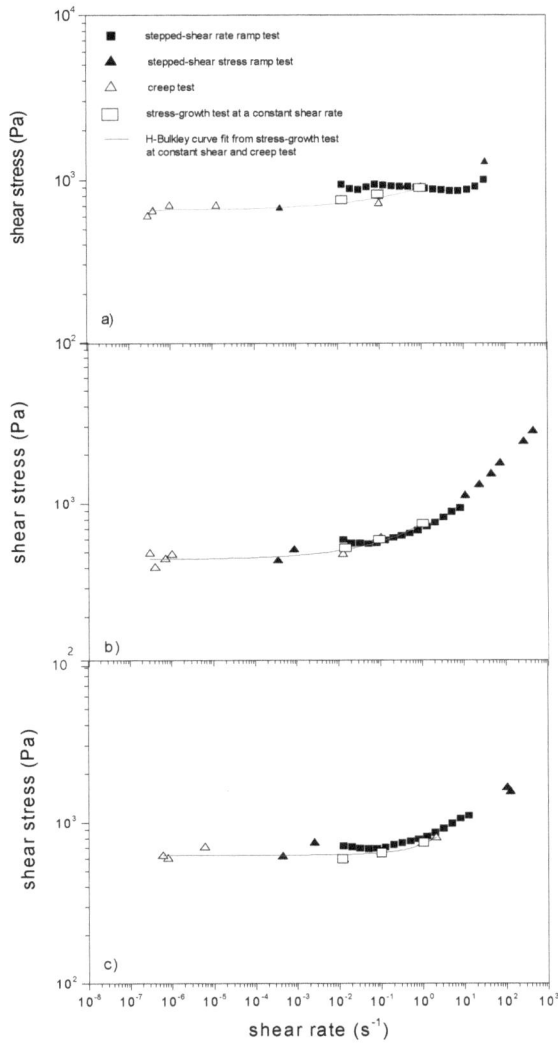

Figure 7. Steady-state flow curves obtained with different rheological tests and fitted to the Herschel-Bulkley model for aluminum complex (**a**), lithium (**b**), and lithium complex (**c**) greases.

As it has been discussed before, the polyurea-based grease exhibits a relatively strong gel-like behaviour in the linear viscoelastic regime and, accordingly, the highest value of τ_0 was obtained for this grease. On the other hand, both calcium complex soap- and polyurea-based greases show the lowest values of the flow index, closely related to the shear-induced structural breakdown [39], which are also in agreement with the values of the amount of overshoot collected in Table 2. In general, lubricating greases with the highest values of the elapsed time at the overshoot reveal the highest values of flow index (Tables 2 and 3). This fact indicates an easier reorientation of the thickener particles in the oily medium, better supporting the shearing action. In consequence, lithium, lithium complex, and aluminum complex greases show a much lower tendency to instabilities during flow.

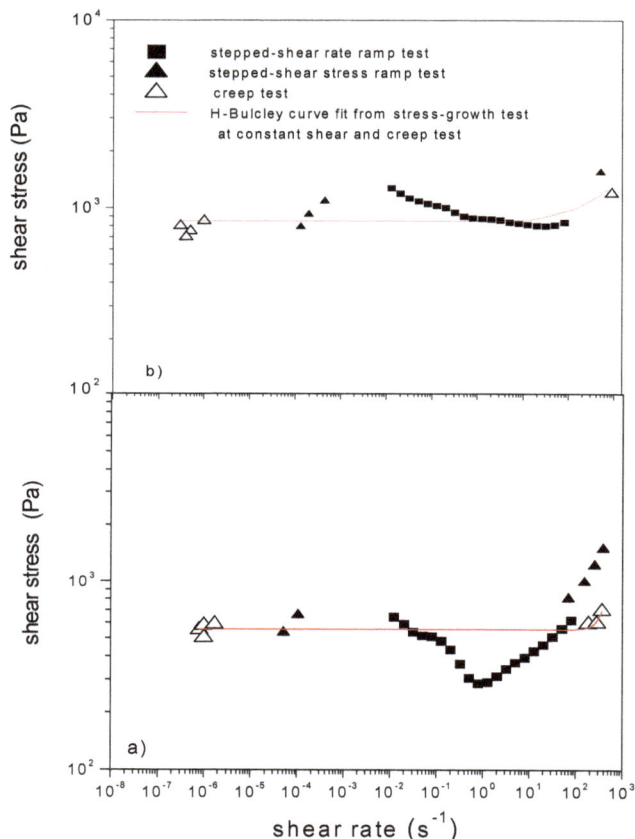

Figure 8. Steady-state flow curves obtained with different rheological tests and fitted to the Herschel-Bulkley model for calcium complex (**a**) and polyurea (**b**) greases.

4. Concluding Remarks

Practical viscous flow curves of some lubricating grease formulations evaluated through stepped-shear rate/stress ramp tests exhibit a non-monotonic evolution of shear stress in a wide range of shear rates, which is mainly the consequence of the non-consecution of the steady conditions. In this sense, long-term transient experiments must be conducted to correctly achieve the steady-state for theses greases. Transient tests at a constant shear rate are affected by the fracture phenomenon in the shear rate range of 0.0125–1 s^{-1}. Creep experiments make it possible to avoid or minimize these adverse situations. Both types of transient tests effectively allow the correction of the "practical" flow curves obtained in shear rate/stress-stepped ramps, noticeably dampening the non-monotonic evolution of the stress in the yielding region when considering the real equilibrium stresses or shear rates. These findings may contribute to better understanding the grease flow dynamics in real situations (bearings, seals, ducts, . . .). On one hand, the suitable rheological evaluation of the steady-state flow is necessary to deduce accurate values of rheological parameters, like those of the commonly employed Herschel-Bulkley model, to be used in simulations. On the other hand, the strong effect of time to achieve the steady-state viscosity values and the correct extension of the yielding flow regime must be incorporated in more complex analytical models in order to predict and/or simulated the grease fluid dynamics in practical situations like those described in recent literature [11–17].

Regarding the composition of greases, in spite of the same NLGI grade (except for the calcium complex grease), important differences between them have been detected from both rheological and microstructural points of view, which differentiate these greases in terms of applications and performance levels. Under the same test conditions, while aluminum complex, lithium complex, and lithium greases exhibit the actual flow behaviour of yielding structured materials, calcium complex and polyurea greases show a clearly apparent minimum in the shear stress vs. shear rate plots. Moreover, both of them exhibit the higher values of amount of overshoot in the entire studied shear rate range. Indeed, the calcium complex grease displays a weaker and not so interconnected structure composed of polydisperse and large particle aggregates, which entails a certain brittle character. This fact justifies the highest relative elasticity in the linear viscoelasticity regime and the dramatic drop in viscosity when the yield stress is reached, characterized by the lowest values of the flow index. On the other hand, the low values of the amount of overshoot and the high values of the elapsed time at the overshoot, together with the highest values of the flow index, reveal that lithium and lithium complex greases have the highest structural stability against shearing. In fact, the fiber-like structural units confer a high level of interactions to these greases, without detriment to the easier reorientation of the soap particles in the oil medium, better supporting the shearing action. Finally, it is worth pointing out the stronger gel-like behavior exhibited by the polyurea-based grease in the linear viscoelasticity regime, which confers to this grease the highest values of the yield stress under steady conditions and stress overshoot in transient experiments. Overall, lubricating greases display a complex yielding steady-state flow behaviour, strong time dependence, and associated flow instability effects, which depend on the microstructural characteristics.

Author Contributions: Conceptualization, M.A.D. and J.M.F.; Methodology, S.S. and C.V.; Formal Analysis, M.A.D., S.S. and C.V; Investigation, M.A.D., S.S., C.V. and J.M.F.; Writing—Original Draft Preparation, M.A.D. and J.M.F..; Writing—Review & Editing, M.A.D. and J.M.F.; Supervision, J.M.F.

Funding: This research received no external funding.

Acknowledgments: Authors gratefully acknowledge Total (France) for kindly providing commercial grease samples.

Conflicts of Interest: The authors declare no conflicts of interest.

References

1. Dawtray, S. Lubricating greases. In *Modern Petroleum Technology*; Hobson, G.D., Pohl, W., Eds.; Applied Science: Essex, UK, 1975.
2. NLGI. *Lubricating Greases Guide*; National Lubricating Grease Institute: Kansas City, MO, USA, 1994.
3. Gow, G. Lubricating grease. In *Chemistry and Technology of Lubricants*, 2nd ed.; Mortier, R.M., Orszulik, S.T., Eds.; Blackie Academic & Professional: London, UK, 1997; pp. 306–319.
4. Mas, R.; Magnin, A. Rheology of colloidal suspensions: Case of Lubricating Greases. *J. Rheol.* **1994**, *38*, 889–908. [CrossRef]
5. Cho, Y.I.; Choi, E.; Kirkland, W.H. The rheology and hydrodynamic analysis of grease flows in a circular pipe. *Tribol. Trans.* **1993**, *36*, 545–554. [CrossRef]
6. Delgado, M.A.; Valencia, C.; Sánchez, M.C.; Franco, J.M.; Gallegos, C. Thermorheological behaviour of a lithium lubricating grease. *Tribol. Lett.* **2006**, *23*, 47–54. [CrossRef]

7. Bondi, A. Theory and Applications. In *Rheology*, 3rd ed.; Eirich, F.R., Ed.; Academic Press: New York, NY, USA, 1960; p. 443.

8. Madiedo, J.M.; Franco, J.M.; Valencia, C.; Gallegos, C. Modeling of the nonlinear rheological behavior of lubricating grease at low shear rates. *J. Tribol.* **2000**, *122*, 590–596. [CrossRef]

9. Yeong, S.K.; Luckhama, P.F.; Tadros, T.F. Steady flow and viscoelastic properties of lubricating grease containing various thickener concentration. *J. Colloid Interface Sci.* **2004**, *274*, 285–293. [CrossRef] [PubMed]

10. Balan, C.; Franco, J.M. Influence of the geometry on the rotational rheometry of lubricating greases. In *The Rheology of Lubricating Greases*; Balan, C., Ed.; ELGI: Amsterdam, The Netherlands, 2000; pp. 43–66.

11. Westerberg, L.G.; Sarkar, C.; Farre-Llados, J.; Lundstrom, T.S.; Hoglund, E. Lubricating grease flow in a double restriction seal geometry: A computational fluid dynamics approach. *Tribol. Lett.* **2017**, *65*, 82. [CrossRef]

12. Westerberg, L.G.; Farre-Llados, J.; Sarkar, C.; Casals-Terre, J. Contaminant particle motion in lubricating grease flow: A computational fluid dynamics approach. *Lubricants* **2018**, *6*, 10. [CrossRef]

13. Yu, R.F.; Li, P.; Chen, W. Study of grease lubricated journal bearing with partial surface texture. *Ind. Lubr. Technol.* **2016**, *68*, 149–157. [CrossRef]

14. Wu, Z.H.; Xu, Y.Q.; Deng, S.E. Analysis of dynamic characteristics of grease-lubricated tapered roller bearings. *Shock Vibr.* **2018**, *2018*, 7183042. [CrossRef]

15. Sarkar, C.; Westerberg, L.G.; Hoglund, E.; Lundstrom, T.S. Numerical simulations of lubricating grease flow in a rectangular channel with and without restrictions. *Tribol. Trans.* **2018**, *61*, 144–156. [CrossRef]

16. Dobrowolski, J.D.; Gawlinski, M.; Paszkowski, M.; Westerberg, L.G.; Hoglund, E. Experimental study of lubricating grease flow inside the gap of a labyrinth seal using microparticle image velocimetry. *Tribol. Trans.* **2018**, *61*, 31–40. [CrossRef]

17. Li, J.X.; Westerberg, L.G.; Hoglund, E.; Lugt, P.M.; Baart, P. Lubricating grease shear flow and boundary layers in a concentric cylinder configuration. *Tribol. Trans.* **2014**, *57*, 1106–1115. [CrossRef]

18. Cyriac, F.; Lugt, P.M.; Bosman, R. On a new method to determine the yield stress in lubricating grease. *Tribol. Trans.* **2015**, *58*, 1021–1030. [CrossRef]

19. Cyriac, F.; Lugt, P.M.; Bosman, R. Yield stress and low-temperature start-up torque of lubricating greases. *Tribol. Lett.* **2016**, *63*, 6. [CrossRef]

20. Møller, P.C.F.; Mewis, J.; Bonn, D. Yield stress and thixotropy: On the difficulty of measuring yield stresses in practice. *Soft Matter* **2006**, *2*, 274–283. [CrossRef]

21. Barrnes, H.A. The Yield Stress-a review or "παντα ρει"-everything flow? *J. Non-Newtonian Fluid Mech.* **1999**, *81*, 133–178. [CrossRef]

22. Magnin, A.; Piau, J.M. Shear rheometry of fluids with a yield stress. *J. Non-Newtonian Fluid Mech.* **1987**, *23*, 91–106. [CrossRef]

23. Hartnett, J.P.; Hu, R.Y.Z. The yield stress - an engineering reality. *J. Rheol.* **1989**, *33*, 671–679. [CrossRef]

24. Almdal, K.; Dyre, J.; Hvidt, S.; Kramer, O. Towards a phenomenological definition of the term "gel". *Polym. Gels Netw.* **1993**, *1*, 5–17. [CrossRef]

25. Sánchez, M.C.; Franco, J.M.; Valencia, C.; Gallegos, C.; Urquiola, F.; Urchegui, R. Atomic force microscopy and thermo-rheological characterization of lubricating greases. *Tribol. Lett.* **2011**, *41*, 463–470. [CrossRef]

26. Delgado, M.A.; Sánchez, M.C.; Valencia, C.; Franco, J.M.; Gallegos, C. Relationship among microstructure, rheology and processing of a lithium lubricating grease. *Chem. Eng. Res. Des.* **2005**, *83*, 1085–1092. [CrossRef]

27. Coussot, P. Slow flows of yield stress fluids: Yielding liquids or flowing solids? *Rheol. Acta* **2018**, *57*, 1–14. [CrossRef]

28. Britton, M.M.; Callagham, P.T. Nuclear magnetic resonance visualization of anomalous flow in cone-and-plate rheometry. *J. Rheol.* **1997**, *41*, 1365. [CrossRef]

29. Coussot, P.; Nguyen, Q.D.; Huynh, H.T.; Bonn, D. Avalanche behavior in yield stress fluids. *Phys. Rev. Lett.* **2002**, *88*, 175–207. [CrossRef] [PubMed]

30. Coussot, P. Yield stress fluid flows: A review of experimental data. *J. Non-Newtonian Fluid Mech.* **2014**, *211*, 31–49. [CrossRef]

31. Ovarlez, G.; Rodts, S.; Chateau, X.; Coussot, P. Phenomenology and physical origin of shear localization and shear banding in complex fluids. *Rheol. Acta* **2009**, *48*, 831–834. [CrossRef]

32. Rubio-Hernandez, F.J.; Paez-Flor, N.M.; Velazquez-Navarro, J.F. Why monotonous and non-monotonous steady-flow curves can be obtained with the same non-Newtonian fluid? A single explanation. *Rheol. Acta* **2018**, *57*, 389–396. [CrossRef]

33. Mewis, J. Thixotropy—A general review. *J. Non-Newtonian Fluid Mech.* **1979**, *6*, 1–20. [CrossRef]

34. Barnes, H.A. Thixotropy—A review. *J. Non-Newtonian Fluid Mech.* **1997**, *70*, 1–33. [CrossRef]

35. Delgado, M.A.; Franco, J.M.; Valencia, C.; Kuhn, E.; Gallegos, C. Transient shear flow of model lithium lubricating greases. *Mech. Time-Depend. Mater.* **2009**, *13*, 63–80. [CrossRef]

36. Papenhuijzen, J.M.P. The role of particle interactions in the rheology of dispersed systems. *Rheol. Acta* **1972**, *11*, 73–88. [CrossRef]

37. Kuhn, E. Analysis of a grease-lubricated contact from an energy point of view. *Int. J. Mater. Prod. Technol.* **2010**, *38*, 5–15. [CrossRef]

38. Coussot, P.; Lenov, A.I.; Piau, J.M. Rheology of concentrated dispersed systems in a low molecular weight matrix. *J. Non-Newtonian Fluid Mech.* **1993**, *46*, 179–217. [CrossRef]

39. Delgado, M.A.; Valencia, C.; Sánchez, M.C.; Franco, J.M.; Gallegos, C. Influence of soap concentration and oil viscosity on the rheology and microstructure of lubricating greases. *Ind. Eng. Chem. Res.* **2006**, *45*, 1902–1910. [CrossRef]

Article

Rheological Characterization of Carbopol® Dispersions in Water and in Water/Glycerol Solutions

Priscilla R. Varges, Camila M. Costa, Bruno S. Fonseca, Mônica F. Naccache and Paulo R. de Souza Mendes *

Department of Mechanical Engineering, Pontifícia Universidade Católica-RJ, Rua Marquês de São Vicente 225, Rio de Janeiro, RJ 22453-900, Brazil; prvarges@puc-rio.br (P.R.V.); camila.moreira.costa@hotmail.com (C.M.C.); brunodasilva.fonseca@gmail.com (B.S.F.); naccache@puc-rio.br (M.F.N.)
* Correspondence: pmendes@puc-rio-br; Tel.:+5521-99982-9653

Received: 15 December 2018; Accepted: 2 January 2019; Published: 4 January 2019

Abstract: The influence of the solvent type on the rheological properties of Carbopol® NF 980 dispersions in water and in water/glycerol solutions is investigated. The material formulation, preparation procedure, common experimental challenges and artifact sources are all addressed. Transient and steady-state experiments were performed. For both solvent types, a clearly thixotropic behavior occurs slightly above the yield stress, where the avalanche effect is observed. For larger stresses, thixotropy is always negligible. Among other findings, it is observed that, for a given Carbopol concentration, the dispersion in the more viscous solvent possesses a lower yield stress and moduli, a larger power-law index, and a longer time to reach steady state.

Keywords: Carbopol; yield stress; thixotropy

---

1. Introduction

Materials like colloidal suspensions, emulsions, foams, gels, and granular materials only flow irreversibly when a finite threshold shear stress—known as the *yield stress*—is exceeded [1–9].

Carbopol®, a trademark owned by Lubrizol Corporation (Wickliffe, OH, USA), is a family of commercial polymers frequently employed in the cosmetics, pharmaceutical, paint, and food industry as a thickening, suspending, dispersing, and stabilizing agent [10]. In research activities, its solutions are frequently employed in flow visualization experiments [11–14], because they are transparent gels that are relatively easy to prepare [15–17].

There are more than 10 grades of Carbopol polymers, which may be subdivided into several categories based on their physical structure and chemical composition, crosslink density, polymerization solvent, type of cross-linking, network electrical charge, and physical appearance [18]. Carbopol is a high molecular weight, hydrophilic, and crosslinked polyacrylic acid polymer. This physical hydrogel presents a three-dimensional polymer network that is swollen by water, and presents temporary, reversible interchain entanglements that are stronger when compared to chemical hydrogels.

Specifically, in the present research we employed Carbopol NF 980, also known as a monograph Carbomer Homopolymer Type C (former Carbomer 940) by U.S. Pharmacopeia/National Formulary (USP/NF) in the United States. Carbopol NF 980 is a synthetic homopolymer, polyacrylic acid crosslinked with allyl sucrose or allyl pentaerythritol, which is polymerized in a co-solvent system. It is supplied as a white and dry powder of primary particles averaging $0.2\,\mu m$ in diameter. It is a weak anionic polyelectrolyte polymer that must be neutralized in order to achieve a high viscosity. Each particle (or network structure) is a mixture of tightly coiled linear polymer chains, which are soluble in polar solvents. The viscosity of Carbopol solutions is not a function of the size of its powder particles [19].

It is well known that this polymer forms a colloidal dispersion when hydrated in water at controlled pH and temperature. Frequently, a Carbopol aqueous dispersion is neutralized with a common base, such as sodium hydroxide (NaOH), converting the acidic polymer into a salt. When neutralized, the polymer presents the ability to absorb and retain water. Polymer chains interconnected by crosslinks begin to hydrate, and partially uncoil due to electrostatic repulsion in order to form irreversible agglomerates [20]. The desired yield stress nature is due to the presence of high molecular weight polyacrylate branched chains that form interchain entanglements which prevent flow at low shear stresses [19,21,22]. During hydration the chains may increase up to 10 times their original diameter [23], and the ionization process leads to a crosslink of the swollen molecules, forming a microgel network with stronger bonds.

In order to better understand the rheological properties of Carbopol solutions, it is fundamental to improve the knowledge of their relation with the material microstructure [24]. Data obtained in rheological measurements are better interpreted when supported by visualization techniques such as microscopy, X-ray, neutron and light scattering, nuclear magnetic resonance (NMR) or differential scanning calorimetry (DSC) [25]. Indeed, Ref. [23] investigated Carbopol viscoelastic properties through direct analysis of the microscopic network structure by scanning electron cryomicroscopy (cryoSEM). Likewise, Ref. [26] studied elasticity through microrheology. On the other hand, Ref. [20] emphasized the need for novel microscopic scale experiments, that would be able to simultaneously describe local dynamics and flow behavior.

Steady and oscillatory simple shear flows provide information on the relationship between the rheological properties and the macromolecular structure. Viscoelastic effects are a result of the interaction between polymer chains that is related to the viscous behavior, and the chains recoiling due to thermal motion, which is related to the elastic behavior [27]. Colloidal dispersions such as Carbopol present a typical viscoelastic behavior at low stresses [28]. On the other hand, elastic effects decrease and eventually tend to become negligible at the higher stresses found in the nonlinear steady state.

The rheological properties of Carbopol dispersions have been extensively investigated [16,23,29–31]. These properties depend on the type and degree of crosslinking, which in turn depend on molecule swelling and medium density. Carbopol dispersion is usually seen as simple model yield stress fluids [32,33]. Their viscoplastic steady state behavior is well represented by the Herschel-Bulkley equation [34], which can accommodate a yield stress and a power-law shear-thinning behavior. However, for some soft glassy materials, other important features cannot be neglected, such as pronounced elastic and thixotropic behavior (especially when the stress is close to the yield stress), transient and initial internal stresses, normal stress differences, and irreversibility of the deformed states [9,32,35–37]. Ref. [38] argue that, at low stresses, the microgels do not move relatively to each other, but are able to deform, resulting in a solid-like elastic behavior. At large stresses, there is relative mobility of the microgels, leading to a liquid-like viscoelastic behavior.

Yield stress materials can be classified according to their microstructure as repulsive-dominated jammed glasses, networked gels with attractive interactions or a combination thereof [39,40]. According to [39], Carbopol gels are classified into the first category.

Ref. [16] describes Carbopol gels as concentrated, percolated, and disordered dispersions with glassy structure. Structural variations occur as the polymer concentration is increased. For $c < 0.035$ wt%, the dispersion obtained possesses no yield stress; for $0.035 < c < 0.12$ wt% a percolated viscoelastic dispersion is obtained; for $0.12 < c < 0.21$ wt% the dispersion obtained is phase-inverted percolated with excess of solvent; and for $c > 0.21$ wt% a closely packed and disordered structure is observed, due to polydispersity.

Using confocal fluorescence microscopy, Ref. [17] observed that the Carbopol microgels are indeed polydisperse. They also observed that, at the same concentration, Carbopol Ultrez 10 and ETD 2050 (Lubrizol Corporation, Wickliffe, OH, USA) present the same mesostructure, despite a difference in particle size. The microgels have complex shape when unconstrained, and possess soft elasticity and capacity to adapt their outer shape and swelling ratio to local space and solvent availability conditions. In addition, the maximum volume concentration can be quite high due to the also high albeit as yet not quantified polydispersity.

For a 1.5 wt% concentration, Ref. [19] observed the existence of highly swollen deformable microgel particles closely packed and in intimate contact. The high viscosity is determined by network cross-link density, which governs particle-particle interaction. On the other hand, dilute dispersions, i.e., dispersions of fully swollen particles with no contact, the particle-solvent interactions result in lower viscosity and elasticity.

Carbopol dispersions show different behavior depending on the solvent [41]. In general, water is used as a solvent. However, co-solvent and anhydrous systems have also been used [42]. Ethanol and isopropanol can be thickened adding Carbopol polymers [43]. In this case, it is crucial to use the appropriate neutralizer, which varies depending upon the alcohol content. In addition, previous works have already demonstrated the possibility of formulating Carbopol systems using different hydrophilic solvents such as polyethylene glycol (PEG) 400, glycerol, silicone, and tetraglycol without neutralization [44].

The choice of the solvent is also important because solvents such as glycerol and propylene glycol can modify hydrogen bond characteristics between water, solvent, and polymer, thereby affecting polymer swelling and the viscoelastic properties [30,41,45].

The present research aims to explore the rheological properties of Carbopol NF 980 dispersions in different formulations. Solvent type, polymer concentration, gel preparation, mechanical properties, and phenomenological behavior are all discussed. We carried out systematic rheological experiments, in transient and steady state regimes, from which we observe that Carbopol gels are not as "simple" as usually assumed.

2. Experimental Protocol

2.1. The Carbopol

Carbopol NF 980 polymer presents a bulk density of 176 Kg/m^3 and in the crosslinked form shows a molecular weight as high as 4.5 billion due to the interlinkage of many polymer chains [46]. Ref. [10] estimated the Carbopol 940 molecular weight between crosslinks as 1.04×10^5, while [19] calculated 5×10^6. In general, all members of the Carbopol family possess a considerably high molecular weight.

2.2. Fluid Preparation

Carbopol aqueous and water/glycerol dispersions were prepared based on the procedures recommended by the manufacturer [43,45,47,48]. Weight concentrations ranging from 0.1 wt% to 0.15 wt% were investigated. All the water/glycerol solutions used as continuous phase were composed of 60% of glycerol and 40% of water, in volume.

Ref. [16] emphasized that a strict protocol for preparation of Carbopol dispersions is fundamental to achieve reproducibility. For the same concentration, he affirms that water properties, reagents involved, dispersion methods, and also the reservoir used during the mixing process can affect the dispersion properties. Therefore, its rheology is a function of concentration, composition, pH, temperature, aging and preparation procedure [31].

A detailed description of the dispersion preparation is now presented. Before preparation, it is important to sift the Carbopol dry powder through a 20-mesh metallic screen to eliminate aggregates which prevent complete hydration, severely affecting the quality of the dispersion.

A 6 L plastic vessel is used to prepare and store the dispersion. Initially, it is filled with a predetermined mass of water obtained from a reverse osmosis system. The vessel with water is placed on a mechanical stirrer equipped with a 3-blade marine impeller positioned close to vessel bottom. The stirrer is turned on at 1200 rpm. A cover is used to minimize solvent evaporation and contamination. Carbopol should be carefully and slowly added to the vessel, approximately halfway between the blade and the vessel wall, to avoid adhesion to solid surfaces.

For Carbopol aqueous dispersions, after polymer addition the agitation is maintained at 1200 rpm for 15 min and then kept at rest for 30 min. On the other hand, for dispersions of Carbopol in water/glycerol solutions, the water/glycerol solution is added after powder addition, with the mixer at 700 rpm. Stirring is applied for 15 min, and then the dispersion is kept at rest.

For both types of dispersion, the 3-blade marine impeller is then replaced by a naval blade and the mixer is set to 150 rpm, to minimize formation of air bubbles. A 18 wt% NaOH aqueous solution is then added to neutralize the dispersion.

Finally, for the aqueous dispersion, the agitation is increased to 300 rpm and maintained uninterruptedly for 5 days, to homogenize the dispersion. In the presence of glycerol, the agitation is kept at 150 rpm for 7 days. This final step is very important since we observed that longer mixing times favor reproducible data as they improve the complete process of hydration and material stability over time [31,49].

2.3. Rheological Measurements

The rheological properties of the Carbopol dispersions are measured using two stress-controlled rheometers, namely the AR-G2 and the DHR-3 by TA Instruments (New Castle, DE, USA). Both are combined motor and transducer (CMT) instruments. Instrument inertia is an important source of artifact, especially for low-viscosity liquids. Therefore, it is important to apply the suitable corrections [50–53].

The rheology of Carbopol dispersions depends very weakly on temperature [23,30]. Nevertheless, all tests were performed at 25 °C. A Peltier system was employed to control the sample temperature. Carbopol dispersions are volatile [15–17]. Therefore, to minimize evaporation and keep constant the shape of the free meniscus, a solvent trap with water was used to create a saturated atmosphere around the sample [54].

In general, the geometry selection is based on the characteristics of the material to be tested. Due to the yield stress, Carbopol dispersions tend to present apparent wall slip at low strain rates [55–57], as usually observed in structured materials. Therefore, roughened surfaces are in order. The plate-plate geometry is usually preferred for dispersions as opposed to the cone-plate geometry, to ensure that the minimum gap throughout the sample is about 10 times the characteristic size of the dispersed phase [54]. In the present research, a 60 mm diameter cross-hatched plate-plate geometry with a 1 mm gap was used in the tests. This geometry was made of titanium, which has the advantage of preventing chemical attacks and minimizing instrument inertia.

A disadvantage of the plate-plate geometry is the flow inhomogeneity, in the sense that the shear rate throughout the sample is not uniform. Rather, it varies linearly with the radial position, and hence for non-Newtonian liquids the viscosity also varies with the radial position. For steady flow, the Weissenberg-Rabinowitsch equation circumvents this problem by giving the shear stress at the rim [58]. A similar treatment was proposed by [59] to evaluate the stress amplitude at the rim for oscillatory flows. For creep (viscosity or shear-rate bifurcation) tests, no correction is needed, because the material behaves as a solid below the yield stress [59].

With the aid of a glass syringe, the sample is positioned at the center of the bottom plate. Carbopol dispersions tend to retain bubbles [15–17]. If bubbles are trapped during loading, they should be removed by suction with a needle syringe. Then the upper plate is positioned slightly above its final position, and the sample excess is trimmed. The upper plate is then brought to the measuring position, the free surface at the rim is checked for irregularities, and the solvent trap is positioned.

Before the beginning of the test, the sample is kept at rest for 30 min to allow for thermal equilibrium and rebuilding of the microstructure which is partially broken during loading. Ref. [60] argues that the resting time essentially eliminates residual stresses.

To verify repeatability, the tests should be repeated at least once, always with a fresh sample.

3. Results and Discussion

3.1. Stress Amplitude Sweep Tests

Typical results for a stress amplitude sweep test are presented in Figure 1. The material is a 0.123 wt% Carbopol aqueous dispersion, and the frequency was kept fixed at 1 Hz. In this figure, the storage modulus G', loss modulus G'', and complex modulus G^* are given as a function of the imposed shear stress amplitude, τ_a. Instrument inertia [61] and flow inhomogeneity [59] effects are corrected for.

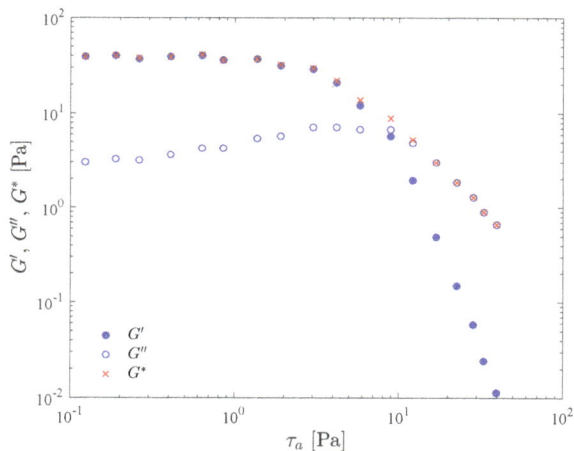

Figure 1. Stress amplitude sweep test results for a 0.123 wt% Carbopol aqueous dispersion.

Figure 1 can be divided into three regions, namely the linear viscoelastic region (LVR), the yielding region, and the non-linear region. The LVR occurs in the low stress amplitude range, and is characterized by constant (i.e., stress-amplitude independent) moduli. It corresponds to the so-called "small-amplitude oscillatory shear" (SAOS) flow regime. Both the yielding and the non-linear region are in the domain of the so-called "large-amplitude oscillatory shear" (LAOS) flow regime.

In the LVR, the imposed stress is small in comparison to the strength of the bonds that sustain the microstructure, so that its integrity is unaffected. The microgels remain in their "cages", i.e., are allowed to deform elastically but do not move significantly relative to each other. This microstructure confers a predominantly elastic behavior to the material, as indicated by a large G' in comparison with G''. The elastic preponderance indicates a low internal dissipation within microgels [16].

At this point it is interesting to emphasize that the definitions of G' and G'' originate from the assumption that both the input and output waves be sinusoidal, i.e., that the stress depends linearly on the shear strain and on the shear rate. In other words, G' and G'' cannot depend on the stress amplitude, by definition. Therefore, in principle the values of G' and G'' shown in Figure 1 would be meaningless beyond the LVR, because they do depend on the stress amplitude. There is no guarantee that the output (strain) wave is sinusoidal outside the LVR, in potential conflict with the basic hypothesis of the theory from which G' and G'' arise. This fact is typically overlooked in the literature, probably because the trends observed for G' and G'' beyond the LVR seem to be in qualitative agreement with the expected behavior.

The reason for the successful performance of G' and G'' outside their domain is the fact that in this case the output waves are not too far from sinusoidal beyond the SAOS regime. This is illustrated in Figure 2. That is, for this Carbopol dispersion at 1 Hz (and seemingly for Carbopol dispersions in general at 1 Hz), the so-called quasilinear large-amplitude oscillatory shear (QL-LAOS) flow regime prevails beyond the LVR. The existence of this regime was recently discovered theoretically by [62] and later confirmed experimentally [63]. The quasilinearity (sinusoidal output wave) occurs because at a given stress amplitude the microscopic state does not change along the cycle (as in SAOS), although it changes with the stress amplitude, in contrast to what is observed in the SAOS regime. The microscopic state remains unchanged along the cycle because the characteristic time needed for changes in the microscopic state is longer than the characteristic time of the flow, namely $1/2\pi$ s in the present case.

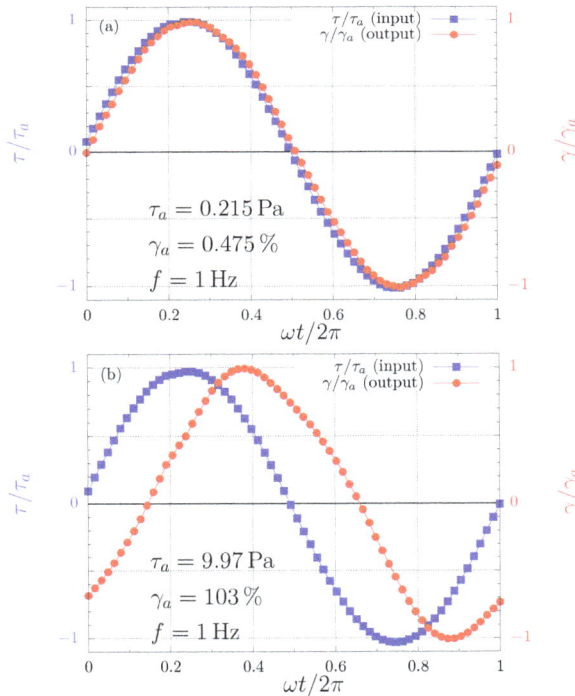

Figure 2. The input and output waves for two different stress amplitudes. (a) $\tau_a = 0.215$ Pa [in the linear viscoelastic region (LVR)]; and (b) $\tau_a = 9.97$ Pa (in the yielding region).

In addition, the characteristic time needed for changes in the microscopic state is expected to decrease as the stress amplitude is increased, i.e., the microstructure is expected to break faster at larger stresses. Therefore, at large enough stress amplitudes the output waves should cease to be sinusoidal, i.e., the flow should no longer be within the QL-LAOS regime. This is because, when the time required for changes in the microscopic state is shorter or comparable with the characteristic time of the flow (namely the cycle period), then the microscopic state changes along the cycle. Since the mechanical behavior is a function of the microscopic state, the mechanical behavior varies along the cycle, rendering the output wave non-sinusoidal. In our experiments, however, we did not attain large enough stress amplitudes to leave significantly the QL-LAOS regime.

The dependence of G' and G'' on the stress amplitude is a result of the onset of a bond breaking process (yielding). The larger the stress amplitude the lower the structuring level, which results in less elasticity, as indicated by the decreasing G' as the stress amplitude is increased. In the yielding region, located in the middle range of stress amplitude and around the crossover point, the stress amplitude is high enough to cause some bond breaking, i.e., some microgels escape from their cages and more relative motion occurs. The frictional forces that arise due to relative motion dissipate more mechanical energy, which explains the increase in G''. Within this range of stress amplitude the structuring level is still high enough to retain a percolated microstructure and a sizable elasticity response, as indicated by the still high values of G' found in this region.

It is tempting to take a characteristic stress amplitude of the yielding region as the yield stress, because it is throughout this region that yielding takes place. An obvious choice of characteristic stress amplitude would be the one at which the G' and G'' curves cross. However, the crossover stress amplitude depends to a great extent on the frequency of oscillation, while the definition of yield stress precludes such a dependence. This frequency dependence occurs because the bond breaking process depends both on the stress intensity and on the time period during which the imposed stress persists, in consonance with the discussion above regarding the characteristic time needed for changes in the microscopic state. The larger the imposed stress the lower the time period required to break the bonds, and vice versa. Therefore, in oscillatory flows, for a given stress amplitude the larger the frequency the shorter the stress persistence. Hence, for larger frequencies (shorter stress persistencies) larger stress amplitudes are required to break the bonds. In view of this discussion it might be argued that the yield stress can be identified as the limit as the frequency approaches zero of the crossover stress amplitude. However, as the frequency is reduced, the output wave will eventually cease to be sinusoidal, and hence G' and G'' will lose their physical meanings.

The non-linear region starts beyond the crossover point, where G'' attains its maximum and starts decreasing, and where G' starts decreasing much faster than G''. The high stress amplitudes found in this region cause a massive bond breakage, so that the microstructure is no longer percolated. This confers a liquid behavior to the material. Moreover, the higher the stress amplitude, the lower the structuring level, leading to a more viscous and less elastic behavior.

Figure 3 shows the complex shear modulus as a function of the stress amplitude for three different dispersions, namely two in water and one in a water/glycerol solution. Comparing the results for the two aqueous dispersions, it is seen that G^* is larger for the more concentrated one throughout the stress amplitude range, as expected. Moreover, the LVR range does not change significantly with concentration, as also observed by [38].

Figure 3 also illustrates the influence of the solvent on the complex modulus. It is seen that the G^* curve for the water/glycerol Carbopol dispersion lies below the curves for the two aqueous dispersions, despite the fact that its Carbopol concentration is the highest, illustrating that the presence of glycerol reduces the complex modulus. Since in the LVR $G^* \approx G'$ for the three dispersions, this G^* reduction indicates that glycerol leads to a less elastic structure. The presence of glycerol slightly increases the loss modulus (not shown), as a consequence of the fact that glycerol is more viscous than water. It is interesting to observe that the 0.1 wt% Carbopol aqueous dispersion and the 0.125 wt% Carbopol dispersion in water/glycerol were designed to possess the same yield stress, namely $\tau_y = 2.7$ Pa.

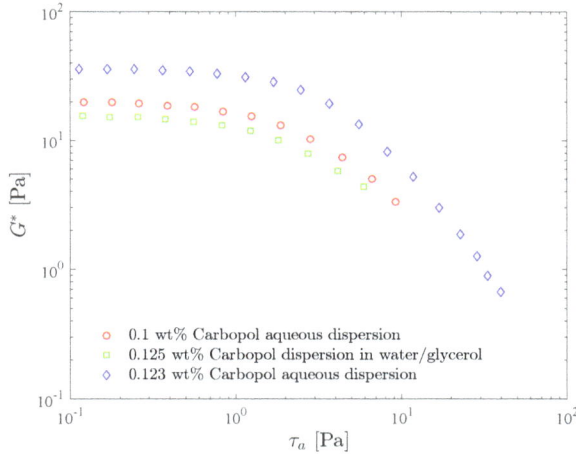

Figure 3. Stress amplitude sweep test resuts for three Carbopol dispersions: effects of the concentration and of the solvent.

3.1.1. Frequency Sweep Tests

Frequency sweep tests are a useful tool to assess the mechanical response of the material microstructure in its quiescent state [23]. It is known that the magnitude of the storage modulus G' can be related to the molecular weight and network density [18]. In this test, the stress amplitude is kept constant at a value low enough to ensure that the flow is kept within the linear viscoelastic regime throughout the range of frequency.

Figure 4 shows the storage modulus G' and the loss modulus G'' as a function of the frequency, for the 0.1 wt% aqueous Carbopol dispersion and for the 0.125 wt% Carbopol dispersion in the water/glicerol solution. It is worth recalling that the composition of these two solutions were chosen such that both possess the same yield stress. Comparison of the curves for the two dispersions illustrates the influence of the solvent type on the moduli.

It is seen in this figure that our data lie within the classic rubbery region (or frequency range) [54], since G' remains essentially constant and much larger than G'' throughout the frequency range, as expected for viscoelastic solids and crosslinked systems [64,65]. The G'' curves, on the other hand, increase with frequency.

Moreover, it is seen in this figure that the G' and G'' curves for the two materials are essentially coincident, except towards the high end of the frequency range, where the curve for G'' the aqueous dispersion lies below the one for the dispersion in water/glycerol, while for the G' curves the reverse is true. This indicates that more mechanical energy is dissipated at larger frequencies, and that the more viscous continuous phase tends to dissipate more energy. Moreover, the fact that the moduli of the two dispersions are essentially coincident suggests that the yield stress and the moduli—which are completely distinct properties—are closely related to the microstructure characteristics at its quiescent state, which is expected to be similar for both dispersions.

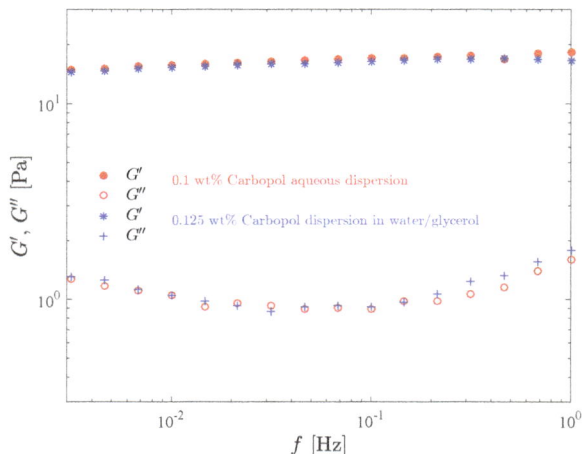

Figure 4. Results of the frequency sweep test for the Carbopol dispersions with the same yield stress.

3.1.2. Time Sweep Tests

The time sweep test consists of observing how the moduli change over time for a constant frequency and stress amplitude. The choice of stress amplitude and frequency should be such as to ensure that the flow is within the linear viscoelastic regime, i.e., the imposed stresses should be negligible when compared to the strength of the microstructure. This test is usually employed to investigate the material stability, i.e., if microscopic changes occur over time. Useful information such as polymer degradation, molecular weight building, cross-linking, solvent evaporation, sedimentation, setting and curing can be obtained with this test. It also allows the determination of the maximum time duration of a test to ensure that none of the just listed effects will interfere in the results.

Figure 5 compares the complex modulus G^* time sweep results for the 0.1 wt% Carbopol aqueous dispersion and for the 0.125 wt% dispersion of Carbopol in water/glycerol solution. Both curves were obtained at 1 Hz and at stress amplitudes well below the yield stress (which is 2.7 Pa), namely 0.08 and 0.1 Pa, respectively.

The results in Figure 5 show that the dispersions under investigation are very stable and free from aging, as indicated by the fact that the G^* curves are essentially horizontal, except for a very short transient behavior probably due to microstructure rebuilding after sample loading. The curve for the water/glycerol dispersion lies above the one for the aqueous dispersion, in accordance with the discussion and results presented above.

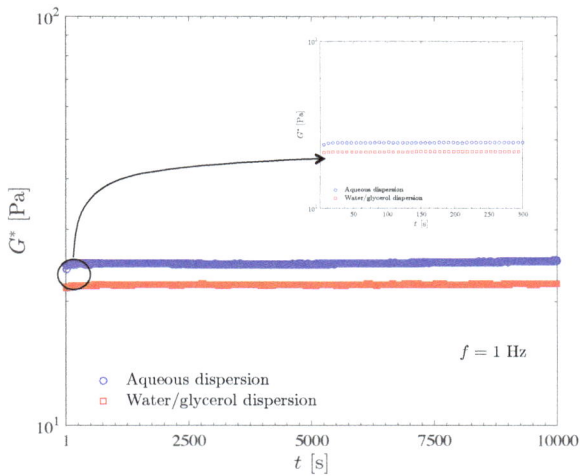

Figure 5. Evaluation of the stability of Carbopol dispersions by means of the time sweep test.

3.2. Constant Shear Rate Tests

Figure 6 shows the transient viscosity of the two Carbopol dispersions that possess the same yield stress. In this test, the sample is loaded on the rheometer and left to rest for a few minutes to allow for the microstructure reconstruction. Then, at time $t = 0$ we impose a constant shear rate, and record the transient shear stress (or viscosity) response. Due to the error caused by the instrument inertia, the data pertaining to the first second of each test were discarded.

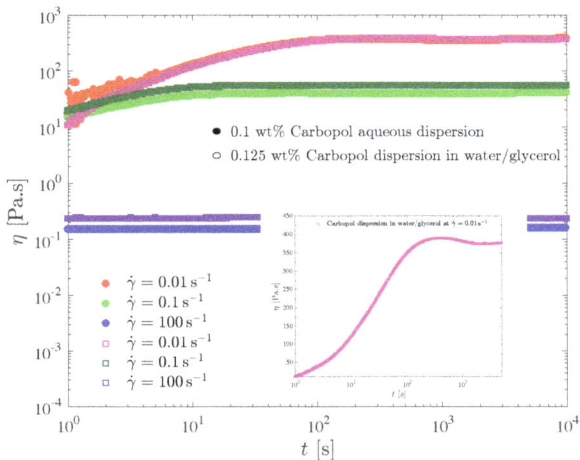

Figure 6. The constant shear rate test for the dispersions with the same yield stress.

For both dispersions and all shear rate values investigated, it is seen in Figure 6 that the viscosity eventually reaches a steady state. The larger the shear rate the lower the steady-state viscosity, because larger shear rates imply larger shear stresses, which in turn result in more intense microstructure breakdown. Consequently, structuring levels pertaining to larger shear rates are lower, and there is a one-to-one relationship between the steady-state viscosity and the structuring level (e.g., [62,66,67]).

Figure 6 also shows that for both dispersions the time required for the steady state to be reached, say t_{ss}, decreases as the shear rate is increased. Specifically, $t_{ss} \approx 100; 10; 0 \, \text{s}$ for $\dot{\gamma} = 0.01; 0.1; 100 \, \text{s}^{-1}$, respectively. These results are in agreement with the generally accepted rule of thumb $t_{ss} \approx 1/\dot{\gamma}$ used to estimate t_{ss}. To understand this trend we first note that the steady-state stresses corresponding to very low shear rates are always very close to the yield stress, while for high shear rates the corresponding shear stresses are well above the yield stress. Therefore, the long times required for the achievement of the steady state at low shear rates mean that the Carbopol dispersions display a sizable thixotropic effect when the stress is very close but above the yield stress. For larger stresses there is no thixotropy, as indicated by the very short times required for the steady state to be achieved.

The shear stress evolution with time is well illustrated by the curve for the 0.125 wt% Carbopol dispersion in glycerol/water at $\dot{\gamma} = 0.01 \, \text{s}^{-1}$, highlighted in the insert of Figure 6. At time $t < 0$, the gel is unstrained and fully structured, and thus it is a viscoelastic solid. At $t = 0$, when the shear rate jump occurs, a corresponding shear stress jump is expected due to an initially purely viscous response of the still unstrained sample. This viscous contribution to the stress is expected to remain fixed while the material remains fully structured, because the shear rate remains fixed in this test. The strain increases linearly, starting from zero at $t = 0$, causing the growth (also from zero) of the elastic contribution to the stress.

At large enough strains (but still below the yield point), the elastic contribution to the stress becomes much larger than the viscous one, and a linear growth of the (total) stress is observed. When the stress reaches the yield stress, the microstructure undergoes a major collapse and ceases to be percolated, and hence the material becomes a viscoelastic liquid and eventually a steady state is reached.

A sizable stress overshoot followed by a slight undershoot is observed in this curve, because in this case the imposed shear rate is small enough to cause stresses just above the yield stress, so that the microstructure does not respond instantaneously to the stress changes (i.e., thixotropic behavior).

For larger imposed shear rate values, most of the features of the stress evolution described here for $\dot{\gamma} = 0.01 \, \text{s}^{-1}$ are expected to be preserved, but occur too fast to be observable in this test.

The influence of the solvent type on the transient viscosity is quite mild. It is observed that for the lower shear rates, slightly longer t_{ss} values are required for Carbopol dispersions in water/glycerol. This suggests that the breakup process is hindered by the higher viscosity of the water/glycerol solution.

3.3. Flow Curves

The effect of the type of solvent and Carbopol concentration on the flow curve is illustrated in Figure 7, where the (steady-state) shear stress is given as a function of the shear rate for the 0.1 wt%, 0.123 wt% and the 0.125 wt% Carbopol aqueous dispersions, and also for the 0.125 wt% and 0.15 wt% Carbopol dispersions in the water/glycerol solution.

All flow curves are obtained by decreasing the shear rate from 100 to $10^{-3} \, 1/\text{s}$. This procedure reduces significantly the time duration of the tests, as compared to imposing ascending shear rate values. It is worth emphasizing the importance of determining a large enough time $t \geq t_{ss}$ at each applied shear rate, to allow steady state to be achieved before moving to the next shear rate value. As mentioned above, the steady-state time can be estimated as $t_{ss} \approx 1/\dot{\gamma}$. In our experiments, we impose each shear rate value for a maximum period of 1000 s. At each 30 s, an averaged torque is recorded. After three consecutive 30-s periods the corresponding values of the averaged torque are compared, and steady state is assumed when they are the same within 0.1%.

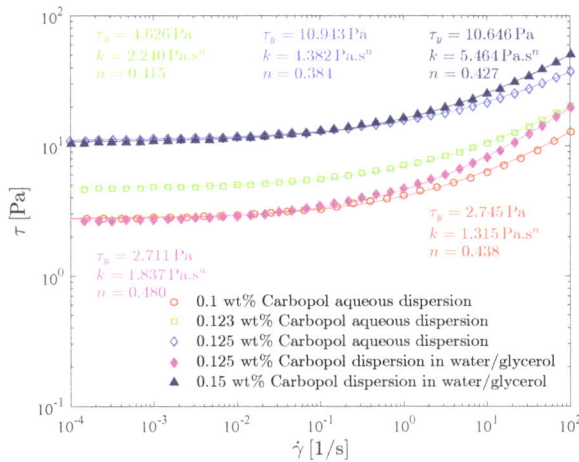

Figure 7. Solvent and polymer concentration influence on flow curve tests: shear stress as a function of shear rate.

The range of shear rate is limited from below by the rheometer torque resolution and from above by sample inertia effects. In addition, the use of cross-hatched parallel plates should be limited to moderate shear stresses, to avoid the occurrence of flow within the protuberances. In this connection, a smooth Couette geometry was used in the range between 3000 and $100\,\mathrm{s}^{-1}$.

All the flow curves given in Figure 7 possess the usual Herschel-Bulkley shape. In addition, all of them are monotonic, indicating that shear banding does not occur [68,69].

The Herschel-Bulkley parameters for all curves are also given in Figure 7. As expected, it is seen that, for a given continuous phase, increasing the Carbopol concentration causes both the yield stress and the consistency index to increase, and the power law index to slightly decrease [70]. Moreover, the values found for the power-law index ranged from 0.38 to 0.48, in good agreement with the observations of [17].

It can be seen in Figure 7 that the flow curve for the 0.125 wt% aqueous dispersion lies well above the one for the 0.125 wt% dispersion in water/glycerol, demonstrating that at the same, Carbopol concentration, the one with a more viscous solvent, possesses a significantly lower viscosity level. In addition, the dispersion in water/glycerol possesses a higher power-law index than the aqueous dispersion, and hence the former is less pseudoplastic than the latter.

Comparing now the flow curves of two dispersions with the same yield stress (0.1 wt% aqueous dispersion and 0.125 wt% dispersion in water/glycerol; and 0.125 wt% aqueous dispersion and 0.15 wt% dispersion in water/glycerol), the dispersions in water/glycerol possess a viscosity level higher than the one of the corresponding aqueous dispersions, as it can be directly observed in Figure 7.

3.4. Creep Tests

The creep test consists of imposing a constant shear stress to an initially unstrained and fully structured sample, and recording the time evolution of the shear rate (or viscosity). It is commonly employed to assess viscoplastic and thixotropic effects of complex fluids, and especially to measure the yield stress [9,71–73]. A number of creep tests are required in order to identify the yield stress: if the imposed stress is lower than the yield stress, then the shear rate approaches zero as a maximum elastic strain is approached, as expected for solids of amorphous microstructure [60,61,74]; and if the imposed stress is above the yield stress then the shear rate eventually attains a constant (steady-state) value [75].

The yield stress obtained from creep tests is the so-called static yield stress, $\tau_{y,s}$, i.e., it is the stress needed to cause irreversible flow on a sample initially unstrained and fully structured [9,71,72]. In contrast, the shear stress obtained from curve fitting to data of a flow curve test is the so-called dynamic yield stress, $\tau_{y,d}$. In general, these two yield stresses are distinct from each other, the dynamic yield stress being smaller than static yield stress [71].

Figure 8 presents creep test results for the 0.1 wt% Carbopol aqueous dispersion and for the 0.125 wt% Carbopol dispersion in water/glycerol. In Figure 8a the shear rate is plotted as a function of time, while in Figure 8b the same data are plotted as a function of the shear strain, $\gamma = \int_0^t \dot{\gamma} dt'$. For all curves shown in Figure 8, the data obtained below one second were discarded due to contamination with the instrument inertia [76,77]. Therefore, for each curve in Figure 8b a different initial strain is observed.

Figure 8 illustrates that the two dispersions possess similar static yield stresses, namely between 4 and 5 Pa for the dispersion in water/glycerol and between 3.5 and 4 Pa for the aqueous dispersion. It is noted that the curve for 4 Pa pertaining to the aqueous dispersion possesses a minimum at about 1400 s (or at a Hencky strain of about 2.5). Therefore, at 4 Pa the sample deformed elastically up to about 1400 s and then the microstructure collapsed, leading to the onset of irreversible flow. This long time delay before the occurrence of yielding is usually referred to as the avalanche effect [78], and is another manifestation of the thixotropic behavior of Carbopol dispersions when the imposed stress is very close but above the yield stress. It is interesting to comment that had we conducted the creep tests for 1000 s only, we would have concluded that the static yield stress of the aqueous dispersion is above 4 Pa, because at this point there was no sign of irreversible flow. On the other hand, if the tests had lasted for two hours, for example, it is possible that the aqueous dispersion would flow at stresses below 3.5 Pa. Therefore, the static yield stress obtained with this test is a function of the time duration of the test. In fact, this is also true for any other method for measuring the yield stress, as extensively discussed in the literature (e.g., [62]).

Ref. [36] observed a similar behavior for a dispersion of Carbopol ETD 2050, and interpreted it as an avalanche-like behavior involving wall slip and transient shear banding. They assumed the Carbopol dispersion as a "simple yield stress fluid", i.e., a non-thixotropic viscoplastic material. A roughened Couette cell was employed and shear-rate controlled experiments were performed. The same behavior was later reported by [79] for shear-stress controlled experiments. On the other hand, the transient shear banding phenomenon was not observed by [80] under similar conditions for a yield stress fluid. Ref. [81] affirm that steady state is very difficult to be achieved under the influence of wall slip. Likewise, they observed that the solid-liquid transition is not reversible upon increasing or decreasing of the applied stresses, due to elastic and thixotropic effects. According to [82], imposing a constant shear stress, the glass fluidization near yielding is very slow, and persistent spatial heterogeneities remain present. Ref. [83] obtained a monotonic flow curve for a material that presented a time delay before yielding [78], and concluded that thixotropy cannot be ignored. More recently, Ref. [84] emphasized that the preparation procedure of Carbopol dispersions is of paramount importance as far as the final mechanical behavior is concerned. Low stirring times lead to large (non-colloidal) microgels and a non-thixotropic behavior. Long stirring times, on the other hand, lead to colloidal microgels and a time-dependent behavior. All this discussion attests that the physical origin of the time delay before yielding is not fully understood.

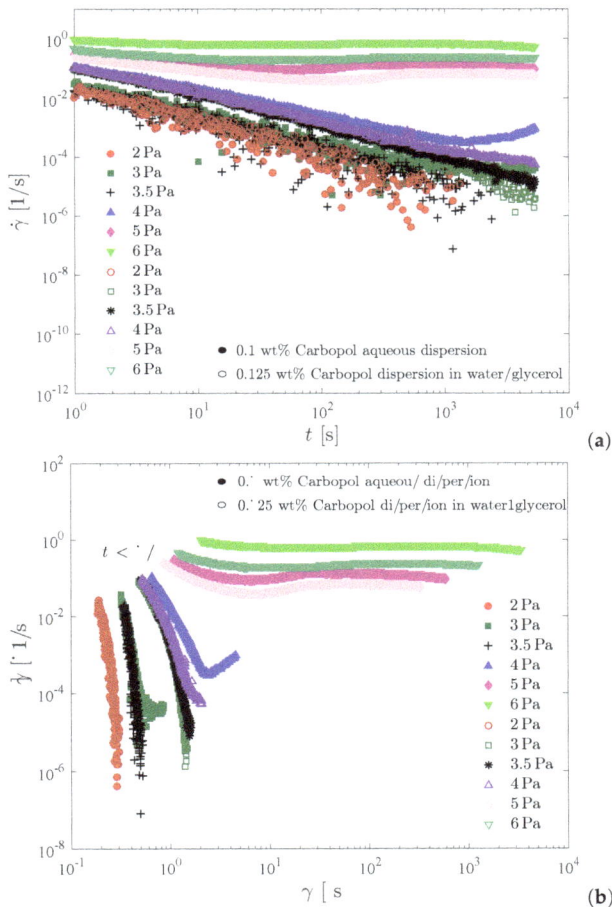

Figure 8. The creep tests. (a) Shear rate *vs.* time; (b) Shear rate *vs.* strain.

3.5. Constant Shear Rate vs. Constant Shear Stress Tests

Figure 9 presents a comparison between results obtained by imposing a constant shear rate, as in Figure 6, and a constant shear stress, as in Figure 8. For each dispersion, the values of imposed stress and shear rate were selected such that the steady-state viscosity was roughly the same both for imposed stress and rate.

For both dispersions considered in Figure 9, a viscosity overshoot is observed when a constant shear stress just above the yield stress is applied. On the other hand, imposing a constant shear rate leads to a much milder and difficult to observe viscosity overshoot, and to a faster achievement of steady state.

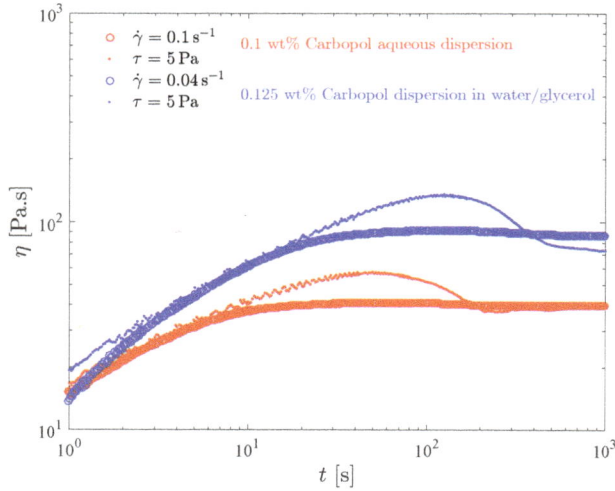

Figure 9. The transient viscosity obtained in constant-stress and constant-rate tests.

This behavior can be explained by observing that the viscosity overshoot that occurs with the application of a constant stress just above the yield stress is due to thixotropy, i.e., due to the time lag that is observed between the application of the stress and the microstructure collapse (yielding). Moreover, the viscosity overshoot is milder when the shear rate is imposed because in this case the transient shear stress is larger, which reduces or eliminates the thixotropic effect.

Creep-Recovery Tests

The creep-recovery test consists of imposing a stress above the yield stress until steady state is achieved, and then imposing a step change to a stress below the yield stress. The shear strain is recorded, and plotted as a function of time. This test is useful to assess the elasticity of the yielded material and also to provide the time required for the microstructure to rebuild. During the first step of the test, namely when a stress above the yield stress is imposed, the structuring level is brought down to a certain extent, while during the second step when the stress is below the yield stress the microstructure rebuilds.

Figure 10 presents the time evolution of the shear strain for the 0.1 wt% Carbopol aqueous dispersion and the 0.125 wt% Carbopol dispersion in water/glycerol. Upon inspection of these figures, it becomes clear that the qualitative behavior of both solutions is the same. The results indicate that the dispersions possess no elasticity whatsoever at the structuring levels that correspond to 5 and 3.5 Pa respectively, as attested by the absence of recoil right after the stress reduction. Another result is that the microstructure of both dispersions rebuild instantaneously, since no flow is observed after the stress reduction.

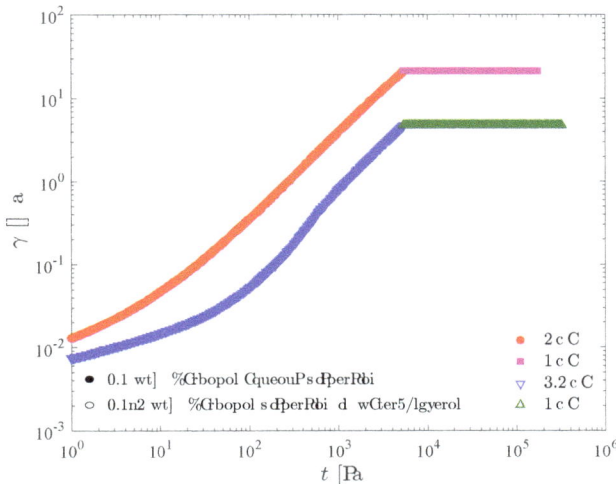

Figure 10. The strain evolution in the creep-recovery test for two Carbopol dispersions.

4. Final Remarks

The rheological properties of Carbopol dispersions in water and in water/glycerol solutions were investigated for concentrations ranging from 0.1 to 0.15 wt%. The preparation procedure and rheological experiments were discussed in detail.

For both formulations, an elasto-viscoplastic behavior was observed as well as a remarkable material stability over time (no aging). The elastic effects are dominant when the microstructure is fully structured, while viscous effects dominate after yielding. At stresses just above the yield stress, thixotropic effects are observed, in contrast to the observed absence of thixotropy at larger stresses. No elasticity was observed in the dispersions while unstructured, and their microstructure rebuilds instantaneously after reduction of the imposed stress to a value below the yield stress.

Comparing aqueous dispersions with dispersions in water/glycerol for the same Carbopol concentration, it is shown that the dispersion in water/glycerol possesses lower yield stress and moduli, while the power-law index is larger (less shear thinning) and the time required for steady state is longer.

Author Contributions: Conceptualization, P.R.V., M.F.N. and P.R.S.M.; Methodology, P.R.V., M.F.N. and P.R.S.M.; Validation, P.R.V., C.M.C. and B.S.F.; Formal Analysis, P.R.V., M.F.N. and P.R.S.M.; Investigation, P.R.V., C.M.C., B.S.F., M.F.N. and P.R.S.M.; Resources, P.R.S.M. and M.F.N.; Data Curation, P.R.V. and C.M.C.; Writing-Original Draft Preparation, P.R.V. and M.F.N.; Writing-Review Editing, P.R.S.M.; Supervision, P.R.S.M.; Project Administration, P.R.S.M.; Funding Acquisition, P.R.S.M. and M.F.N.

Funding: This research received no external funding.

Acknowledgments: The authors are indebted to Petrobras, Equinor, CNPq, CAPES, FAPERJ, FINEP, and MCT for the financial support to the Group of Rheology at PUC-Rio.

Conflicts of Interest: The authors declare no conflict of interest.

References

1. Barnes, H.A.; Walters, K. The yield stress myth? *Rheol. Acta* **1985**, *24*, 323–326. [CrossRef]
2. Barnes, H.A. The yield stress—A review or 'panta rei'—Everything flows? *J. Non-Newton. Fluid Mech.* **1999**, *81*, 133–178. [CrossRef]
3. Astarita, G. The Engineering Reality of The Yield Stress. *J. Rheol.* **1990**, *34*, 275–277. [CrossRef]
4. Hartnett, J.P.; Hu, R.Y.Z. The yield stress—An engineering reality. *J. Rheol.* **1989**, *33*, 671–679. [CrossRef]
5. Evans, I.D. On the nature of the yield stress. *J. Rheol.* **1992**, *36*, 1313–1316. [CrossRef]
6. Nguyen, Q.D.; Boger, D.V. Measuring the flow properties of yield stress fluids. *Annu. Rev. Fluid Mech.* **1992**, *24*, 47–88. [CrossRef]
7. Møller, P.C.F.; Mewis, J.; Bonn, D. Yield stress and thixotropy: On the difficulty of measuring yield stresses in practice. *Soft Matter* **2006**, *2*, 274–283. [CrossRef]
8. Moller, P.; Fall, A.; Bonn, D. Origin of apparent viscosity in yield stress fluids below yielding. *EPL (Europhys. Lett.)* **2009**, *87*, 38004. [CrossRef]
9. Balmforth, N.J.; Frigaard, I.A.; Ovarlez, G. Yielding to Stress: Recent Developments in Viscoplastic Fluid Mechanics. *Annu. Rev. Fluid Mech.* **2014**, *46*, 121–146. [CrossRef]
10. Carnali, J.O.; Naser, M.S. The use of dilute solution viscometry to characterize the network properties of carbopol microgels. *Colloid Polym. Sci.* **1992**, *270*, 183–193. [CrossRef]
11. Magnin, A.; Piau, J. Cone-and-plate rheometry of yield stress fluids. Study of an aqueous gel. *J. Non-Newton. Fluid Mech.* **1990**, *36*, 85–108. [CrossRef]
12. Alba, K.; Taghavi, S.M.; Bruyn, J.R.; Frigaard, I. Incomplete fluid–fluid displacement of yield-stress fluids. Part 2: Highly inclined pipes. *J. Non-Newton. Fluid Mech.* **2013**, *201*, 80–93. [CrossRef]
13. Coussot, P. Yield stress fluid flows: A review of experimental data. *J. Non-Newton. Fluid Mech.* **2014**, *211*, 31–49. [CrossRef]
14. Jørgensen, L.; Le Merrer, M.; Delanoë-Ayari, H.; Barentin, C. Yield stress and elasticity influence on surface tension measurements. *Soft Matter* **2015**, *11*, 5111–5121. [CrossRef] [PubMed]
15. Roberts, G.P.; Barnes, H.A. New measurements of the flow-curves for Carbopol dispersions without slip artefacts. *Rheol. Acta* **2001**, *40*, 499–503. [CrossRef]
16. Piau, J.M. Carbopolgels: Elastoviscoplastic and slippery glasses made of individual swollen sponges. Meso- and macroscopic properties, constitutive equations and scaling laws. *J. Non-Newton. Fluid Mech.* **2007**, *144*, 1–29. [CrossRef]
17. Gutowski, I.A.; Lee, D.; de Bruyn, J.R. Scaling and mesostructure of Carbopol dispersions. *Rheol. Acta* **2012**, *51*, 441–450. [CrossRef]
18. Ahmed, E.M. Hydrogel: Preparation, characterization, and applications: A review. *J. Adv. Res.* **2015**, *6*, 105–121. [CrossRef]
19. Taylor, N.W.; Bagley, E.B. Dispersions or Solutions? A Mechanism for Certain Thickening Agents. *J. Appl. Polym. Sci.* **1974**, *18*, 2747–2761. [CrossRef]
20. Putz, A.; Burghelea, T. The solid-fluid transition in a yield stress shear thinning physical gel. *Rheol. Acta* **2009**, *48*, 673–689. [CrossRef]
21. Brodnyan, J.G.; Kelley, E.L. The Rheology of Polyelectrolytes. I. Flow Curves of Concentrated Poly(acrylic Acid) Solutions. *Trans. Soc. Rheol.* **1961**, *5*, 205–220. [CrossRef]
22. Fischer, W.H.; Bauer, W.H.; Wiberley, S.E. Yield Stresses and Flow Properties of Carboxypolymethylene Water Systems. *Trans. Soc. Rheol.* **1961**, *5*, 221–235. [CrossRef]
23. Kim, J.Y.; Song, J.Y.; Lee, E.J.; Park, S.K. Rheological properties and microstructures of Carbopol gel network system. *Colloid Polym. Sci.* **2003**, *281*, 614–623. [CrossRef]
24. Bonn, D.; Denn, M.M.; Berthier, L.; Divoux, T.; Manneville, S. Yield stress materials in soft condensed matter. *Rev. Mod. Phys.* **2017**, *89*, 035005. [CrossRef]
25. Larson, R.G. *The Structure and Rheology of Complex Fluids*; Oxford University Press Inc.: New York, NY, USA, 1999.

26. Oppong, F.K.; Rubatat, L.; Frisken, B.J.; Bailey, A.E.; de Bruyn, J.R. Microrheology and structure of a yield-stress polymer gel. *Phys. Rev. E* **2006**, *73*, 041405. [CrossRef] [PubMed]

27. Yoshimura, A.S.; Prud'homme, R.K. Response of an elastic Bingham fluid to oscillatory shear. *Rheol. Acta* **1987**, *26*, 428–436. [CrossRef]

28. Al-Hadithi, T.S.R.; Barnes, H.A.; Walters, K. The relationship between the linear (oscillatory) and nonlinear (steady-state) flow properties of a series of polymer and colloidal systems. *Colloid Polym. Sci.* **1992**, *270*, 40–46. [CrossRef]

29. Barry, B.; Meyer, M. The rheological properties of Carbopol gels I. Continuous shear and creep properties of Carbopol gels. *Int. J. Pharm.* **1979**, *2*, 1–25. [CrossRef]

30. Islam, M.T.; Rodríguez-Hornedo, N.; Ciotti, S.; Ackermann, C. Rheological Characterization of Topical Carbomer Gels Neutralized to Different pH. *Pharm. Res.* **2004**, *21*, 1192–1199. [CrossRef]

31. Di Giuseppe, E.; Corbi, F.; Funiciello, F.; Massmeyer, A.; Santimano, T.N.; Rosenau, M.; Davaille, A. Characterization of Carbopol hydrogel rheology for experimental tectonics and geodynamics. *Tectonophysics* **2015**, *642*, 29–45. [CrossRef]

32. Moller, P.; Fall, A.; Chikkadi, V.; Derks, D.; Bonn, D. An attempt to categorize yield stress fluid behaviour. *Philos. Trans. R. Soc. Lond. A Math. Phys. Eng. Sci.* **2009**, *367*, 5139–5155. [CrossRef] [PubMed]

33. Coussot, P.; Tocquer, L.; Lanos, C.; Ovarlez, G. Macroscopic vs. local rheology of yield stress fluids. *J. Non-Newton. Fluid Mech.* **2009**, *158*, 85–90. [CrossRef]

34. Herschel, V.W.H.; Bulkley, R. Konsistenzmessungen von Gummi-Benzollösungen. *Colloid Polym. Sci.* **1926**, *39*, 291–300. [CrossRef]

35. Tabuteau, H.; Coussot, P.; de Bruyn, J.R. Drag force on a sphere in steady motion through a yieldstress fluid. *J. Rheol.* **2007**, *51*, 125–137. [CrossRef]

36. Divoux, T.; Tamarii, D.; Barentin, C.; Manneville, S. Transient Shear Banding in a Simple Yield Stress Fluid. *Phys. Rev. Lett.* **2010**, *104*, 208301. [CrossRef] [PubMed]

37. Weber, E.; Moyers-González, M.; Burghelea, T.I. Thermorheological properties of a Carbopol gel under shear. *J. Non-Newton. Fluid Mech.* **2012**, *183–184*, 14–24. [CrossRef]

38. Ketz, R.J.; Prud'homme, R.K.; Graessley, W.W. Rheology of concentrated microgel solutions. *Rheol. Acta* **1988**, *27*, 531–539. [CrossRef]

39. Nelson, A.Z.; Ewoldt, R.H. Design of yield-stress fluids: A rheology-to-structure inverse problem. *Soft Matter* **2017**, *13*, 7578–7594. [CrossRef]

40. Nelson, A.Z.; Bras, R.E.; Liu, J.; Ewoldt, R.H. Extending yield-stress fluid paradigms. *J. Rheol.* **2018**, *62*, 357–369. [CrossRef]

41. Chu, J.S.; Yu, D.M.; Amidon, G.L.; Weiner, N.D.; Goldberg, A.H. Viscoelastic Properties of Polyacrylic Acid Gels in Mixed Solvents. *Pharm. Res.* **1992**, *9*, 1659–1663. [CrossRef]

42. Proniuk, S.; Blanchard, J. Anhydrous Carbopol polymer gels for the topical delivery of oxygen/water sensitive compounds. *Pharm. Dev Technol.* **2002**, *7*, 249–255. [CrossRef] [PubMed]

43. Corporation, T.L. *Neutralizing Carbopol and Pemulen Polymers in Aqueous and Hydroalcoholic Systems*; Technical Report TDS-237; The Lubrizol Corporation: Wickliffe, OH, USA, 2009.

44. Bonacucina, G.; Cespi, M.; Misici-Falzi, M.; Palmieri, G.F. Rheological evaluation of silicon/carbopol hydrophilic gel systems as a vehicle for delivery of water insoluble drugs. *AAPS J.* **2008**, *10*, 84–91. [CrossRef] [PubMed]

45. Noveon, I. *Formulating Topical Properties*; Technical Report Bulletin 14; Noveon Inc.: Cleveland, OH, USA, 2002.

46. Corporation, T.L. *Molecular Weight of Carbopol and Pemulen Polymers*; Technical Report TDS-222; The Lubrizol Corporation: Wickliffe, OH, USA, 2007.

47. Corporation, T.L. *Dispersion Techniques for Carbopol Polymers*; Technical Report TDS-103; The Lubrizol Corporation: Wickliffe, OH, USA, 2007.

48. Corporation, T.L. *Formulating Hydroalcoholic Gels with Carbopol Polymers*; Technical Report TDS-255; The Lubrizol Corporation: Wickliffe, OH, USA, 2009.

49. Sikorski, D.; Tabuteau, H.; de Bruyn, J. Motion and shape of bubbles rising through a yield-stress fluid. *J. Non-Newton. Fluid Mech.* **2009**, *159*, 10–16. [CrossRef]

50. Krieger, I.M. Bingham Award Lecture—1989: The role of instrument inertia in controlled-stress rheometers. *J. Rheol.* **1990**, *34*, 471–483. [CrossRef]
51. Baravian, C.; Quemada, D. Using instrumental inertia in controlled stress rheometry. *Rheol. Acta* **1998**, *37*, 223–233. [CrossRef]
52. Baravian, C.; Benbelkacem, G.; Caton, F. Unsteady rheometry: Can we characterize weak gels with a controlled stress rheometer? *Rheol. Acta* **2007**, *46*, 577–581. [CrossRef]
53. Läuger, J.; Stettin, H. Effects of instrument and fluid inertia in oscillatory shear in rotational rheometers. *J. Rheol.* **2016**, *60*, 393–406. [CrossRef]
54. Barnes, H.A. *A Handbook of Elementary Rheology*; University of Wales, Institute of Non-Newtonian Fluid Mechanics: Wales, UK, 2000.
55. Buscall, R.; McGowan, J.I.; Morton-Jones, A.J. The rheology of concentrated dispersions of weakly attracting colloidal particles with and without wall slip. *J. Rheol.* **1993**, *37*, 621–641. [CrossRef]
56. Barnes, H.A. A review of the slip (wall depletion) of polymer solutions, emulsions and particle suspensions in viscometers: Its cause, character, and cure. *J. Non-Newton. Fluid Mech.* **1995**, *56*, 221–251. [CrossRef]
57. Barnes, H.A. Measuring the viscosity of large-particle (and flocculated) suspensions—A note on the necessary gap size of rotational viscometers. *J. Non-Newton. Fluid Mech.* **2000**, *94*, 213–217. [CrossRef]
58. Rabinowitsch, B. Über die Viskosität und Elastizität von Solen. *Z. Phys. Chem.* **1929**, *A145*, 1–26. [CrossRef]
59. De Souza Mendes, P.R.; Alicke, A.A.; Thompson, R.L. Parallel-plate geometry correction for transient rheometric experiments. *Appl. Rheol.* **2014**, *24*, 52721.
60. Lidon, P.; Villa, L.; Manneville, S. Power-law creep and residual stresses in a carbopol gel. *Rheol. Acta* **2017**, *56*, 307–323. [CrossRef]
61. Dimitriou, C.J.; Ewoldt, R.H.; McKinley, G.H. Describing and prescribing the constitutive response of yield stress fluids using large amplitude oscillatory shear stress (LAOStress). *J. Rheol.* **2013**, *57*, 27–70. [CrossRef]
62. De Souza Mendes, P.R.; Thompson, R.L. A unified approach to model elasto-viscoplastic thixotropic yield-stress materials and apparent-yield-stress fluids. *Rheol. Acta* **2013**, *52*, 673–694. [CrossRef]
63. De Souza Mendes, P.R.; Thompson, R.L.; Alicke, A.A.; Leite, R.T. The quasilinear large-amplitude viscoelastic regime and its significance in the rheological characterization of soft matter. *J. Rheol.* **2014**, *58*, 537–561. [CrossRef]
64. Raghavan, S.R.; Chen, L.A.; McDowell, C.; Khan, S.A.; Hwang, R.; White, S. Rheological study of crosslinking and gelation in chlorobutyl elastomer systems. *Polymer* **1996**, *37*, 5869–5875. [CrossRef]
65. Kocen, R.; Gasik, M.; Novak, S. Viscoelastic behaviour of hydrogel-based composites for tissue engineering under mechanical load. *Biomed. Mater.* **2017**, *12*, 025004. [CrossRef]
66. De Souza Mendes, P.R. Modeling the thixotropic behavior of structured fluids. *J. Non-Newton. Fluid Mech.* **2009**, *164*, 66–75. [CrossRef]
67. de Souza Mendes, P.R. Thixotropic elasto-viscoplastic model for structured fluids. *Soft Matter* **2011**, *7*, 2471–2483. [CrossRef]
68. Quemada, D.; Berli, C. Describing the Flow Curve of Shear-Banding Fluids Through a Structural Minimal Model. *arXiv* **2009**, arXiv:0903.0808.
69. Jain, A.; Singh, R.; Kushwaha, L.; Shankar, V.; Joshi, Y.M. Transient start-up dynamics and shear banding in aging soft glassy materials: Rate-controlled flow field. *arXiv* **2018**, arXiv:1801.07088.
70. Hassan, M.A.; Pathak, M.; Khan, M.K. Thermorheological Characterization of Elastoviscoplastic Carbopol Ultrez 20 Gel. *J. Eng. Mater. Technol.* **2015**, *137*, 031002. [CrossRef]
71. Cheng, D.C.H. Yield stress: A time-dependent property and how to measure it. *Rheol. Acta* **1985**, *25*, 542–554. [CrossRef]
72. Mujumdar, A.; Beris, A.N.; Metzner, A.B. Transient phenomena in thixotropic systems. *J. Non-Newton. Fluid Mech.* **2002**, *102*, 157–178. [CrossRef]
73. de Souza Mendes, P.R.; Thompson, R.L. A critical overview of elasto-viscoplastic thixotropic modeling. *J. Non-Newton. Fluid Mech.* **2012**, *187-188*, 8–15. [CrossRef]
74. Da C. Andrade, E.N. On the viscous flow in metals, and allied phenomena. *Proc. R. Soc. Lond. A Math. Phys. Eng. Sci.* **1910**, *84*, 1–12. [CrossRef]

75. Coussot, P.; Nguyen, Q.D.; Huynh, H.T.; Bonn, D. Viscosity bifurcation in thixotropic, yielding fluids. *J. Rheol.* **2002**, *46*, 573–589. [CrossRef]

76. Ferry, J.D. *Viscoelastic Properties of Polymers*; Wiley: New York, NY, USA, 1980.

77. Ewoldt, R.H.; McKinley, G.H. Creep ringing in rheometry or how to deal with oft-discarded data in step stress tests! *Rheol. Bull.* **2007**, *76*, 22–24.

78. Coussot, P.; Nguyen, Q.; Huynh, H.; Bonn, D. Avalanche Behavior in Yield Stress Fluids. *Phys. Rev. Lett.* **2002**, *88*, 175501. [CrossRef]

79. Divoux, T.; Barentin, C.; Manneville, S. From stress-induced fluidization processes to Herschel-Bulkley behaviour in simple yield stress fluids. *Soft Matter* **2011**, *7*, 8409–8418. [CrossRef]

80. Vasu, K.S.; Krishnaswamy, R.; Sampath, S.; Sood, A.K. Yield stress, thixotropy and shear banding in a dilute aqueous suspension of few layer graphene oxide platelets. *Soft Matter* **2013**, *9*, 5874–5882. [CrossRef]

81. Poumaere, A.; Moyers-González, M.; Castelain, C.; Burghelea, T. Unsteady laminar flows of a Carbopol® gel in the presence of wall slip. *J. Non-Newton. Fluid Mech.* **2014**, *205*, 28–40. [CrossRef]

82. Chaudhuri, P.; Horbach, J. Onset of flow in a confined colloidal glass under an imposed shear stress. *Phys. Rev. E* **2013**, *88*, 040301. [CrossRef] [PubMed]

83. Cheng, P.; Burroughs, M.C.; Leal, L.G.; Helgeson, M.E. Distinguishing shear banding from shear thinning in flows with a shear stress gradient. *Rheol. Acta* **2017**, *56*, 1007–1032. [CrossRef]

84. Dinkgreve, M.; Fazilati, M.; Denn, M.; Bonn, D. Carbopol: From a simple to a thixotropic yield stress fluid. *J. Rheol.* **2018**, *62*, 773–780. [CrossRef]

Review

Rheology of Natural Hydraulic Lime Grouts for Conservation of Stone Masonry—Influence of Compositional and Processing Parameters

Luis G. Baltazar [1,*], Fernando M.A. Henriques [1] and Maria Teresa Cidade [2]

[1] Faculdade de Ciências e Tecnologia, Departamento de Engenharia Civil, Universidade NOVA de Lisboa, 2829-516 Caparica, Portugal; fh@fct.unl.pt

[2] Faculdade de Ciências e Tecnologia, Departamento de Ciência dos Materiais e CENIMAT/I3N, Universidade NOVA de Lisboa, 2829-516 Caparica, Portugal; mtc@fct.unl.pt

* Correspondence: luis.baltazar@fct.unl.pt; Tel.: +351-212-948-300

Received: 13 December 2018; Accepted: 16 January 2019; Published: 18 January 2019

Abstract: This review provides an overview of the recent progress in the field of the rheology of grouts for historic masonry consolidation. During the last two decades, significant research has been devoted on the grouting technique for stone masonry consolidation but most results are scattered by scientific papers, congress communications, and thesis. This paper compiles and briefly demonstrates the effect of several intrinsic and extrinsic parameters, such as admixtures, additions, pressure, temperature, and measuring instrumentation, on the rheological performance of natural hydraulic lime-based grouts.

Keywords: grout; rheology; natural hydraulic lime; masonry; consolidation

---

1. Introduction

Grouting is generally used as a means of changing or improving the masonry's load bearing capacity to vertical and horizontal actions. A commonly used method for grouting is the grout injection, whereby the grout material, i.e., the suspension of binder particles in an aqueous medium, is forced by pressure into voids and fractures of stone masonry wall (Figure 1). It has been largely used in Europe, since the beginning of the twentieth century for consolidation of stone masonry walls, particularly in seismic areas [1–6].

Grouting is an invisible retrofitting technique, which is an advantage when working on historic buildings. On the other hand, it is an irreversible technique, which means that a badly designed grout can lead to regrettable consolidation failures [7,8]. Cement based grouts are the most widely used reinforcing material in concrete structures. However, it is not suitable for consolidation works of stone masonry buildings and the main reasons are the mechanical strength and rigidity (higher modulus of elasticity) of cement, which results in poor compatibility with the masonry characteristics in mechanical, physical, and chemical aspects. In this sense, the natural hydraulic lime (NHL) is today the most commonly used material for grouting operations in historic masonry buildings because of its moderate mechanical strength, water vapor permeability and chemical compatibility with traditional materials found in old masonry walls [7,9].

Depending on the prevailing masonry conditions, grouts with different characteristics must be designed [10–12]. Although, most grout' characteristics like penetrability, rheology, mechanical strength, stability, and durability are characteristics that need measurement techniques and standards, many of which have yet to be undertaken. Nevertheless, when it comes to penetrability and rheology of natural hydraulic lime-based grouts, several research works have been done in the last decades [13–19]. Among the various grouts characteristics, fluidity appears as one of the most determinant characteristics in the grouting performance, since it is important to ensure that the grout can flow and fill well the cracks and voids in the masonry core.

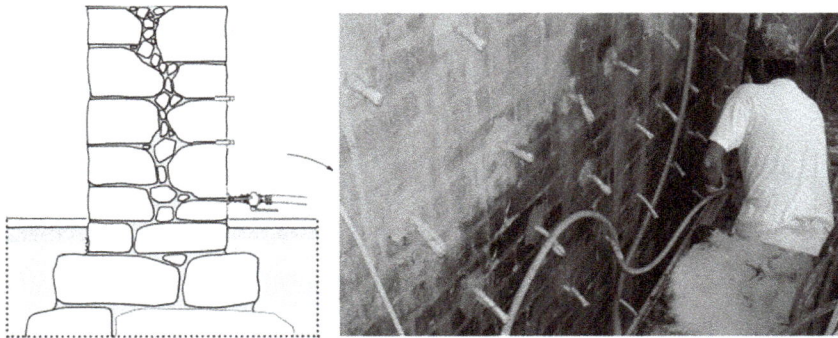

Figure 1. Detail of the grout injection on a stone masonry wall.

NHL-based grout is a suspension of particles in water and the NHL is the element that gives the grout its binder character. However, different elements, currently used by the cement and concrete industries, can be added to the grout, to either optimize its performance or lower the cost. In fact, it is the suspended particles that have an important contribution on the rheological behavior of the grout [20,21]. The particles will influence the flow properties and set a limit to void size that can be penetrated. Notwithstanding, apart from the penetration issues due to particle size, understanding and controlling the rheology of the grout is essential for a successful grouting intervention. The knowledge will, under the prevailing conditions, facilitate the choice of additive, admixtures, grouting pressure, and temperature. The present paper reviews the current knowledge concerning the measurements of the rheological properties of NHL-based grouts, with emphasis on a contribution of different factors, additives, and admixtures on the rheology of injection grouts.

2. Rheology of Natural Hydraulic Lime Grouts

Rheology plays an important role in injection grouts since it provides valuable data about the influence of composition, temperature, pressure, resting time, among others, on flowability of grouts. The rheological behavior of NHL-based grouts is a difficult task since it can be considered a complex system due to simultaneous interactions between the two phases and also between the particles themselves. Moreover, the fresh behavior is also influenced by the hydration of the NHL; despite considering that the hydration of lime during the dormant period is practically stationary, it will inevitably lead to changes in the rheological properties with time.

2.1. Yield Stress

The interactions between the particles in a suspension (like a grout) result in a yield stress value. The presence of yield stress means that under static conditions, the grout essentially acts as a solid and will continue acting as a solid until the stress reaches the shear force needed to overcome the internal bonding between the particles. This yield stress (also called static yield stress) can thus be regarded as the property that represents the transition between solid and liquid behavior [22,23]. This behavior can also be characterized by the flow curves, resulting from the relation between shear stress (τ) and shear rate ($\dot{\gamma}$) under simple steady shear; so the yield stress is equal to the intersection point on the stress axis, as shown in Figure 2.

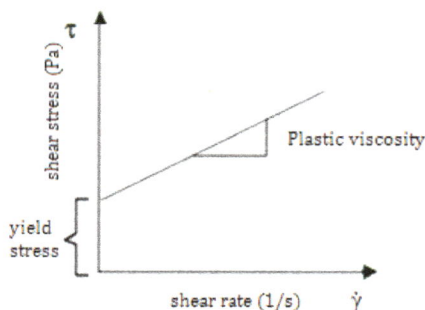

Figure 2. Example of a flow curve of a fluid with yield stress.

In addition to a yield stress, cementitious grout compositions display another peculiarity, which is the change of its properties during time, namely due to hydration. Therefore, from a practical point of view, the grout fluidity is not only a function of the instantaneous shear rate but also presents time-dependent behavior. Shaking or shearing the grouts causes a gradual breakdown of its microstructure, which is recovered when the grout is at rest. A reversible and time-dependent microstructure defines thixotropy that will be discussed below. Nevertheless, this behavior causes several challenges in defining the yield stress as a constant property since different measuring protocols, history of shear, and type of geometry lead to different yield stress results [24,25]. Thus, to solve these issues, two yield stress values, namely static and dynamic yield stress, were proposed [26,27]. The static yield stress has already been defined above, while the dynamic yield stress can be seen as the yield stress when the grout is subjected to shear and is in a fully broken down state. Rahman et al. [28] measured the yield stress of cement-based grout considering the effect of thixotropy and hydration. In the same work, it was shown that there exists a critical shear rate range, below which there is a transition from the dynamic to the static yield stress, which should be taken into account for grout design. It should be noted, however, that despite the importance of yield stress in cementitious suspensions design, no standard methods are yet available to determine the yield stress of grouts.

2.2. Thixotropy

Thixotropy is a gradual decrease of the viscosity under constant shear stress followed by a gradual recovery of structure when the stress is removed [29,30]. Thixotropy should not be confused with shear thinning. When a material is shear thinning it changes the microstructure instantly while in a thixotropic material the microstructure does change (by breaking down or building up) and such changes take time [29,30]. It can be said that thixotropy is due to the structure degradation resulting from rupturing flocs or linked particles when the grout is sheared. When the shearing stress is removed, the grout microstructure rebuilds and is eventually restored to its original condition [31,32]. According to Billberg [30] with today's knowledge of microstructural changes it is probably safe to say that shear thinning materials are also thixotropic since it always takes time, even limited, to create the re-grouping of the microstructural elements to result in shear thinning.

A quantitative measurement of thixotropy can be performed in several ways. The most apparent characteristic of a thixotropic system is the hysteresis loop, which is formed by the up-and down-curves of the flow curve [23,33]. If the grout is thixotropic, the resulting two curves (up and down curves) do not coincide, as shown in Figure 3. The degree of thixotropic behavior can be quantified by the area of the hysteresis loop, which indicates a breakdown of structure that does not reform immediately when the stress is removed or reduced.

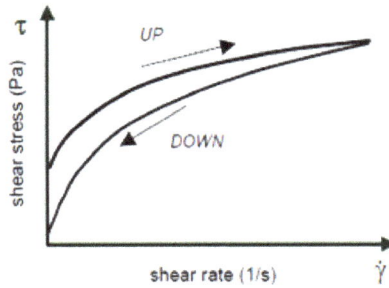

Figure 3. Illustration of thixotropic behavior.

2.3. Rheological Models

The injection grouts are often referred to as non-Newtonian, thixotropic, and in possession of a yield stress [14,34–36]. This non-Newtonian behavior can be attributed to mechanisms in which the shear stress orients the suspended binder particles in opposition to the randomizing effects of Brownian motion [14,37–39]. The rheological behavior of cementitious suspensions is often approximated by a rheological model, incorporating various rheological parameters. The typical model used for NHL grouts have to take into account the yield stress. The yield stress value will limit the penetration distance that the grout will reach at a certain injection pressure. A review of the literature shows that several mathematical models have been proposed for the behavior of the injection grouts. Some of the more widely used models include Bingham, Modified Bingham, Casson, Herschel Bulkley, and Power Law, as shown in Table 1.

Table 1. Rheological models used for describing the flow curve of injection grouts.

Model	Equation	Description	Parameters
Bingham	$\tau = \tau_0 + \eta_p \dot{\gamma}$	Yield and linear	τ_0 = yield stress, η_p = plastic viscosity
Modified Bingham	$\tau = \tau_0 + \eta_p \dot{\gamma} + c\dot{\gamma}^2$	Yield and nonlinear	τ_0 = yield stress, η_p = plastic viscosity, c = constant
Herschel-Bulkley	$\tau = \tau_0 + k\dot{\gamma}^n$	Shear thinning	τ_0 = yield stress, k = consistency, n = power law index
Power law	$\tau = k\dot{\gamma}^n$	Shear thinning	k = consistency; n = power law index
Casson	$\sqrt{\tau} = \sqrt{\tau_0} + \sqrt{\eta \cdot \dot{\gamma}}$	Linear between the square root of shear stress and the square root of the shear rate	τ_0 = yield stress, η = viscosity

The simplest model including a yield stress is the Bingham model which is a two parameter model widely used in the injection grouts. The grout behavior has been modeled with Bingham model by several authors [40–44]. However, for injection grouts that exhibit a pronounced shear thinning behavior, and according to several studies [14,20,45,46], the modified Bingham model or the Herchel–Bulkley leads to better fittings especially at very low shear rates. The Hershel–Bulkley model is an extension of the Bingham model to include shear rate dependence by replacing the plastic viscosity term (in the Bingham model) with the power law expression, where k is the consistency (Pa.s^n) and n is the flow behavior index (dimensionless) [47]. Non-linear relations between shear stress and shear rate can be described by the power law model. However, care should be taken in the use of this model outside the range of the data used to define it. For instance, the power law model fails at high shear rates, where the viscosity approaches a constant value. This weakness of the power law model can be rectified by the use of other models which can fit different parts of the flow curve.

The Casson model is a structure based model traditionally used to describe the flow of viscoelastic fluids [48]. This model has a more gradual transition from Newtonian to the yield region. Taking into account the testimonies of various authors [14,18,49,50] the geometry and morphology of the flow channels in the core of the masonry that is to be consolidated is very difficult to define, so the

41

development of sophisticated rheological models may not be justified when most of the time the Bingham model is the model used because the cementitious suspensions in general follows this equation fairly well and the two parameters in the Bingham model, yielding stress and viscosity, can be determined. In this sense, it can be concluded that researchers have not agreed upon a rheological model which satisfactorily describes the flow of injection grouts. Many researchers [25,51,52] are in effect using the Bingham model due to its simplicity.

It should be mentioned, however, that there are some practical difficulties in using theses classical rheological models especially in computational simulations of viscoplastic materials, due to their intrinsic singularities [53]. All rheological models presented in Table 1 are discontinuous, which means that for flow field of viscoplastic suspensions it is often required to develop numerical techniques to track down yielded/unyielded regions in flow fields. In this sense, regularized versions of the original rheological models have been proposed and are often used for the simulations of viscoplastic flows, such as the Bingham–Papanastasiou model (equation 1) proposed by Papanastasiou [54].

$$\tau = \eta_p \dot{\gamma} + \tau_0 \left[1 - exp\left(-m.\dot{\gamma} \right) \right], \tag{1}$$

where m is a non-rheological parameter acting on yield stress term.

This regularized model rendered the original discontinuous Bingham viscoplastic model as a purely viscous one by introducing into a continuation parameter, which facilitates the solution process and is valid for all rates of deformation.

3. Rheological Measurements Apparatus

The rheological characterization of injection grouts can be challenging because of the need for suitable devices as well as measurement procedures and data analysis appropriate to each grout composition [27,55]. Below, a summary is given of each general method along with descriptions of common measurement devices and geometries.

3.1. Viscometer

The viscometers measure the grout apparent viscosity as a function of rotation speeds by driving a measurement tool (called spindle), immersed in the test sample (Figure 4a). According to Hackley and Ferraris [56] the viscous drag of the sample against the spindle causes the spring to deflect, and this deflection is correlated with torque. The calculated shear rate depends on the rotation speed, the tool geometry and the size and shape of the sample container. Furthermore, the conversion factors are needed to calculate shear stress vs. shear rate curve and are typically pre-calibrated for specific tool and container geometries [57,58]. Despite some limitations that are often pointed to viscometers (for example related to the range of shear rates, accuracy in measurements and inability to perform some types of measurements) these devices can be used to measure the typical rheological properties of injection grouts. Previous studies [14,59] where the effect of mixing procedures on some fresh properties of hydraulic grout, including the rheological ones, were conducted using a Brookfield viscometer (Figure 4b).

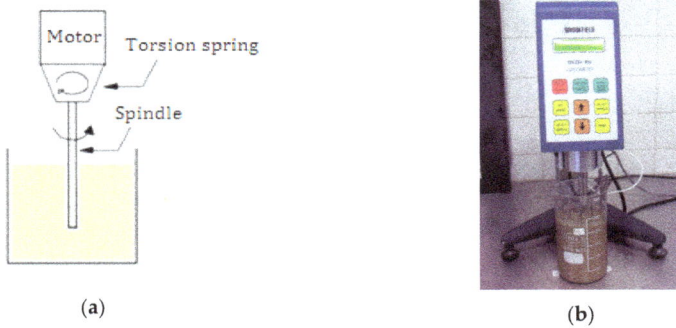

Figure 4. Viscometer: (**a**) Schematic diagram of a Brookfield-type viscometer; (**b**) Brookfield viscometer LV DV-II + PRO with spindle immersed in the sample.

3.2. Rotational Rheometer

These kind of devices (Figure 5a) are of higher precision when compared with viscometers and are some of the most commonly used devices for measuring the rheological properties of cementitious pastes and grouts. The basic rotational system consists of four parts: (i) a measurement tool with a well-defined geometry, (ii) a device to apply a constant torque or rotation speed to the tool over a wide range of shear stress or shear rate values, (iii) a device to determine the stress or shear rate response and (iv) the temperature unit control for the test sample and tool. Depending on the design specifications, rheometers may also include built-in corrections or compensations for inertia and temperature fluctuations during measurements. The measurement of the forces and torques acting on the geometry yields the stresses, and the ratio of shear stress to the shear rate (which is related with the rotation speed) gives the apparent viscosity.

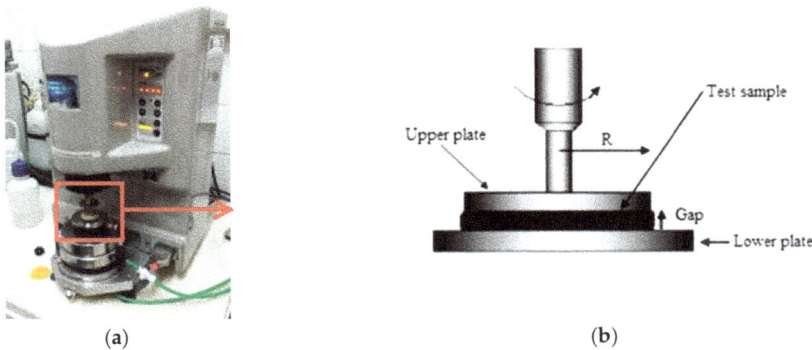

Figure 5. Rheometer: (**a**) Rotational rheometer; (**b**) Illustration of parallel plates geometry.

Most rheometers are based on the relative rotation about a common axis of three alternative tool geometries: concentric cylinder, cone and plate, or parallel plates. In concentric cylinder geometry, either the inner, outer, or both cylinders may rotate, depending on instrument design. The test material is maintained in the annulus between the cylinder surfaces. The large surface area of this geometry improves sensitivity when measuring samples with low viscosity. Furthermore, it allows good thermal control and when this geometry is used in conjunction with solvent traps, the sample evaporation can be minimized.

On the other hand, a concentric cylinder generally requires greater sample volumes than the others geometries. The cone and plate geometry consists of an inverted cone in near contact with a lower plate. The geometry advantages are that only a small sample is required, the shear rate is constant all over the gap, and the shear fracture is minimized because of the small free surface. The parallel plate geometry (see Figure 5b) can be considered a simplified version of the cone and plate, having an angle of 0°. The test sample is constrained in the narrow gap between the two surfaces. An advantage of the parallel-disk geometry is that the gap height can easily be changed, even without reloading the sample. One disadvantage when compared to the cone and plate geometry is the fact that the shear rate varies with the radius of the plate. However, the parallel plate geometry is more suitable than the cone and plate for large particles, since the gap is fixed in the cone and plate and is very small (the order of tens of microns in the top of the truncated cone) [60]. Measuring grouts with large particle sizes can then be problematic because of the limited small gap size of the cone and plate geometry.

3.3. Ultrasound Velocity Profiling Method

The in-line rheological instruments appear as a response to the ever-increasing demand for the development of rheometers capable of dealing with complex fluids. The in-line measurement techniques that combine the ultrasound velocity profiling (UVP) method with the pressure difference (PD) measurements (also known as UVP+PD) belongs to the non-invasive rheological measurement techniques, which has been investigated and applied on the characterization of cement grouts [43,61,62].

The UVP+PD measurement method is based on the emission of pulsed ultrasound bursts and echo reception. This method determines the relative time lags of the echoes received between successive emitted pulses [63,64]. The time lags are related to the speed of the moving fluid. The UVP+PD method contrasts with conventional off-line rheometers, in which a data fitting is required to determine the rheological behavior and the velocity profiles. One of the greatest advantages of the UVP+PD method is the determination of the true rheological properties using the non-model approach [62]. This means that common unreliable determination of rheological properties due to the influence on the microstructure of the measuring geometry can be avoided with this non-invasive measurement method. It was evidenced by Håkansson and Rahman [65] that the UVP+PD technique can also be used for continuous monitoring of grout concentrations, which will confirm the water/cement ratio of the grout and thereby act as a quality control. The potential of UVP+PD as rheometric method for measuring the rheological properties of cement-based grouts has been studied and demonstrated in several works carried out at the Royal Institute of Technology in Sweden [43,62,63,65].

3.4. Marsh Cone and Slump Test

The Marsh cone is a simple instrument (see Figure 6) that initially started to be used in the oil industry to measure the flow performance of drill muds in the field. The instrument is currently used in the construction industry to empirically measure fluidity of consolidation grouts or pastes for preplaced-aggregate concrete as specified in the standard ASTM C939:02 [66]. Based on this standard from the time it takes a certain volume to flow out of the cone, the flow properties can be estimated. In order to improve the physical significance of the Marsh cone test some authors suggest that the grout fluidity (especially in very fluid grouts) should be evaluated using a modified Marsh cone having an outlet diameter of 5 mm [17,50]. However, the only property that can be measured with the Marsh cone is the fluidity, which is an empirical measurement since the physical parameter "viscosity" is not in fact determined. In this sense, mathematical models were proposed by Baltazar & Henriques [67] that allow predicting the rheological properties of the grout just by performing the Marsh cone test on field. The proposed models can be very useful to streamline the grout design methodology since these models are able to calculate a physical parameter (e.g., viscosity) instead of an empirical one (e.g., fluidity).

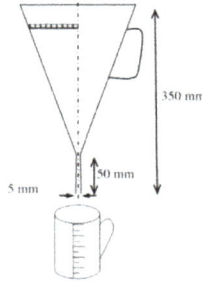

Figure 6. Illustration of the Marsh cone apparatus.

Despite the Marsh cone has been extensively used to make rapid measurements, its accuracy still raises some controversy. However, the growing development of computational fluid dynamics, has prompted several attempts of numerical and analytical simulation of Marsh cone flows [25,42,68,69]. Nowadays, it is already possible to predict the flow of some fluids through the Marsh cone as an alternative to the experiments. Notwithstanding, according to Sadrizadeh et al. [70] limited investigations with numerical and analytical calculations have been done, thus further investigation is needed to develop an easy, fast, and reliable fluid viscosity analysis in terms of numerical and analytical calculations of the Marsh cone.

The slump test is another common method to quantify the fluidity of cementitious mixtures. This measurement technique is used extensively for the evaluation of self compacting concrete due to the low yield stress of such materials. The modus operandi is the following: the cone is laid on a horizontal glass surface. After the careful placing of the grout into the cone, to avoid bubble formation and grout overflow, the cone is vertically lifted (Figure 7a). After lifting the cone, the grout flows by gravity and the slump occurs. The grout will flow while the local stress is higher than the material yield stress. The spread diameter is measured when the flow stops. Each spread diameter value is the mean of two measurements along two perpendicular directions (Figure 7b). The spread at stoppage is directly linked to the yield value, which may even be calculated. In fact, the relation between yield value and slump has been studied by Hu [71] or Christensen [72]. More recently Roussel et al. [73] proposed another approach, the mini-cone test, which is very suitable for yield stress measurements of cement pastes and grouts. From a practical point of view, it is desirable to perform "in site" measurements, whereby the significant rheological properties are measured during the grouting operations. In this sense, the research work should be guided in this direction.

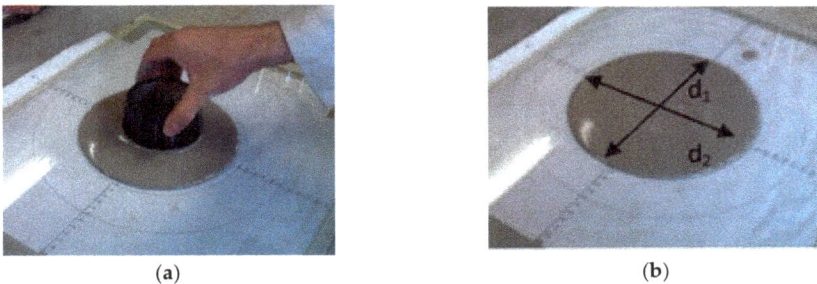

(a)   (b)

Figure 7. Slump test, (**a**) Lift the cone; (**b**) Spread sample at stoppage.

3.5. Important Features Affecting the Rheological Measurements of Grouts

During rheological measurements many disturbing effects may arise. They often reflect the changes that are happening in the microstructure of the sample. For instance, particle suspension sedimentation and migration of particles can significantly alter the stress distribution and thus the measured torque [60]. Other disturbing effects are experimental problems pertaining to the geometry type. For instance, when using a smooth metallic shearing surface, wall slip can occur (the called slippage).

A substantial source of problems may arise due to the presence of a wall, since a wall modifies the particle arrangement. This phenomenon called particle depletion involves a decrease in particle concentration close to the wall, which leads to the development of a lubricated fluid layer close to the solid boundary and to the slipping of the bulk. This is consequence of several factors, such as hydrodynamic, viscoelastic, chemical and gravitacional forces acting on the disperse phase [74]. Depending on the grout properties, there may be interactions between the metallic surface of the geometry and the constituents of the suspension. This is particularly problematic for particle suspensions like injection grouts; the presence of a solid boundary may alter the local structure within the liquid, thus producing local viscosity, and may lead to an under-evaluation of apparent viscosity and an improper evaluation of yield stress, as demonstrated in [75].

Other important factors that influence the rheological measurement is the lack of reaching a state of equilibrium of shear stress at each shear rate (steady state). During measurements this is one of the main problems in obtaining precise data on grout parameters like the yield stress and plastic viscosity. This can be a problem when a linear change in rate of shear is applied. This would not be a problem when testing Newtonian liquid, but when testing a very thixotropic and non-Newtonian suspension like NHL-grouts, it can influence the measurement significantly [76,77].

4. Intrinsic and Extrinsic Factors That Influence the Grouts' Rheology

NHL-based grouts are complex suspensions and there are several intrinsic and extrinsic factors that can influence the rheological behavior of injection grouts, such as: chemical admixtures, mineral additions, particle size distribution, particles shape, volume fraction of particles, temperature, pressure, etc. Different factors and materials that have been reported in several research works, which should be taken into account in the grout design, are briefly outlined in this review in order to highlight their contribution on the rheology of the injection grouts.

4.1. High Range Water Reducer

High range water reducers (HRWRs) like superplasticizers have a significant influence on the rheology of the grout, since they are capable of reducing the water contents by 30% or, for the same water content, will improve the fluidity of the grout [78]. The most available HRWR is based upon the polycarboxylates, which are high molecular weight polymers. The HRWR will disperse the binder agglomerates into primary particles and give them a negatively charged surface leading to repulsion between the particles, which causes a reduction of yield stress and plastic viscosity values [20,21,35]. Different works have shown that the HRWRs effect is only pronounced to dosages around 1% of the binder weight and their efficiency is also a function of their own chemical composition [18,30].

Nevertheless, other authors [79] suggested that some HRWR, when used at higher dosages, have the reverse effect. A higher dosage worsens the fluidity and consequently the injecatbility of the mixture [79–81]. Furthermore, it must be highlighted that the HRWR effect is dependent on the moment at which the HRWR is added during the mixing stage and its action is limited (ranging from 30–60 min). Baltazar et al. [59] concluded that adding the HRWR 10 min after the beginning of the mixing improves the grout fluidity. Other studies [51,82,83] for cement-based systems corroborate this result, showing that the delay in the addition of HRWR improves the effectiveness of the particles dispersing, when compared to an addition without delay. As the delay addition leads to a lower amount of HRWR being intercalated in diverse hydration products [84], a higher amount of HRWR will be available for an effective dispersion of binder particles.

4.2. Water

Water is one of the main elements that assures fluidity of the grout and enables the hydration of the NHL components. The determination of proper water dosage, i.e., the water/solids ratio (w/s), is an important aspect. The w/s ratio has a similar behavior of the HRWR since high water content gives improved flow and injectability [85,86]. However, simple addition of water to make the grout more fluid is an inappropriate decision because a higher w/s ratio decreases the mechanical properties of the hardened grout and will increase the shrinkage deformation as well as the free water amount that might cause instability of the grout.

Grouts for consolidation of old masonry buildings have a w/s ratio in the range of 0.5-1.5 [4], which is dependent on the presence of admixtures (like HRWR) and/or additives. In the same field, Bras [14] studied the optimization of hydraulic lime grout with a w/b ratio in the range 0.6–0.8. In the case of permeation grouting, the typical water/cement ratios are 0.6 to 1.0 [61,62].

4.3. Silica Fume

Silica fume is an ultrafine powder that works as pozzolan and is a by-product of the silicon metal production using electric arc furnaces. The addition of silica fume will contribute to the formation of additional C–S–H (calcium silicate hydrate), and it is expected that the small and spherical silica fume particles will fill the voids between binder particles and produce a ball-bearing effect [87–89].

On the other hand, the addition of silica fume leads to some difficulties regarding the workability of the cementitious materials, requiring the presence of HRWR to minimize these problems. According to Kadri et al. [90], the influence of silica fume on workability is a complex issue since the silica fume increases the volume concentration of the solid phase and also the specific surface; besides, it should be mentioned that silica fume, HRWR and binder constituents will interact with each other as a function of its concentrations.

This complex contribution was confirmed in previous studies, for instance Baltazar et al. [21] concluded that silica fume has an adverse effect on the rheological properties of grouts; the yield stress increases exponentially with the increasing of the content of silica fume and the use of HRWR is indispensable. A similar behavior was also observed on the results of Park et al. [91] that reported that flow resistance of cementitious-based suspensions increases with increasing silica fume dosage. Moreover, Vikan et al. [92] and Zhang & Han [93] concluded that the effect of silica fume on the rheology of cementitious materials depends on the dispersion ability of the HRWR used. Despite some rheological issues, the use of silica fume will enhance the stability which leads to less risk for grout settlement [14,27].

Knowing that a reduced workability is associated with a high water demand when silica fume is used, a recent research work [94] put forward a pretreatment with a polydimethylsiloxane solution for ordinary silica fume in order to obtain a silica fume with hydrophobic behavior to be used in injection grouts. Therefore, this study found that the majority of fluidity and workability problems of injection grouts may be easily overcome with hydrophobic silica fume. The idea can be explained by the combined effect of hydrophobicity and the spherical shape of the silica fume particles, which cause a pure ball-bearing action between the bigger and elongated NHL particles.

Different dosages of silica fume (0, 10, 20, and 30% as replacement of NHL in weight percentage) were tested, revealing that the rheological properties of grouts containing hydrophobic silica fume are significantly improved [94]. Silica fume has a strong impact on the shear thinning behavior of NHL grouts and the presence of hydrophobic silica fume leads to a more Newtonian behavior. The grouts containing 10 wt% of hydrophobic silica fume showed the best rheological performance. Additionally, hydrophobic silica fume has made the hardened grout hydrophobic as well, which significantly contributed to a higher durability of grouts. Therefore, this result should be considered in a consolidation intervention, whenever problems of rising dampness must be also solved. This way, both problems can be overcome with a lower number of physical interventions on the building.

4.4. Fly Ash

Fly ash is a by-product of coal burning from thermal electric power plants and is widely used in the concrete manufacture. It is a very fine powder that can react with calcium hydroxide ($Ca(OH)_2$) or in other words it has a pozzolanic reactivity and consequently it must be used with hydraulic binders such as cement and hydraulic lime, which produce $Ca(OH)_2$. Fly ashes have been used in cementitious-based systems since they improve the durability [95] and contribute to mechanical strength development by both the pozzolanic and filler effect [96,97].

Besides that, the replacement of a certain amount of binder by fly ash will reduce the costs and the environmental impacts [98]. Mirza et al. [99] reported the improvement of stability and the reduction of drying shrinkage in cementitious grouts proportioned by fly ash. A reduced shrinkage of injection grouts is desirable since it interfaces between the grout and the original materials of the masonry.

Sonebi [100] demonstrated that the small size and spherical shape of fly ash particles increase the grout density and reduces the yield stress values due to the ball bearing effect between the binder particles, which reduces the friction forces and consequently reduces the shear stress needed to start the flow. Similar trends of higher fluidity in the presence of fly ash were observed by Baltazar et al. [34].

4.5. Ambient Temperature

In general, a decrease in temperature leads to an increase in viscosity and vice versa, approximately following the Arrhenius relationship [22]. This is also true for hydraulic grouts and seems to be equally true for the yield stress. Bras [14], Baltazar et al. [20] and Jorne et al. [35] have focused on the effect of temperature (ranging from 5 °C to 40 °C) on the rheological properties of hydraulic grouts. From the results presented by these authors, it is clear that the rheological behavior of the hydraulic grout is a strong function of temperature. Yield stress is slightly influenced by temperatures between 5 °C and 20 °C and tends to a higher value with temperature increase. Plastic viscosity decreases between 5 °C and 20 °C but shows an incremental increase between 20 °C and 40 °C, which means a workability loss. The decrease of viscosity with the increase of temperature can be attributed to an increase in the Brownian motion of the particles, which partially weakens the interactions between agglomerates and keeps the particles away from each other. Based on the conclusion of these authors, the ambient temperature of 20 °C is the one that leads to the best grout rheological behavior and consequently the most suitable for performing the masonry injection [14,20,35].

As a consequence of the temperature dependence it is obvious that the rheological properties must be measured at the temperature that will be prevailing in the application site. Moreover, it should be noted that it is not only the ambient temperature that influences the rheology but also the temperature rise due to hydration reactions and shearing induced in high-shear mixers.

4.6. Nano-Silica

There have been progressive developments in the field of nanotechnology, which enabled the manufacture of nanoscale materials (e.g., nanosilica, carbon nanotubes, etc.) that can be incorporated in a cementitious systems [101]. With more studies on the behavior of cement-based systems incorporating nanoscale particles, a better understanding of such composites can be gained and contribute to producing cementitious materials with improved overall performance. Nanoparticles of silica can fill the spaces between/within layers of C–S–H, acting as a nanofiller.

Furthermore, the pozzolanic reaction of nanosilica with calcium hydroxide produces secondary C–S–H, resulting in a higher densification of the matrix, which improves the strength and durability of the material [102]. Other studies [103–105] also reported that the inclusion of nanosilica modifies fresh and hardened properties of cementitious materials compared to conventional mineral additions (at microscale). For example, relative to silica fume, nanosilica shortened the setting time of hydraulic binder mixture and reduced bleeding water and segregation in a more meaningful way. Baltazar et al. [106] investigated the effect of nanosilica dosages from 0 to 3.5 wt%, by mass of binder, on NHL-based grouts prepared with w/s of 0.5 and a HRWR dosage ranging from 0.8 to 1.6 wt%. In the same study they reported that the yield stress increased considerably when the nanosilica dosage increased; however, the effect on plastic viscosity was less pronounced. It was also observed that higher nanosilica dosage (>1.5 wt%) significantly affects the rheological performance of grouts, which compromises the expected injection capacity; thus, it is recommended that the use of nanosilica is always accompanied by the incorporation of a superplasticizer. Senff et al. [103] also concluded that the fluidity of cement pastes and mortars is reduced when incorporating high nanosilica dosages.

Moreover, it is believed that the reduced particle size of nanosilica (even at lower dosages) can contribute to overcome most of the grout's penetrability problems (like the so-called blockage phenomenon). It is known that grout's penetrability problems are due to the grain size characteristics of the solid phase of the grout [4,17,50]. Unfortunately, the uses of these materials to get a specific grading of the solid phase is difficult to implement and supervise on site, which in addition to higher difficulties in the mixing procedure, are real setbacks to the use of these materials. To make these grouts injectable a specific mixing procedure is essential and the use of the ultrasonic mixing procedure is crucial. An ultrasonic mixing time of around ten min (the first five with only the nanoparticles and the other five with NHL included) was set ideal to deflocculate the grain elements and obtain the appropriate fluidity of the mixtures [106].

4.7. Injection Pressure and Resting Time

A proper choice of the injection pressure is of great importance in grouting operations since it is the injection pressure that guarantees the sufficient shear rates within the injection pipes in order to cause a reduction of the grout viscosity. Nevertheless, very high pressure should not be adopted to avoid an excessive outward pressure loading, which could endanger the stability of the structure or cause fewer attached stones to blow from the surface. Thus, the pressure must be limited to a few bars in order to avoid any damage with the weakened masonry. It can be found in the literature [5,107,108] that grout injection for old masonry consolidation should be made with a pressure in the range of 0.2–1.5 bar.

The results presented by Baltazar et al. [20] clearly show the influence of injection pressure on the rheological properties of grouts, using a rotational rheometer equipped with a high pressure cell. The analysis of these results reveals that an increase of the yield stress as well as of the viscosity occurs with the increase of pressure. Moreover, this study also shows that almost no difference exists between the rheological properties of the grout at pressure of 0.5 bar (above the atmospheric pressure) and the one at atmospheric pressure. The biggest difference in rheological parameters was observed for pressures above 0.5 bar. Thus, considering these conclusions and the fact that the injection pressure should be limited to a few bars to avoid the masonry disruption, the injection pressure around 0.5 bar is recommended for the grouting operation of stone masonry [20,109]. These results emphasize how harmful it is to increase the injection pressure to overcome some injectability difficulties, which is caused by several phenomena such as grout blockage. It is worth highlight that the phenomenon behind the blocking is mainly due to particles or agglomerates that get stuck at constrictions [110,111]. Either the solid particle size is too big to allow them to go through the pores or there is segregation. The binder particles deposit themselves at the opening of the pores, producing a "cork" that hinders the injection. Moreover, an experimental program presented by Binda [112] showed that the injection pressure of grout (although it is based on another binder than NHL) should not be higher than 0.6 bar, to optimize the diffusion and penetration of the grout.

As previously mentioned, NHL-based grouts can be considered as thixotropic materials, since they show a shear thinning and time dependent behavior. During shearing of hydraulic lime grout, the weak interparticle bonds are broken by the mechanical stress and the network among them breaks down into separate agglomerates (structural breakdown). If the grout is at rest, the particles will start to flocculate into agglomerates again (structural build-up), leading to a loss of workability. In earlier investigations [14,20] NHL-based grouts were sheared using a rotational rheometer and parallel-plates at a constant shear rate of 1 s^{-1} after different resting times and the initial shear stress was plotted as a function of resting time. The initial shear stress can illustrate the intensity of interactions between particles agglomerates in the grout before shearing takes place (solid-like particle structure). Furthermore, the rate of increase of the initial shear stress, also known as the flocculation rate [14], was used in thixotropic evaluation and it is a measure of the forces acting between particles, namely the electrostatic forces and Van der Waals attractive forces [60]. Alongside the results of different works [14,20,35], it is clear that the initial shear stress rate increases linearly with resting time, which means more flocculation, which leads to workability loss according to the PFI-theory [36]. Moreover, the grout flocculation rate is of particularly relevance during a grouting operation, since it may reduce the injectability up to 59% [35]. Taking these statements into account it is recommended not to exceed 10 min and 60 min of resting time (before injection) for grouts without HRWR and with HRWR, respectively.

Nevertheless, it is important to note that the grout flocculation can be a way to solve some issues after injection. For example, when grouts are at rest in the masonry core, gravity can promote sedimentation of grout particles at rest. Thus, a higher flocculation rate can be useful since it will increase the interparticles bonds (structural build-up) which can be sufficient to prevent the particles from settling [113]. This may seem somewhat contradictory considering what has been said previously but from a practical point of view it is desirable to have grout with low flocculation rate before its injection and with a high flocculation rate after its application. Some research works [25,113] were found on this subject but in the field of self compacting concrete. Summarizing, care should be taken during the whole grouting operation, namely to avoid stops during injection and to restrict the resting time between the mixing and the injection in order to prevent flocculation, which causes a considerable reduction on grout injectability and, consequently, compromises the efficiency of the consolidation operation.

5. Conclusions

Despite the widespread use of NHL-based grouts for structural consolidation and repair of historic masonry structures, research works on the subject are still insufficient. The lack of information and fundamental understanding of the mechanics underlying the effect of some constituents and factors emphasizes the importance of continuing to study this subject in order to contribute to the optimization of the grout injection technique and to avoid new damage in historic masonry as a consequence of bad practices. In this sense, some important aspects need further study, for example: (i) to study how the slippage phenomenon, which is frequently observed in the rheological measurements, influences the flow and penetration length of the grout in practice; (ii) since significant properties are continuously changing during the grouting operation, it would also be interesting and desirable to perform "in-line" measurements (i.e., under real injection conditions) in order to provide a fuller assessment of the rheological performance of grouts. Nevertheless, and considering the review made, the following aspects stand out:

- NHL-based injection grouts are often referred to as non-Newtonian, thixotropic, and possess a yield stress. This non-Newtonian behavior can be attributed to mechanisms in which the shear stress orients the suspended binder particles in opposition to the randomizing effects of the Brownian motion.
- A non linear relationship between shear stress and shear rate implies that a shear thinning relationship like a Herschel–Bulkley model should be adequate. However, in most grouting applications a simple model like the Bingham one is used as it contains the fundamental properties, yield stress, and plastic viscosity.
- Due to the complex nature of hydraulic binders, the rheological characterization of NHL-based grouts is challenging. Therefore, it is important that standard protocols are developed in order to allow the results comparison. Brookfield viscometers or rotational rheometers have been successfully used to characterize the rheology of the injection grouts. However, several details

should be taken into account, such as the type of geometry and other phenomena that may affect the accuracy of the measurements, such as wall slip and segregation.

- The incorporation of HRWR will cause a reduction of yield stress and viscosity values. Several studies have shown that their effect is only pronounced to an amount around 1% of the binder weight. Nevertheless, a higher value worsens the fluidity and consequently the injectability of the mixture. Furthermore, it must be highlighted that the HRWR effect is dependent on the moment that the HRWR is added to during the mixing stage.

- High water content improves the flow and injectability of the grouts. However, a simple addition of water to make the grout more fluid is inappropriate. Grouts for consolidation of old masonry buildings should have a w/s ratio of around 0.5–1.5, the dosage of water depending on the presence of HRWR and ultra fine materials such as silica fume.

- The use of fly ash additions, besides improving the mechanical properties, has a significant influence on the rheology of the grouts. For instance, the small size and spherical shape of fly ash particles increases the grout density and reduces the yield stress values due to the ball bearing effect, which reduces the friction forces and consequently reduces the shear stress needed to start the flow.

- The addition of silica fume leads to some difficulties regarding the rheological properties of grouts and the use of HRWR is indispensable. However, a pretreatment with a polydimethylsiloxane applied on ordinary silica fume was a proposed solution to mitigate the rheological disadvantages of this material.

- The reduced particle size of nanosilica (even at lower dosages) may contribute to overcome most of grout's penetrability problems (like the so-called blockage phenomenon). Nevertheless, the uses of these materials require a different mixing procedure.

- An ambient temperature of 20 °C is the one that leads to the best grout rheological behavior and consequently the most suitable for performing the masonry injection. The injection pressure around 0.5 bar is recommended for grouting operation. Moreover, a low resting time (less 10 min) is desired, especially in the cases of grouts with low HRWR amount.

Funding: This work is funded by National Funds through Portuguese Foundation for Science and Technology (FCT/MCTES), Reference UID/CTM/50025/2013 and Fundo Europeu de Desenvolvimento Regional (FEDER) funds through the COMPETE 2020 Programme under the project number POCI-01-0145-FEDER-007688.

Conflicts of Interest: The authors declare no conflict of interest.

References

1. Ashurst, J. Methods of repairing and consolidating stone buildings, Chapter I. In *Conservation of Building and Decorative Stone*; Butterworth-Heinemann: Oxford, UK, 1990; Volume 2, ISBN 0750612770.
2. Penelis, G.; Karaveziroglou, M.; Papayanni, J. Grouts for Repairing and Strengthening Old Masonry Structural. In *Structural Repair and Maintenance of Historical Buildings*; Brebbia, C.A., Ed.; Computational Mechanics Publications: Southampton, UK, 1989; pp. 179–188.
3. Baronio, G. Criteria and methods for the optimal choice of grouts according to the characteristics of masonries, effectiveness of injection techniques for retrofitting of stone and brick masonry walls in seismic areas. In Proceedings of the International Workshop CNR-GNDT, Milano, Italy, 30–31 March 1992.
4. Miltiadou, A.E. Contribution à L'étude des Coulis Hydrauliques Pour la Réparation et le Renforcement des Structures et des Monuments Historiques en Maçonnerie. Ph.D. Thesis, l'École Nationale des Ponts et Chaussées, Champs-sur-Marne, France, 1990.
5. Binda, L.; Modena, C.; Baronio, G.; Abbaneo, S. Repair an investigation technique for stone masonry walls. *Constr. Build. Mater.* **1997**, *11*, 133–142. [CrossRef]
6. Carocci, C.F. Guidelines for the safety and preservation of historical centres in seismic areas. In Proceedings of the 3rd International Seminar on Structural Analysis of Historical Constructions, Guimarães, Portugal, 7–9 November 2001; Lourenço, P.B., Roca, P., Eds.; pp. 145–166.

7.   Binda, L.; Baronio, G.; Tiraboschi, C.; Tedeschi, C. Experimental research for the choice of adequate materials for the reconstruction of the Cathedral of Noto. *Constr. Build. Mater.* **2003**, *7*, 629–639. [CrossRef]
8.   Collepardi, M. Degradation and restoration of masonry walls of historical buildings. *Mater. Struct.* **1990**, *23*, 81–102. [CrossRef]
9.   Toumbakari, E.-E.; Van Gemert, D. Injection grouts for ancient masonry: Strength properties and microstructural evidence. In *Compatible Materials for the Protection of European Cultural Heritage*; Biscontin, G., Moropoulou, A.I., Erdik, M., Rodrigues, J.D., Eds.; Technical Chamber of Commerce: Athens, Greece, 1998; pp. 191–200.
10. Binda, L.; Penazzi, D.; Saisi, A. Historic masonry buildings: Necessity of a classification of structures and masonries for the adequate choice of analytical models. In Proceedings of the VI International Symposium Computer Methods in Structural Masonry-STRUMAS, Rome, Italy, 22–24 September 2003; pp. 168–173.
11. Binda, L.; Saisi, A.; (State of the Art of Research on Historic Structures in Italy. Department of Structural Engineering, Politecnico di Milão, Italy). Personal communication, 2001.
12. Binda, L.; Saisi, A.; Tedeschi, C. Compatibility of materials used for repair of masonry buildings: Research and applications. *Fract. Fail. Natl. Build. Stone* **2006**, 167–182. [CrossRef]
13. Binda, L.; Moderna, C.; Baroni, G.; Gelmi, A. Experimental qualification of injection admixtures use for repair and strengthening of stone masonry walls. In Proceedings of the 10th International Brick & Block Masonry Conference, Calgary, AB, Canada, 5–7 July 1994; pp. 539–548.
14. Bras, A. Grout Optimization for Masonry Consolidation. Ph.D. Thesis, Universidade Nova de Lisboa, Lisbon, Portugal, 2011.
15. Tassios, T.P.; Miltiadou-Fezans, A. Stability of hydraulic grouts for masonry strengthening. *Mater. Struct.* **2013**, *46*, 1631–1652.
16. Kalagri, A.; Miltiadou-Fezans, A.; Vintzileou, E. Design and evaluation of hydraulic lime grouts for the strengthening of stone masonry historic structures. *Mater. Struct.* **2010**, *43*, 1135–1146. [CrossRef]
17. Jorne, F.; Henriques, F.M.A.; Baltazar, L.G. Injection capacity of hydraulic lime grouts in different porous media. *Mater. Struct.* **2014**, *48*, 2211–2233. [CrossRef]
18. Jorne, F.; Henriques, F.M.A.; Baltazar, L.G. Evaluation of consolidation of different porous media with hydraulic lime grout injection. *J. Cult. Herit.* **2015**, *16*, 438–451. [CrossRef]
19. Luso, E.; Lourenço, P.B. Experimental characterization of commercial lime based grouts for stone masonry consolidation. *Constr. Build. Mater.* **2016**, *102*, 216–225. [CrossRef]
20. Baltazar, L.G.; Henriques, F.M.A.; Jorne, F.; Cidade, M.T. Combined effect of superplasticizer, silica fume and temperature in the performance of natural hydraulic lime grouts. *Constr. Build. Mater.* **2014**, *50*, 584–597. [CrossRef]
21. Baltazar, L.G.; Henriques, F.M.A.; Jorne, F.; Cidade, M.T. The use of rheology in the study of the composition effects on the fresh behaviour of hydraulic lime grouts for injection of masonry walls. *Rheol. Acta* **2013**, *52*, 127–138. [CrossRef]
22. Barnes, H.A.; Hutton, J.F.; Walters, K. *An Introduction to Rheology*; Rheology Series; Elsevier: Amsterdam, The Netherland, 1989; ISBN 9780444871404.
23. Barnes, H.A. Thixotropy—A review. *J. Non-Newton. Fluid Mech.* **1997**, *70*, 1–33. [CrossRef]
24. Nguyen, Q.D.; Boger, D.V. Measuring the flow properties of yield stress fluids. *Annu. Rev. Fluid Mech.* **1992**, *24*, 47–88. [CrossRef]
25. Nguyen, V.H.; Rémond, S.; Gallias, J.L.; Bigas, J.P.; Muller, P. Flow of Herschel–Bulkley fluids through the Marsh cone. *J. Non-Newton. Fluid Mech.* **2006**, *139*, 128–134. [CrossRef]
26. James, A.E.; Williams, D.J.A.; Williams, P.R. Direct measurement of static yield properties of cohesive suspensions. *Rheol. Acta* **1987**, *26*, 437–446. [CrossRef]
27. Hakansson, U. Rheology of Fresh Cement-Based Grouts. Ph.D. Thesis, The Royal Institute of Technology, Stockholm, Sweden, 1993.
28. Rahman, M.; Wiklund, J.; Kotzé, R.; Håkansson, U. Yield Stress of Cement Grouts. *Tunn. Undergr. Space Technol.* **2017**, *61*, 50–60. [CrossRef]
29. Larson, R.G. *The Structure and Rheology of Complex Fluids*; Oxford University Press: New York, NY, USA, 1999; ISBN 978-0195121971.

30. Billberg, P. Form Pressure Generated by Self-Compacting Concrete—Influence of Thixotropy and Structural Behaviour at Rest. Ph.D. Thesis, School of Architecture and the Built Environment, Division of Concrete Structures Royal Institute of Technology, Stockholm, Sweden, 2006.
31. Mewis, J. Thixotropy—A general review. *J. Non-Newton. Fluid Mech.* **1979**, *6*. [CrossRef]
32. Barnes, H.A. *A Handbook of Elementary Rheology*; Institute of Non-Newtonian Fluid Mechanics, University of Wales: Wales, UK, 2000; ISBN 9780953803200.
33. Cheng, D. Thixotropy. *Int. J. Cosmet. Sci.* **1987**, *9*, 151–191. [CrossRef] [PubMed]
34. Baltazar, L.G.; Henriques, F.M.A.; Cidade, M.T. Contribution to the design of hydraulic lime-based grouts for masonry consolidation. *J. Civ. Eng. Manag.* **2015**, *21*, 698–709. [CrossRef]
35. Jorne, F.; Henriques, F.M.A.; Baltazar, L.G. Influence of superplasticizer, temperature, resting time and injection pressure on hydraulic lime grout injectability. Correlation analysis between fresh grout parameters and grout injectability. *J. Build. Eng.* **2015**, *4*, 140–151. [CrossRef]
36. Wallevik, J. Rheological properties of cement paste: Thixotropic behavior and structural breakdown. *Cem. Concr. Res.* **2009**, *39*, 14–29. [CrossRef]
37. Krieger, I.M.; Dougherty, I.J. A mechanism for non-Newtonian flow in suspension of rigid spheres. *J. Rheol.* **1959**, *3*, 137–152. [CrossRef]
38. Coussot, P.; Ngugen, O.D.; Huynh, H.T.; Bonn, D. Viscosity bifurcation in thixotropic, yielding fluids. *J. Rheol.* **2002**, *46*, 573–589. [CrossRef]
39. Wallevik, J. *Introductions to Rheology of Fresh Concrete*; Innovation Center Iceland: Reyjavik, Iceland, 2009.
40. Wallner, M.; (Propagation of sedimentation stable cement paste in jointed rock. Rock Mechanics and Waterways Construction, University of Achen, BRD). Personal communication, 1976.
41. Abdali, S.S.; Mitsoulis, E.; Markatos, N.C. Entry and exit flows of Bingham fluids. *J. Rheol.* **1992**, *36*, 389–407. [CrossRef]
42. Håkansson, U.; Hässler, L.; Stille, H. Rheological properties of microfine cement grouts. *Tunn. Undergr. Space Technol.* **1992**, *7*, 453–458. [CrossRef]
43. Rahman, M. Rheology of Cement Grout: Ultrasound Based in-Line Measurement Technique and Grouting Design Parameters. Ph.D. Thesis, Royal Institute of Technology, Stockholm, Sweden, 2015.
44. Amadei, B.; Savage, W.Z. An analytical solution for transient flow of Bingham viscolastic materials in rock fractures. *Int. J. Rock Mech. Min. Sci.* **2001**, *38*, 285–296. [CrossRef]
45. Baltazar, L.G.; Henriques, F.M.A.; Cidade, M.T. Experimental study and modeling of rheological and mechanical properties of NHL grouts. *J. Mater. Civ. Eng.* **2015**, *27*. [CrossRef]
46. Baltazar, L.G.; Henriques, F.M.A.; Miguel, D.; Cidade, M.T. Effects of hydrophobic additives on the rheology of hydraulic grouts. In Proceedings of the Iberian Meeting of Rheology (Ibereo 2017), Valencia, Spain, 6–8 September 2017.
47. Larrard, F.D.; Ferraris, C.F.; Sedran, T. Fresh concrete: A Herschel-Bulkley material. *Mater. Struct.* **1998**, *31*, 494–498. [CrossRef]
48. Casson, W. A flow equation for pigment-oil suspensions of the printing ink type. In *Rheology of Disperse Systems*; Mill, C.C., Ed.; Pergamon: London, UK, 1959.
49. Hassler, L. Grouting of Rock-Simulation and Classification. Ph.D. Thesis, Department of Soil and Rock Mechanics, Royal Institute of Technology, Stockholm, Sweden, 1991.
50. Miltiadou-Fezans, A.; Tassios, T.P. Penetrability of hydraulic grouts. *Mater. Struct.* **2013**, *46*, 1653–1671. [CrossRef]
51. Aiad, I. Influence of time addition of superplasticizers on the rheological properties of fresh cement pastes. *Cem. Concr. Res.* **2003**, *33*, 1229–1234. [CrossRef]
52. Axelsson, M.; Gustafson, G. A robust method to determine the shear strength of cement-based injection grouts in the field. *Tunn. Undergr. Space Technol.* **2006**, *21*, 499–503. [CrossRef]
53. Papanastasiou, T.C.; Boudouvis, A.G. Flows of viscoplastic materials: Models and computations. *Comput. Struct.* **1997**, 677–694. [CrossRef]
54. Papanastasiou, T.C. Flows of materials with yield. *J. Rheol.* **1987**, *31*, 385–404. [CrossRef]
55. Baltazar, L.G.; Henriques, F.M.A.; Cidade, M.T. Grouts with improved durability for masonry consolidation: An experimental study with non-standard specimens. *Key Eng. Mater.* **2017**, *747*, 480–487. [CrossRef]

56. Hackley, V.A.; Ferraris, C.H.; Guide to Rheological Nomenclature: Measurements in Ceramic Particulate Systems. National Institute of Standards and Technology, U.S. Department of Commerce, Special Publication 946. Personal communication, 2001.

57. Brookfield Engineering Labs Inc.; Viscometers, Rheometers & Texture Analyzers for Laboratory and Process Applications. Brookfield Eng. Labs, 4–35. Personal communication, 2008.

58. Chhabra, R.P.; Richardson, J.F. *Non-Newtonian Flow and Applied Rheology*, 2nd ed.; Elesvier: Oxford, UK, 2008; ISBN 9780750685320.

59. Baltazar, L.G.; Henriques, F.M.A.; Jorne, F. Optimisation of flow behaviour and stability of superplasticized fresh hydraulic lime grouts through design of experiments. *Constr. Build. Mater.* **2012**, *35*, 838–845. [CrossRef]

60. Mewis, J.; Wagner, N.J. *Colloidal Suspension Rheology*; Cambridge University Press: Cambridge, UK, 2012; ISBN 978-0521515993.

61. Wiklund, J.; Rahman, M.; Håkansson, U. In-Line rheometry of micro cement based grouts—A promising new industrial application of the ultrasound based UVP+PD method. *Appl. Rheol.* **2012**, *22*, 42783. [CrossRef]

62. Rahman, M.; Håkansson, U.; Wiklund, J. In-Line Rheological measurements of cement grouts: Effects of water/cement ratio and hydration. *Tunn. Undergr. Space Technol.* **2015**, *45*, 34–42. [CrossRef]

63. Takeda, Y. Velocity profile measurement by ultrasound Doppler shift method. *Int. J. Heat Fluid Flow* **1986**, *7*, 313–318. [CrossRef]

64. Wiklund, J.A.; Stading, M.; Pettersson, A.J.; Rasmuson, A. A comparative study of UVP and LDA techniques for pulp suspensions in pipe flow. *AIChE J.* **2006**, *52*, 484–495. [CrossRef]

65. Håkansson, U.; Rahman, M. Rheological properties of cement based grouts using the UVP-PD method. In Proceedings of the Nordic Symposium of Rock Grouting, Helsinki, Finland, 5 November 2009.

66. ASTM C939. *Standard Test Method of Flow of Grout for Preplaced-Aggregate Concrete (Flow Cone Method)*; ASTM International: West Conshohocken, PA, USA, 2002.

67. Baltazar, L.G.; Henriques, F.M.A. Rheology of grouts for masonry injection. *Key Eng. Mater.* **2015**, *624*, 283–290. [CrossRef]

68. Le Roy, R. The Marsh Cone as a viscometer: Theoretical analysis and practical limits. *Mater. Struct.* **2004**, *38*, 25–30. [CrossRef]

69. Roussel, N.; Le Roy, R. The Marsh cone: A test or a rheological apparatus? *Cem. Concr. Res.* **2005**, *35*, 823–830. [CrossRef]

70. Sadrizadeh, S.; Ghafar, A.N.; Halilovic, A.; Håkansson, U. Numerical, experimental and analytical studies on fluid flow through a Marsh funnel. *J. Appl. Fluid Mech.* **2017**, *10*, 1501–1507. [CrossRef]

71. Hu, C. Rheologie des Be'tons Fluides (Rheology of Fluid Concretes). Ph.D. Thesis, Ecole des Ponts ParisTech, Paris, France, 1995.

72. Christensen, G. Modelling the Flow of Fresh Concrete: The Slump Test. Ph.D. Thesis, Princeton University, Princeton, NJ, USA, 1991.

73. Roussel, N.; Stefani, C.; Leroy, R. From mini-cone test to Abrams cone test: Measurement of cement-based materials yield stress using slump tests. *Cem. Concr. Res.* **2005**, *35*, 817–822. [CrossRef]

74. Barnes, H.A. A review of the slip (wall depletion) of polymer solutions, emulsions and particle suspensions in viscometers: Its cause, character, and cure. *J. Non-Newtin. Fluid Mech.* **1995**, *56*, 221–231. [CrossRef]

75. Baltazar, L.G.; Henriques, F.M.A.; Cidade, M.T. Rheological characterization of injection grouts using rotational rheometry. In *Advances in Rheology Research*; Nova Science Publishers: New York, NY, USA, 2017; pp. 13–42, ISBN 978-1-53612-876-5.

76. Macosko, C. *Rheology Principles, Measurements, and Applications*, 1st ed.; VCH: New York, NY, USA, 1994; ISBN 978-0471185758.

77. Morrison, F. *Understanding Rheology*; Oxford University Press: Oxford, UK, 2001; ISBN 9780195141665.

78. Atkinson, R.; Schuller, M. Evaluation of injecatble cementitious grouts for repair and retrofit of masonry. In *Masonry: Design and construction, Problems and Repair*; Melander, J.M., Lauersdorf, L.R., Eds.; American Society for Testing Materials: Philadelphia, PA, USA, 1993.

79. Chandra, S.; Van Rickstal, F.; Van Gemert, D. Evaluation of cement grouts for consolidation injection of ancient masonry. In Proceedings of the Nordic Concrete Research Meeting, Goteborg, Sweden, 17–19 August 1993.

80. Banfill, P.F.G. Additivity effects in the rheology of fresh concrete containing water-reducing admixtures. *Constr. Build. Mater.* **2011**, *25*, 2955–2960. [CrossRef]

81. Flatt, R.J. Interparticle Forces and Superplasticizers in Cement Suspensions. Ph.D. Thesis, École Polytechnique Fédérale de Lausanne, Lausanne, Switzerland, 1999.
82. Chandra, S.; Bjornstrom, J. Influence of superplasticizer type and dosage on the slump loss of portland cement mortars—Part II. *Cem. Concr. Compos.* **2002**, *32*, 1613–1619. [CrossRef]
83. Fernàndez-Altable, V.; Casanova, I. Influence of mixing sequence and superplasticiser dosage on the rheological response of cement pastes at different temperatures. *Cem. Concr. Res.* **2006**, *36*, 1222–1230. [CrossRef]
84. Flatt, R.J.; Houst, F.Y. A simplified view on chemical effects perturbing the action of superplasticizers. *Cem. Concr. Res.* **2001**, *31*, 1169–1176. [CrossRef]
85. Eriksson, M.; Friedrich, M.; Vorschulze, C. Variations in the rheology and penetrability of cement-based grouts—An experimental study. *Cem. Concr. Res.* **2004**, *34*, 1111–1119. [CrossRef]
86. Rosquoët, F.; Alexis, A.; Khelidj, A.; Phelipot, A. Experimental study of cement grout. *Cem. Concr. Res.* **2003**, *33*, 713–722. [CrossRef]
87. Shannag, M.J. High strength concrete containing natural pozzolan and silica fume. *Cem. Concr. Compos.* **2000**, *22*, 399–406. [CrossRef]
88. Shannag, M.J. High-performance cementitious grouts for structural repair. *Cem. Concr. Res.* **2002**, *32*, 803–808. [CrossRef]
89. Shihada, S.; Arafa, M. Effects of silica fume, ultrafine and mixing sequences on properties of ultra high performance concrete. *Asian J. Mater. Sci.* **2010**, *2*, 137–146. [CrossRef]
90. Kadri, E.H.; Aggoun, S.; De Schutter, G. Interaction between C_3A, silica fume and naphthalene sulphonate superplasticiser in high performance concrete. *Constr. Build. Mater.* **2009**, *23*, 3124–3128. [CrossRef]
91. Park, C.K.; Noh, M.H.; Park, T.H. Rheological properties of cementitious materials containing mineral admixtures. *Cem. Concr. Res.* **2009**, *35*, 842–849. [CrossRef]
92. Vikan, H.; Justnes, H.; Winnefeld, F.; Rigib, F. Correlating cement characteristics with rheology of paste. *Cem. Concr. Res.* **2007**, *37*, 1502–1511. [CrossRef]
93. Zhang, X.; Han, J. The effect of ultra–fine admixture on the rheological property of cement paste. *Cem. Concr. Res.* **2000**, *30*, 827–830. [CrossRef]
94. Baltazar, L.G.; Henriques, F.M.A.; Douglas, R.; Cidade, M.T. Experimental characterization of injection grouts incorporating hydrophobic silica fume. *J. Mater. Civ. Eng.* **2017**, *29*, 04017167. [CrossRef]
95. ACI Committee 226. Use of fly ash in concrete. *ACI Mater. J.* **1987**, *84*, 381–409. [CrossRef]
96. Papadakis, V.G. Effect of fly ash on Portland cement systems. Part 1: Low-calcium fly ash. *Cem. Concr. Res.* **1999**, *29*, 1727–1736. [CrossRef]
97. Poon, C.S.; Kou, S.C.; Lam, L.; Lin, Z.S. Activation of fly ash/cement systems using calcium sulfate anhydrite (CaSO4). *Cem. Concr. Res.* **2001**, *31*, 873–881. [CrossRef]
98. Ozcan, T.; Zaimoglu, A.S.; Hinislioglu, S.; Altunb, S. Taguchi approach for optimization of the bleeding on cement-based grouts. *Tunn. Undergr. Space Technol.* **2005**, *20*, 167–173. [CrossRef]
99. Mirza, J.; Mirza, M.S.; Roy, V.; Saleh, K. Basic Rheological and Mechanical Properties of High-Volume Fly Ash Grouts. *Constr. Build. Mater.* **2002**, *16*, 353–363. [CrossRef]
100. Sonebi, M. Experimental design to optimize high-volume of fly ash grout in the presence of Welan Gum and superplasticizeri. *Mater. Struct.* **2002**, *35*, 373–380. [CrossRef]
101. Sobolev, K.; Gutiérrez, M.F. How nanotechnology can change the concrete world. *Am. Ceram. Soc. Bull.* **2005**, *84*, 14–18. [CrossRef]
102. Collepardi, S.; Borsoi, A.; Ogoumah Olagot, J.J.; Troli, R.; Collepardi, M.; Cursio, A.Q. Influence of nano-sized mineral additions on performance of SCC. In Proceedings of the 6th International Congress, Global Construction, Ultimate Concrete Opportunities, Dundee, UK, 5–7 July 2005.
103. Senff, L.; Labrincha, J.A.; Ferreira, V.M.; Hotza, D.; Repette, W.L. Effect of nano-silica on rheology and fresh properties of cement pastes and mortars. *Constr. Build. Mater.* **2009**, *23*, 2487–2491. [CrossRef]
104. Björnström, J.; Martinelli, A.; Matic, A.; Börjesson, L.; Panas, I. Accelerating effects of colloidal nano-silica for beneficial calcium–silicate–hydrate formation in cement. *Chem. Phys. Lett.* **2004**, *392*, 242–248. [CrossRef]
105. Qing, Y.; Zenan, Z.; Deyu, K.; Rongshen, C. Influence of nano-SiO_2 addition on properties of hardened cement paste as compared with silica fume. *Constr. Build. Mater.* **2007**, *21*, 539–545. [CrossRef]

Fluids **2019**, *4*, 13

106. Baltazar, L.G.; Henriques, F.M.A.; Gouveia, T.; Cidade, M.T. Rheological properties of injection grouts incorporating nano-silica. In Proceedings of the Annual European Rheological Conference (AERC 2018), Sorrento, Italy, 17–20 April 2018.

107. Keersmaekers, R.; Schueremans, L.; Van Rickstal, F.; Van Gemert, D.; Knapen, M.; Posen, D. NDT—Control of injection of an appropriate grout mixture for the consolidation of the columns foundations of our lady's basilica at Tongeren. In Proceedings of the 4th International Conference on Structural Analysis of Historical Constructions, New Delhi, India, 6–7 November 2006; ISBN 972-8692-27-7.

108. Corradi, M.; Tedeschi, C.; Binda, L.; Borri, A. Experimental evaluation of shear and compression strength of masonry wall before and after reinforcement: Deep repointing. *Constr. Build. Mater.* **2008**, *22*, 463–472. [CrossRef]

109. Valluzzi, M.R. Requirements for the choice of mortar and grouts for consolidation of three-leaf stone masonry walls. In Proceedings of the Workshop Repair Mortars for Historic Masonry, Delft, The Netherlands, 26–28 January 2005.

110. Draganović, A.; Stille, H. Filtration and penetrability of cement-based grout: Study performed with a short Slot. *Tunn. Undergr. Space Technol.* **2011**, *26*, 548–559. [CrossRef]

111. Eklund, D.; Håkan, S. Penetrability due to filtration tendency of cement-based grouts. *Tunn. Undergr. Space Technol.* **2008**, *23*, 389–398. [CrossRef]

112. Binda, L. Strengthening and durability of decayed brick-masonry repair by injection. In Proceedings of the 5th North American Masonry Conference, Champaign, IL, USA, 3–6 June 1990; Volume IV, pp. 839–852.

113. Roussel, N. A theoretical frame to study stability of fresh concrete. *Mater. Struct.* **2006**, *39*, 81–91. [CrossRef]

Review

Rheological Behavior of Fresh Cement Pastes

Francisco-José Rubio-Hernández

Departamento de Física Aplicada II, Universidad de Málaga, 29071 Málaga, Spain; fjrubio@uma.es;
Tel.: +34-951-952-296

Received: 7 November 2018; Accepted: 7 December 2018; Published: 11 December 2018

Abstract: Rheology of a concrete is mainly controlled by the rheological behavior of its cement paste. This is the main practical reason for the extensive research activity observed during 70 years in this research subfield. In this brief review, some areas of the research on the rheological behavior of fresh cement pastes (mixture method influence, microstructure analysis, mineral additions influence, chemical additives influence, blended cements behavior, viscoelastic behavior, flow models, and flow behavior analysis with alternative methods) are examined.

Keywords: cement pastes; rheology; particle suspensions

1. Introduction

Concrete can be dealt with as a dispersion of gravel (around 20 mm diameter) in mortar, and mortar can be dealt with as a dispersion of sand (around 1 mm diameter) in cement paste. So, cement paste that is a dispersion of cement particles (around 10 μm) in water, is the fluid media where aggregate particles (sand and gravel) are dispersed.

Interest in gaining information about the rheological behavior of fresh cement pastes (RBFCP) is mainly due to the role played by this phase in concrete and mortar formulations. Tattersall firstly pointed to this field of research in 1955 [1], enhancing the fact that this is a non-Newtonian fluid. More precisely, he assumed a relationship between thixotropic behavior of the cement paste and the performance of the vibration process commonly used when the concrete is placed. Although a direct connection between cement paste and concrete rheology can be questioned [2] due to the effect of the shape and grain size distribution of coarse aggregates, the subject left open the search for possible relationships between both materials [3]. In this line, Ferraris and Gaidis [4] studied the rheology of the paste with parallel plate geometry (Figure 1) in an attempt to simulate the effect of coarse aggregate on the rheological behavior of pastes.

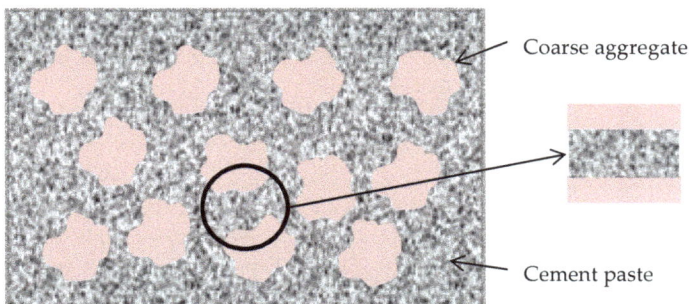

Figure 1. Cement paste flow in concrete is directly influenced by the action of coarse aggregate [4].

Interest in the study of fresh cement paste rheological behavior is founded in the assumption that concrete rheology is mainly determined by cement paste rheology [4]. Considering that the hydration reaction of cement starts at the first contact between cement and water, and continues when the rheological studies are usually made (10–60 min after), it is reasonably expected that rheometric work with fresh cement pastes faces a variety of challenges. Even if it is assumed that during the dormant period the hydration of cement is basically stopped, additional problems associated with rheological measurements of cement pastes can be listed. Banfill [5,6] pointed to possible reasons for these problems. Concretely, (a) discrepancies in time-dependent viscosity curves are usually reported, that can be due to variable competition between shear-induced breakdown and by-hydration build-up of structure when the design of hysteresis loops is modified; (b) the rheological results depend on the mixing and handling protocol, probably due to the decrease of the yield stress with vigorous mixing; (c) slippage of pastes at the walls of the rotor can certainly be overcome with roughened surfaces, but it is not possible to determine to which extension; and (d) particle sedimentation during measurement can be neglected when helical impeller or similar are used, or when the water/cement ratio is kept below 0.4 with conventional geometries. All these problems can be overcome with a good experimental design, however plug flow due to the existence of a range of stresses in the bulk of the material, some of them with values lower than yield stress, is described as an unsolved problem [5,6].

It can be affirmed that the experimental study of the rheological behavior of fresh cement pastes is a difficult task [5,6]. Generally, this is due to several factors and conditions that have certain influence on the response of the material. We can include physical factors like water/cement ratio or the morphology of cement grains, mineralogical factors like cement composition, chemical factors like structural modifications due to hydration processes, mixing conditions like the type of stirrer, or measurement conditions like experimental procedures.

The first interest for the study of RBFCP was initially practical; i.e., it was motivated by the necessity to control and predict the rheological behavior of the concrete obtained when sand and gravel are added to the cement paste. But, also, this study is interesting from a fundamental point of view. This is because cement pastes are particle dispersions featured by a decisive hydration chemical activity, which generates products of the reaction that form a gel phase interconnecting the core of cement particles. It is easy to understand that a material with such a brand mark had been wide and deeply studied. Different aspects of these particle suspensions have attracted the interest of researchers. Some of them will be briefly described in this review:

a) Mixture method influence.
b) Microstructure analysis.
c) Mineral additions influence.
d) Chemical additives influence.
e) Blended cements behavior.
f) Viscoelastic behavior.
g) Flow models.
h) Flow behavior analysis with alternative methods.

2. Mixture Method Influence

The purpose of cement paste mixing processes is to reduce the size of particle clusters in order to make effective the wetting of each individual cement particle. Additionally, specific interest in the study of the mixture method influence on cement paste rheology is justified by the fact that cement paste forms part of the medium where aggregates are dispersed. Certainly, when concrete flows cement paste is mixed similarly to that which occurs in a laboratory stirrer. Therefore, it is of capital importance to determine the influence of mixing procedure on cement paste rheology, because it has been clearly demonstrated that the mixing procedure to obtain the cement paste has a direct influence on its rheological behaviour. This is probably due to the fact that different water–cement contacts (in number and quality) can be obtained when the mixture is agitated or mixed following different methods. Consequently, several research studies have been undertaken with the objective of achieving an optimal mixing protocol, specifically referring to the reproducibility and repeatability of the rheological results. In this line, Jones and Taylor [7] obtained rheological reproducible results applying vibration to cement pastes that had been previously mixed by hand. Yang and Jennings [8] investigated the effect of low and high mixing procedures on the peak stress value, which was obtained when a constant shear rate was applied to samples just after they were at rest along different time intervals. These authors observed that peak stress values were directly related to the size of agglomerates, and concluded that the lack of hydration of the inner particles of the agglomerates is a determinant source of microstructural defects. Additionally, they proposed a new empirical parameter, the limiting gap, as a measure of the degree of mixing. As the shear stress increases when the gap reduces, the limiting gap was defined as the gap that separates the two plates of plate–plate geometries just when the shear stress was three times higher than the shear stress measured at high gaps. Although some correlation between limiting gap and microstructure was induced, the utility of this parameter is still open to discussion. Williams et al. [9] used, in a very detailed work, plastic viscosity and the area enclosed by the up and down curves of the hysteresis loop to compare the effect of several mixing methods (hand-, paddle-, high-shear-, and concrete-mixed) on the degree of structure remaining in the cement paste. Results confirmed, as is reasonably expected, that the structure be comparatively much more broken down when the mixed method was more vigorous. Additionally, they obtained similar effects for concrete- and high-shear mixed pastes; a result that suggests that, in concrete, cement paste is highly sheared by aggregates (see Figure 1). Therefore, it could be concluded that standard initially used to prepare cement paste samples [10] is not adequate to simulate the ball-milling effect of aggregates in concrete [8]. Then, another standard [11] for the preparation of cement pastes was introduced. The capability of both standards to influence on structural and rheological properties of cement pastes was evaluated by Han and Ferron [12]. These authors obtained an unexpected result. They observed that the mixing intensity had much more influence on the yield stress and the plastic viscosity for pastes formulated with chemical additives and lower particle concentration (higher water/cement ratio). Their results were clearly counterintuitive because, while a decrease in both rheological parameters was expected with the increase of the mixing intensity, the opposite result was found using the ASTM C1738 protocol [11]. The discrepancy was especially high when plasticizers form part of the cement paste formulation. Han and Ferron [12] pointed out that if, certainly, the vigorous mixing breaks particle agglomerates, it also accelerates the cement hydration. Then, a hindering effect of the plasticizer agent due to the high velocity of mixing, which leads to a decrease in the extent of the steric diffuse double layer that surrounds cement particle and promotes particle agglomeration, was claimed [12,13].

The main conclusion of this research area is that the mixing protocol applied to cement pastes preparation is not a trivial issue. It should be intensive and lengthy enough to break all particle agglomerates, but limited by the possibility of provoking the opposite effect, especially when chemical additives form part of the cement paste formulation.

3. Microstructure Analysis

Breakdown of structure due to shear and rebuilding of structure-at-rest are phenomena of technical interest in cement paste applications. The contact of cement pastes with water gives place to a process of coagulation starting from a completely dispersed state. A gel layer of calcium silicate hydrate (C-S-H) progressively surrounds cement particles, the process being very slow during the dormant or induction period [14]. This phase initiates around 10 min after the first contact between cement and water. During the induction period, fresh cement pastes can be properly analyzed from a rheological point of view [15]. It is worth distinguishing when fresh cement pastes are studied between reversible and irreversible microstructure evolutions. Reversible evolution of the cement paste microstructure (breaking or building) can only be claimed when colloidal interactions between cement particles are taken into account. Hydration of cement particles is always connected to irreversible processes. Fortunately, during the dormant period of the cement hydration, this reaction can be slow enough to assume that chemical changes in cement pastes are negligible. However, it is important to consider that these effects are not fully absent during rheological measurements. Hydration products generate particle flocs that, finally, lead to the formation of clusters or 3D networks formed by aggregates of flocculated cement particles [16]. If we consider only reversible evolution of the microstructure, the picture can be described as follows. Initially, when the network immobilizes the liquid phase, cement pastes behave in a solid-like way, and only when the stress achieves a threshold value, or yield stress, due to the action of shear, does the structural breakdown begin. The increasing of shear rate results in further deflocculation, i.e., the apparent viscosity will decrease with shear. This is the basis for the concrete vibration technique that is applied after placing, which makes that previous stiff material can after flow easily. Therefore, lower water/cement ratios can be used in concrete formulations, and higher strength of the set concrete can, consequently, be obtained. So, it is justified why an understanding of both, reversible (thixotropy) and irreversible (setting) breakdown of structure of cement pastes, is of major importance: simply due to the role-played by cement pastes when a vibratory force is applied in the placing of concrete [1,17].

First studies of the time evolution of the microstructure of Portland cement pastes [1] showed that an exponential decrease of the torque with time, when a step-up in shear rate is applied to samples, fits reasonably well experimental results,

$$M = M_e + (M_o - M_e)e^{-kt} \tag{1}$$

In Equation (1), M is the torque at time t, k is the rate for the stress decay, and M_o and M_e are the initial and the equilibrium torque values, respectively (see Figure 2). Tattersall's results [1] showed that k increases with the angular velocity of the rotor. The area under the torque-time curve (Figure 1) can be used as a measure of pastes workability [17] because it is directly related to the net energy input or work done by the rotor to (i) overcome viscous forces, (ii) breaking the structure, and (iii) maintaining broken the structure [1].

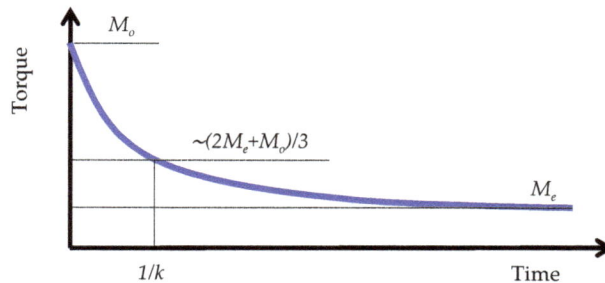

Figure 2. Schematic plot of the evolution with time of the torque after a step-up in angular velocity according with Tattersall [1].

Bouras et al. [18] found that the sum of two exponentials fitted better stress decay with time due to breakdown microstructure, when pastes contain some viscosity-modifying admixture (polysaccharide). They justified their result as due to the existence of two relaxation times for the rebuilding of microstructure. The first one is related to particle–particle interconnections as usual, and the second one refers to the alignment and disentanglement under flow of admixture molecules. Then, using Cheng and Evans' general thixotropic microstructural model [19], the kinetics equation for the structural parameter λ should be expressed as,

$$\frac{d\lambda}{dt} = \frac{1}{T_a} + \frac{1}{T_b} - \lambda\dot{\gamma} \tag{2}$$

In Equation (2), T_a is the characteristic time for cement particles aggregation, T_b is the characteristic time for the admixture entanglement, and $\dot{\gamma}$ is the shear rate. However, when Bouras et al. [18] analyzed the microstructure rebuilt process assumed that, experimentally, stress increases exponentially with time, avoiding the use of Equation (2) that predicts a linear increase of the stress with time.

Irreversible time-evolution of the microstructure is observed during the setting of cement pastes due to the hydration chemical reaction is the dominant effect. It has been observed that in this case the increase in the viscosity with time, which is partially attributed to shear-induced aggregation, is opposite to that which could be expected, faster with increasing organic additives concentration [20]. These processes are dominated by the hydration of cement particles and fall out of the scope of the rheological analysis proposed in this review. However, it is worth distinguishing between the irreversible, or aging due to hydration, and the reversible part, or thixotropy, of the transient behavior of cement pastes, as mentioned above. This is important to avoid misinterpretations of the experimental results. For example, Roussel [21] proposed a thixotropic model for cement pastes. He started from the assumption of the experimental results summarized in Equation (1) although he attributed it to Lapasin et al. [22] instead of Tattersal [1]. Then, Roussel used Cheng and Evans' general microstructure thixotropic model [19] and assumed that the material restructuration occurs in a natural way at an unique constant rate $1/T$, although maintaining the idea that the destructuration is proportional to the existing structure and to the shear rate,

$$\frac{d\lambda}{dt} = \frac{1}{T} - \alpha\lambda\dot{\gamma} \tag{3}$$

Roussel assumption predicts, as Bouras et al. [18], unrealistic linear increase of the microstructure-at-rest. When shear rate is zero,

$$\frac{d\lambda}{dt} = \frac{1}{T} \tag{4}$$

Solving Equation (4),

$$\lambda = \frac{t}{T} + \lambda_o \tag{5}$$

However, experimental tests have demonstrated that the microstructure-at-rest increases exponentially until a maximum (reversible) structure is achieved [18,23]. This result is opposite to which Equation (5) predicts [21].

Very recently, Ma et al. [24] have proposed a kinetics equation for the rebuilt process, which expresses restoration of an equilibrium state from a non-equilibrium condition, i.e.,

$$\frac{d\lambda}{dt} = -\frac{1}{T}(\lambda - \lambda_e) \tag{6}$$

In Equation (6) $\lambda_e > \lambda$ represents the structure of the paste at equilibrium. Effectively, solving Equation (6) a most realistic exponential evolution of the microstructure towards the rest equilibrium state is obtained (Figure 3),

$$\lambda = \lambda_e + (\lambda_o - \lambda_e)e^{-t/T} \tag{7}$$

It is worth noting that experimental results by Ma et al. [24] supported the existence of two structural levels (particle flocs and C-S-H nucleation) in cement pastes, which were previously pointed out by Roussel et al. [25].

Summarizing, fresh cement pastes are concentrated particle suspensions in which particle-particle interaction is not only governed by colloidal and hydrodynamic forces, but also steric (admixtures) and products of the hydration reaction (C-S-H) play a specific role for the development of microstructures. These aspects of cement particle interactions make the study of the microstructural evolution of cement pastes with time (thixotropy) especially interesting.

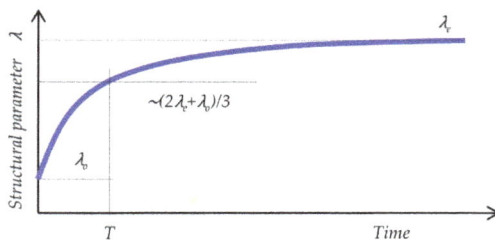

Figure 3. Schematic plot of the evolution with time of the atructure at rest according with Ma et al. [24].

4. Mineral Additions Influence

The flow behavior of cement pastes is characterized by a variety of rheological parameters. It is a task of major importance to determine the influence of the cement properties, and of mineral and chemical additions on these parameters. So, a lot of work has been done with the aim to determine correlations between different geometric, physical and chemical modifications of cement pastes with their rheological properties.

Some results corresponding to this research sub-area refer to the yield stress value. For example, it has been observed that the yield stress increases with the specific surface area of cement particles and the water/cement ratio [26]. Ivanov et al. [27] obtained an increasing dependence of the yield stress and the plastic viscosity with a variety of factors; concretely, they observed that the yield stress is more dependent according to the following succession, superplasticizer > addition of silica fume > water/cement ratio; while the plastic viscosity is dependent in the following succession, water/cement ratio > superplasticizer > addition of silica fume. Nehdi et al. [28] observed that the replacement of cement with limestone filler slightly increased the yield stress of cement paste with added plasticizer, but its plastic viscosity decreased. Latter, Rubio-Hernández et al. [23] observed that when the limestone filler concentration is lower than 3% weight the opposite effect appears, i.e., the yield stress decreases and the plastic viscosity increases.

It is reasonably expected that the different factors on which the rheological behavior of cement pastes is dependent, interact among them giving place to complex dependencies. Then, it is necessary to define combinations of them. So, Wong and Kwan [29] considered the effect of the excess water to solid surface area ratio on the rheological properties of cement pastes. In this way, water content, packing density and solid surface area effects were simultaneously considered. They found that yield stress and plastic viscosity versus excess water/solid surface area ratio curves overlap when this parameter was higher than 0.05 μm, while at lower values the yield stress and the plastic viscosity curves separate, distinguishing between pastes with and without silica fume, i.e., the rheology of cement pastes also depends on silica fume content at the lower water/solid surface area ratios.

One of the technical applications pursued by means of mineral additions is the design of highly concentrated suspensions with moderate yield stress and plastic viscosity values [30]. To this end, advantages resulting from the use of bimodal suspensions have been considered. This strategy can be useful to improve the performance of, for example, extruded materials [31]. For example, bimodal suspensions of cement and clay particles showed higher yield stress values than those obtained for the original cement suspension, despite the maximum packing fraction practically did not vary [32]. This apparently contradictory result was justified as due to the water adsorption of clays, which eliminates lubricant water phase and increases the effective size of clay particles [33]. Fine mineral additions like limestone, silica fume, fly ash, etc. increase maximum packing fraction with good lubricant effect, and the resulting cement pastes can serve as the base for self-compacting concrete design [33]. However, not only the particle size but also the geometric shape have a determinant effect on the cement paste and, consequently, on the concrete fluidity. So, when fly ash particles of spherical shape substituted amorphous cement particles, the viscosity or consistency of fresh cement pastes decreases, and even lower yield stress values can be obtained although the water/binder ratio decreases [34]. Nevertheless, high dosages of fly ash can lead to fluidity loss due to its high specific surface area, which provokes higher water demand and admixture consumption. The combined effects of spherical shape and high polydispersity of fly ash added to cement pastes can enhance the results before pointed out, i.e., for a given solid volume fraction, the plastic viscosity increases and the yield stress decreases with particle polydispersity [35]. In the last sense, the main objective pursued adding fine mineral additions with a high polydispersity index is to reduce, as low as possible, water/binder ratio in order to obtain optimal conditions for the production of high-strength and high-performance concretes [36]. Sometimes, mineral additions are used in combination with chemical additives (superplasticizers) [37]. The objective is to reduce water demand maintaining workability and strength at hardened state. Not all mineral additions can be used to this end. For example, this can be achieved using fly ash but not limestone [38] or nano-$CaCO_3$ [39].

Summarizing, mineral additions are used to improve hardened properties of cementitious materials maintaining or refining fresh state behaviour. Two main types of mineral additions are used, with and without pozzolanic activity. The first type will be specifically considered in part 6 when advances in blended cement are reviewed. The main characteristics of mineral additions without pozzolanic activity that have an influence on the rheology of cement pastes are size, shape and particle polydispersity. In both cases, combined action of mineral additions with plasticizers gives place to cement paste formulations optimal for high-performance concretes design.

5. Chemical Admixtures Influence

There are two most important effects that the presence of chemical additives has on the rheology of fresh cement pastes,

a) the increase of the duration of the dormant period [40], because hydration reaction of cement slows down [41], allowing considering cement pastes as a chemically stable suspension during longer time intervals [42],

b) and the modification of its viscosity [43,44].

The type of chemical additive is determinant for the observation of different rheological behaviors. Specifically, polysaccharide gums increase the viscosity of pastes, enhance rebuilt-up kinetics at rest, and increase the yield stress [18], the effect being much more significant at low than at high shear rates [43]. So, the cement paste will be stable against sedimentation at rest and resistive to solid–liquid separation, but will flow easily when, for example, being pumped.

Sometimes, counterintuitive effects, like a decrease of the viscosity of cement paste when the viscosity of the interstitial liquid phase is increased due to the addition of a chemical admixture, are also observed. This phenomenon has been justified by the lubricant action of the liquid phase with respect to solid particles [45], which is higher when the admixture viscosity is higher [46].

Whatever the case is, the performance of a chemical additive is determined by its compatibility with cements, i.e., the adsorption capability of each type of cement particles [44,47], and temperature [48]. Bonen and Sarkar [49] analyzed the adsorption capacity of sodium salt of polynaphthalene sulfonate superplasticizer by different cement types, and concluded that cement fineness and superplasticizer molecular weight are the main factors determining the adsorption capacity of pastes. Bessaies-Bey et al. [50] studied polyacrylamide adsorption on cement particles, and observed the formation of polymer micro-gels that not only adsorb on particles but also bridge them increasing consequently the yield stress of the cement paste. Then, Mukhopadhyay and Jang [51] proposed a rheological method to quantify cement-admixture incompatibilities. They measured the time evolution of the yield stress and plastic viscosity of pastes and justified, combined with heat of hydration evolution data, that incompatibility of the admixture corresponded to a rate of change of yield stress lower than 14 Pa/h and a rate of change of plastic viscosity lower than 0.02 Pas/h.

Two main types of chemical admixtures can be distinguished, viscosity-modifying admixtures, which increase the viscosity of cement pastes [43,44], and superplasticizers, which disperse cement particles [52,53].

In order to design adequate formulations of cementitious materials, the optimal dosage of chemical admixtures must be determined [54–56]. Sometimes, this does not coincide with that recommended by the producer [57] or is clearly dependent on physical conditions of the cement paste, as temperature [58]. Moreover, incompatibilities before cited between cement and chemical admixtures should be well established to avoid problems in fluidity [58]. In fact, the same chemical admixture can show a different performance when added to different cement types [59]. It appears that polycarboxylate type superplasticizer have the best compatibility with a wider variety of cements [60]. So, the study of this kind of superplasticizers has extended to consider a variety of molecular conformations due to the incorporation of different hydrophobic groups to the polycarboxylate molecule [61].

Another aspect that must be considered when several chemical admixtures are simultaneously used in the formulation of the same cement paste is its synergic effect [62–65]. For example, it was observed that a polysaccharide viscosity-modifying agent (Welan Gum) has a higher influence on rheological properties of cement pastes formulated with an ester polycarboxylate superplasticizer than when it was formulated with an ether polycarboxylate [66]. Nanoclay enhances or modifies [24] thixotropy behavior of cement pastes with polycarboxylate ether superplasticizer [67]. Hydroxypropyl-methyl cellulose ether [68] and polyacrylic acid [69] acts opposite to the dispersive capability of polycarboxylate type superplasticizers. The presence of other mineral additions, as borax, in order to retard cement paste setting must be analyzed in depth due to the negative effect that can induce in the adsorption of superplasticizer molecules onto the particle surface [70]. On the other hand, specific plasticizer and viscosity-modifying admixture concentrations are necessary to achieve the best rheological performance of cement pastes [30]. Although polycarboxylates are more effective than lignosulfonates, i.e., a lower concentration of polycarboxylates is necessary to obtain the same yield stress and consistency reductions, when they are combined, lignosulfonates adsorption onto the cement particle surface is dominant [71], although limited by the amount of C_3A in the composition of the clinker from which the cement powder is obtained [72].

Viscosity-modifying admixture of starch type can give place to different viscous behavior at low (shear-thinning) and high (shear-thickening) shear rates. These opposite behaviours have been ascribed to disentanglement and alignment of admixture molecules at low shear, and the increase of repulsive interparticle forces at high shear [73].

Summarizing, two main types of chemical admixtures are used to modify cement paste rheological behavior, viscosity-modifying admixtures and superplasticizers. The first are used to improve stability of pastes against sedimentation and bleeding, while superplasticizers increase the dormant period and allow for reduction of water content, which leads to higher strength when pastes set. Synergic effects must be carefully considered when different chemical admixtures are tested.

6. Blended Cements Behavior

Although the term blended cements generally refers to materials obtained when the ligand phase is obtained with water added to the mixture of cement and another material in different states of aggregation [74–76], here we will limit the definition to materials that are obtained when cement is partially substituted by another solid phase. This last can or cannot show pozzolanic activity. The term "binder" is used for the mixture cement + solid addition. It is widely accepted that particle morphology [77] and pozzolanic activity [78] of binders are the main features that determine their influence on the rheological behaviour of blend cement pastes.

Limestone filler and silica fume are two examples of solid substitutions without pozzolanic activity. It has been observed that the substitution of cement by limestone filler, maintaining constant the water/binder ratio, increases the yield stress and decreases the plastic viscosity of pastes [28]. Then, the stability at rest of cement pastes is enhanced, although the increase of the segregation of the phases can be an undesirable consequence. It has been also observed that the particle morphology of limestone has much more influence on the rheological behavior of pastes than its chemical composition [79].

Activated kaolinite (metakaolin) is a widely used example of the solid phase with pozzolanic activity [80]. The yield stress of blended pastes increases when the amount of activated kaolin increases [81] or the pozzolanic activation of kaolinite is enhanced [82]. Even then it can give to cement paste adequate characteristics for use in self-compacting concrete formulation [83]. Fly ash is another binder extensively used due to its pozzolanic activity. The origin of fly ash can be the combustion of fuel [34] or different biomasses [84]. In both cases, the regular shape of particles is a characteristic demanded for the optimal rheological behavior of pastes.

The use of natural pozzolans (mixtures of volcanic ash and pumice powder) as partial substitution of Portland cement has been shown to be an economic and very useful alternative in volcanic zones. Again, the yield stress increases and the plastic viscosity decreases with cement substitution. This variation of the rheological parameters has been justified by using a model that treats fresh volcanic cement pastes as suspensions of particles in a fluid phase formed by water and the gel resulting from the chemical hydration reaction of cement [78].

Blended cements contribute to the reduction of CO_2 emissions that result from cement production, and help to conserve the environment when waste solid materials are used. These are two important reasons that justify the interest for this research area that can be added to the increase of concrete durability thanks to improving the resistance to salts, freeze-thaw and carbonation effects.

7. Viscoelastic Behaviour

The objective of the rheological studies of cement pastes is to obtain rheological parameters with some practical meaning. This has traditionally been the case with viscosity, yield stress, and plastic viscosity, which are related to workability, fluidity and resistance to segregation, respectively. However, these rheological parameters account only for the viscous behaviour of the pastes. Additionally, time dependence of viscosity (thixotropy) is used to quantify microstructure changes, which is information useful to avoid, for example, cold joints in sequential casting applications. Moreover, viscoelastic studies are considered when useful knowledge for practical applications is obtained, i.e., on workability [57], microstructural evolution [85,86], and pumping [87].

Creep-recovery tests have been used for cement pastes to determine with much more precision the yield stress value. This is the stress value that limits the transition from viscoelastic solid-like to liquid-like behaviour [88,89].

Small-amplitude oscillatory shear applied to cement pastes has been shown to be the main evolution of the microstructure of the paste occuring just after the first contact of the cement with water [85,86]; the linear viscoelastic behaviour is limited by lower shear strain values when the water/cement ratio increases [88]. Moreover, as is reasonably expected, the storage modulus also decreases when the water/cement ratio increases [90]. Large-amplitude oscillatory shear will be of interest when the meaning for the results is proposed.

8. Flow Models

As cement pastes need, in general, for a threshold shear stress to be surpassed in order to observe flow, viscoplastic models [26,57,91–97] have been largely used to describe shear-stress-shear-rate or steady flow curves (Table 1). Some confusion results from the use of ramp flow curves (non-steady) to fit or describe new viscoplastic models [98]. Whatever the case, the influence of cement paste composition (cement type, water/cement ratio, type and concentration of different additions and admixtures, etc.) on the value of the model parameters, has been the subject of a large number of publications [2,27,59,99–101]. For example, Jones and Taylor [7] proposed a model based on Robertson–Stiff's to relate the flow curve of cement pastes to water/cement ratio,

$$\tau = (aw + b)\left\{\dot{\gamma} + (cw + d)\right\}^{\left(\frac{r}{w} + se^{-w}\right)} \tag{8}$$

where w is the water/cement ratio, and a, b, c, d, r, and s are constants to be determined. However, this 6-parameter model is capable for describing only qualitatively shear-stress-shear-rate-water/cement-ratio curves [7].

Table 1. Viscoplastic models used for describing steady flow curve of cement pastes.

Model Name	Equation	Reference
Bingham	$\tau = \tau_y + \eta_p \dot{\gamma}$	[90]
Modified Bingham	$\tau = \tau_y + \eta_p \dot{\gamma} + c\dot{\gamma}^2$	[91]
Herschel–Bulkley	$\tau = \tau_y + K\dot{\gamma}^n$	[92]
Robertson–Stiff	$\tau = A(C + \dot{\gamma})^B$	[93]
Karam	$\tau = A exp(k\phi)(\dot{\gamma}_o - \dot{\gamma})$	[94]
Casson	$\tau = \tau_y + \eta_\infty \dot{\gamma} + 2\sqrt{\tau_y \eta_\infty}\sqrt{\dot{\gamma}}$	[95]
Modified Casson	$\tau^m = \tau_y^m + \eta_\infty \dot{\gamma}^m$	[96]
Papo–Piani	$\tau = \tau_y + \eta_p \dot{\gamma} + K\dot{\gamma}^n$	[57]
Vom Berg	$\dot{\gamma} = B sinh\left(\frac{\tau - \tau_y}{A}\right)$	[97]

The responses of cementitious materials to flow after rest state cycles have been also modeled [102].

9. Flow Behavior Analysis with Alternative Methods

It has been recognized the influence of the measurement device on the results of rheological testing. A study on concentric cylindrical geometries [103] lead to acknowledge that the surfaces in contact with cement pastes must be roughness to avoid wall-slip phenomena, and the gap between rotor and stator must be large enough to guarantee laminar and homogeneous flow of the fluid. The use of cone-plate geometry to obtain rheological data is not appropriate due to the size of cement particles (10–100 μm) despite, surprisingly, some authors affirming the reproducibility of the results when the gap was the same order than particle size [7].

The sedimentation of cement grains is one of the most important problems that must be avoided to obtain valid results. With the aim to reduce its negative impact on rheological measurements, alternatives to rotational rheometers have been explored. For example, the turning-tube viscometer [104] avoids cement grain sedimentation, although rheometric measurements are relative, i.e., it does not supply absolute or fundamental rheological parameters.

Squeeze flow has been also used to characterize rheologically cement pastes [105]. However, it is necessary to assume a variety of simplifications in the experimental procedure (infinite volume of the sample, neglecting buoyancy force on the top plate, and neglecting of stress due to friction) to obtain results with some physical meaning. Despite squeeze and shear experiments results did not coincide [104], it is a valuable technique to mimic the flow conditions experienced by the cement paste in the inner granular space of concretes [106].

The inclined plane test has been shown to be a valuable way to infer rheological characteristics of cement pastes [107]. Shear stress-shear rate curves show reasonable agreement with those obtained with conventional rotational rheometers but what is most remarkable is that they allow us to show directly that the yield stress increases with the time that the sample is at rest before the test was made.

Extrusion is used in the formation of cement pastes. To determine the conditions that maintain constant the shape and cohesion of the extruder, specific studies complementary to rotational rheometry must be undertaken. Another reason for the use of this technique is the possibility to test materials with very low water/binder ratios [108]. In this respect, the presence of fly ash reduces the value of the extrusion load due to the lower size and the spherical shape of fly ash particles gives place to a lubricant effect [31].

Flow measurements at high pressure conditions can be made with specific cells that can be coupled to conventional rotational rheometers. They are designed to simulate pumping processes [109].

10. Conclusions

Interest in gaining information about the rheological behavior of fresh cement pastes (RBFCP) is mainly due to the role played by this phase in concrete and mortar formulations. Moreover, as cement pastes are particle dispersions featured by a decisive hydration chemical activity, which generates products from the reaction that form a gel phase interconnecting the core of cement particles, academic interest in the study of this material has also given rise to a variety of research sub-fields on RBFCP. Briefly, the state of the art can be summarized as follows:

a)   The mixing protocol applied to cement paste preparation is not a trivial issue. It should be intensive and lengthy enough to break all particle agglomerates, but limited by the possibility of provoking the opposite effect, especially when chemical additives form part of the cement paste formulation.

b)   Fresh cement pastes are concentrated particle suspensions in which particle–particle interactions are governed by colloidal, steric, and hydrodynamic forces, and also by products of the hydration reaction (C–S–H).

a)   The combined action of mineral additions with plasticizers gives rise to cement paste formulations optimal for high-performance concrete design.

d)   Viscosity-modifying admixtures and superplasticizers are used to improve the stability of pastes against sedimentation and bleeding, and to increase the duration of the dormant period and allow for a reduction in water content. Synergic effects must be carefully considered when different chemical admixtures are jointly tested.

e)   Studies on blended cements are justified by the benefits on the environment and the possibility to develop new and improved concrete formulations.

f)   Viscoelastic studies on cement pastes need to be properly justified, giving practical meaning to the rheological parameters thus obtained.

g)   Cement paste is a viscoplastic material. In this case, the rheological parameters commonly analyzed, the yield stress and the plastic viscosity, are related to practical uses. This is because steady viscous flow studies have traditionally been undertaken.

h)   Sedimentation and wall slip are two error sources in rheological tests. Then, new experimental methods must be developed that, in addition, can inform us about other flow cement pastes characteristics.

Funding: This research received no external funding.

Conflicts of Interest: The author declares no conflicts of interest.

References

1.   Tattersall, G.H. The rheology of Portland cement pastes. *Br. J. Appl. Phys.* **1955**, *6*, 165–167. [CrossRef]

2.   Tattersall, G.H.; Banfill, P.F.G. *The Rheology of Fresh Concrete*; Pitman Advanced Publishing Program: Boston, MA, USA, 1983; ISBN 978-0273085584.

3.   Struble, L.; Szecsy, R.; Lei, W.G.; Sun, G.K. Rheology of cement paste and concrete. *Cem. Concr. Aggr.* **1998**, *20*, 269–277. [CrossRef]

4.   Ferraris, C.F.; Gaidis, J.M. Connection between the Rheology of concrete and Rheology of cement paste. *ACI Mater. J.* **1992**, *89*, 388–393. [CrossRef]

5.   Banfill, P.F.G. The Rheology of cement paste: Progress since 1973. In *Properties of Fresh Concrete, Proceedings of the International RILEM Colloquium, Leeds, UK, 22–24 March 1973*; Wierig, H.J., Ed.; Taylor & Francis: New York, NY, USA, 1990; pp. 3–9.

6.   Banfill, P.F.G. The Rheology of fresh cement and concrete-A review. In Proceedings of the 11th International Cement Chemistry Congress, Durban, UK, 11–16 May 2003.

7.   Jones, T.E.R.; Taylor, S. A mathematical model relating the flow curve of a cement paste to its water/cement ratio. *Mag. Concr. Res.* **1977**, *29*, 207–212. [CrossRef]

8. Yang, M.; Jennings, H.M. Influences of mixing methods on the microstructure and rheological behavior of cement paste. *Adv. Cem. Based Mater.* **1995**, *2*, 70–78. [CrossRef]
9. Williams, D.A.; Saak, A.W.; Jennings, H.M. The influence of mixing on the rheology of fresh cement paste. *Cem. Concr. Res.* **1999**, *29*, 1491–1496. [CrossRef]
10. ASTM C305-14. *Standard Practice for Mechanical Mixing of Hydraulic Cement Pastes and Mortars of Plastic Consistency*; ASTM International: West Conshohocken, PA, USA, 2014.
11. ASTM C1738/C1738M-18. *Standard Practice for High-Shear Mixing of Hydraulic Cement Pastes*; ASTM International: West Conshohocken, PA, USA, 2018.
12. Han, D.; Ferron, R.D. Effect of mixing method on microstructure and rheology of cement paste. *Constr. Build. Mater.* **2015**, *93*, 278–288. [CrossRef]
13. Han, D.; Ferron, R.D. Influence of high mixing intensity on rheology, hydration, and microstructure of fresh state cement paste. *Cem. Concr. Res.* **2016**, *84*, 95–106. [CrossRef]
14. RILEM Committee 68-MMH. The hydratium of tricalcium silicate. *Mater. Struct.* **1984**, *17*, 457–468. [CrossRef]
15. Lei, W.G.; Struble, L.J. Microstructure and flow behavior of fresh cement paste. *J. Am. Ceram. Soc.* **1997**, *80*, 2021–2028. [CrossRef]
16. Jiang, W.; Roy, D.M. Microstructure and flow behavior of fresh cement paste. In *Flow and Microstructure of Dense Suspensions*; Struble, L.J., Zukoski, C.F., Maitland, G.C., Eds.; MRS Online Proceedings Library Archive: Warrendale, PA, USA, 1993; Volume 289, pp. 161–166.
17. Nessim, A.A.; Wajda, R.L. The rheology of cement pastes and fresh mortars. *Mag. Concr. Res.* **1965**, *17*, 59–68. [CrossRef]
18. Bouras, R.; Chaouche, M.; Kaci, S. Influence of viscosity-modifying admixtures on the thixotropic behaviour of cement pastes. *Appl. Rheol.* **2008**, *18*, 1–8. [CrossRef]
19. Cheng, D.H.; Evans, F. Phenomenological characterization of the rheological behaviour of inelastic reversible thixotropic and antithixotropic fluids. *Br. J. Appl. Phys.* **1965**, *16*, 1599–1617. [CrossRef]
20. Otsubo, Y.; Miyai, S.; Umeya, K. Time-dependent flow of cement pastes. *Cem. Concr. Res.* **1980**, *10*, 631–638. [CrossRef]
21. Roussel, N. Steady and transient flow behaviour of fresh cement pastes. *Cem. Concr. Res.* **2005**, *35*, 1656–1664. [CrossRef]
22. Lapasin, R.; Longo, V.; Rajgelj, S. Thixotropic behaviour of cement pastes. *Cem. Concr. Res.* **1979**, *9*, 309–318. [CrossRef]
23. Rubio-Hernández, F.J.; Morales-Alcalde, J.M.; Gómez-Merino, A.I. Limestone filler/cement ratio effect on the flow behaviour of a SCC cement paste. *Adv. Cem. Res.* **2013**, *25*, 262–272. [CrossRef]
24. Ma, S.; Qian, Y.; Kawashima, S. Experimental and modeling study on the non-linear structural buil-up of fresh cement pastes incorporating viscosity modifying admixtures. *Cem. Concr. Res.* **2018**, *108*, 1–9. [CrossRef]
25. Roussel, N.; Ovarlez, G.; Garrault, S.; Brumaud, C. The origins of thixotropy of fresh cement pastes. *Cem. Concr. Res.* **2012**, *42*, 148–157. [CrossRef]
26. Vom Berg, W. Inflkuence of specific surface and concentration of solids upon the flow behaviour of cement pastes. *Mag. Concr. Res.* **1979**, *31*, 211–216. [CrossRef]
27. Ivanov, Y.P.; Roshavelov, T.T. Flow behaviour of modified cement pastes. *Cem. Concr. Res.* **1993**, *23*, 803–810. [CrossRef]
28. Nehdi, M.; Mindess, S.; Aïtcin, P.C. Statistical modeling on the microfiller effect on the rheology of composite cement pastes. *Adv. Cem. Res.* **1997**, *9*, 37–46. [CrossRef]
29. Wong, H.H.C.; Kwan, A.K.H. Rheology of cement paste: Role of excess water to solid surface area ratio. *J. Mater. Civ. Eng.* **2008**, *20*, 189–197. [CrossRef]
30. Martins, R.M.; Bombard, A.J.F. Rheology of fresh cement paste with superplasticizer and nanosilica admixtures studied by response surface methodology. *Mater. Struct.* **2012**, *45*, 905–921. [CrossRef]
31. Micaelli, F.; Lanos, C.; Levita, G. Rheology and extrusion of cement-fly ashes pastes. In Proceedings of the XVth International Congress on Rheology, the Society of Rheology 80th Annual Meeting, Monterey, CA, USA, 3–8 August 2008; Co, A., Leal, L.G., Colby, R.H., Giacomm, A.J., Eds.; Amerivan Institute of Physics: College Park, MD, USA, 2008; pp. 665–667.
32. Tregger, N.A.; Pakula, M.E.; Shah, S.P. Influence of clays on the rheology of cement pastes. *Cem. Concr. Res.* **2010**, *40*, 384–391. [CrossRef]

33. Diamantonis, N.; Marinos, I.; Katsiotis, M.S.; Sakellariou, A.; Papathanasiou, A.; Kaloidas, V.; Katsioti, M. Investigations about the influence of fine additives on the viscosity of cement paste for self-compacting concrete. *Constr. Build. Mater.* **2010**, *24*, 1518–1522. [CrossRef]

34. Rubio-Hernández, F.J.; Cerezo-Aizpún, I.; Velázquez-Navarro, J.F. Mineral additives geometry influence in cement pastes flow. *Adv. Cem. Res.* **2011**, *23*, 55–60. [CrossRef]

35. Bentz, D.P.; Ferraris, C.F.; Galler, M.A.; Hansen, A.S.; Guynn, J.M. Influence of particle size distributions on yield stress and viscosity of cement-fly ash pastes. *Cem. Concr. Res.* **2012**, *42*, 404–409. [CrossRef]

36. Kwan, A.K.H.; Chen, J.J. Roles of packing density and water film thickness in rheology and strength of cement paste. *J. Adv. Concr. Technol.* **2012**, *10*, 332–344. [CrossRef]

37. Stefancic, M.; Mladenovic, A.; Bellotto, M.; Jereb, V.; Zavrsnik, L. Particle packing and rheology of cement pastes at different replacement levels of cement by α-Al_2O_3 submicron particles. *Constr. Build. Mater.* **2017**, *139*, 256–266. [CrossRef]

38. Burgos-Montes, O.; Alonso, M.M.; Puertas, F. Viscosity and water demand of limestone- and fly ash-blended cement pastes in the presence of superplasticisers. *Constr. Build. Mater.* **2013**, *48*, 417–423. [CrossRef]

39. Sun, R.; Zhao, Z.; Huang, D.; Xin, G.; Wei, S.; Ge, Z. Effect of fly ash and nano-$CaCO_3$ on the viscosity of cement paste. *Appl. Mech. Mater.* **2013**, *357–360*, 968–971. [CrossRef]

40. Simard, M.A.; Nkinamubanzi, P.C.; Jolicoeur, C.; Perraton, D.; Aïtcin, P.C. Calorimetry, rheology and compressive strength of superplasticized cement pastes. *Cem. Concr. Res.* **1993**, *23*, 939–950. [CrossRef]

41. Ltifi, M.; Guefrech, A.; Mounanga, P. Effects of sodium tripolyphosphate on the rheology and hydration rate of Portland cement pastes. *Adv. Cem. Res.* **2012**, *24*, 325–335. [CrossRef]

42. Mikanovic, N.; Jolicoeur, C. Influence of superplasticizers on the rheology and stability of limestone and cement pastes. *Cem. Concr. Res.* **2008**, *38*, 907–919. [CrossRef]

43. Ghio, V.A.; Monteiro, P.J.M.; Demsetz, L.A. The rheology of fresh cement paste containing polysaccharide gums. *Cem. Concr. Res.* **1994**, *24*, 243–249. [CrossRef]

44. Lachemi, M.; Hossain, K.M.A.; Lambros, V.; Nkinamubanzi, P.C.; Bouzoubaa, N. Performance of new viscosity modifying admixtures in enhancing the rheological properties of cement paste. *Cem. Concr. Res.* **2004**, *34*, 185–193. [CrossRef]

45. Hot, J.; Besdsaies-Bey, H.; Brumaud, C.; Duc, M.; Castella, C.; Roussel, N. Adsorbing polymers and viscosity of cement pastes. *Cem. Concr. Res.* **2014**, *63*, 12–19. [CrossRef]

46. Lombois-Burger, H.; Colombet, P.; Halary, J.L.; Van Damme, H. On the frictional contribution to the viscosity of cement and silica pastes in the presence of adsorbing and non adsorbing polymers. *Cem. Concr. Res.* **2008**, *38*, 1306–1314. [CrossRef]

47. Colombo, A.; Geiker, M.R.; Justnes, H.; Lauten, R.A.; De Weerdt, K. On the effect of calcium lignosulfonate on the rheology and setting time of cement paste. *Cem. Concr. Res.* **2017**, *100*, 435–444. [CrossRef]

48. Vicar, H. Influence of temperature, cement and plasticizer type on the rheology of paste. In Proceedings of the Second International Symposium on Design, Performance and Use of Self-Consolidating Concrete, Beijing, China, 5–7 June 2009.

49. Bonen, D.; Sarkar, S.L. The superplasticizer adsorption capacity of cement pastes, pore solution composition, and parameters affecting flow loss. *Cem. Concr. Res.* **1995**, *25*, 1423–1434. [CrossRef]

50. Bessaies-Bey, H.; Baumann, R.; Schmitz, M.; Radler, M.; Roussel, N. Effect of polyacrylamide on rheology of fresh cement pastes. *Cem. Concr. Res.* **2015**, *76*, 98–106. [CrossRef]

51. Mukhopadhyay, A.K.; Jang, S. Predicting cement-admixture incompatibilities with cement paste rheology. *Transp. Res. Rec. J Transp. Res. B* **2012**, *2290*, 19–29. [CrossRef]

52. Struble, L.; Sun, G.K. Viscosity of Portyland cement paste as a function of concentration. *Adv. Cem. Based Mater.* **1995**, *2*, 62–69. [CrossRef]

53. Houst, Y.F.; Flatt, R.J.; Bowen, P.; Hofmann, H.; Mäder, U.; Widmer, J.; Sulser, U.; Bürge, T.A. Influence of superplasticizer adsorption on the rheology of cement paste. In Proceedings of the International Conference "The Role of Chemical Admixtures in High Performance Concrete", Monterrey, Mexico, 21–26 March 1999; Cabrera, J.G., Rivera-Villareal, R., Eds.; RILEM Publications S.A.R.L.: Cachan, France, 1999; pp. 387–402.

54. Jayasree, C.; Gettu, R. Experimental study of the flow behaviour of superplasticized cement paste. *Mater. Struct.* **2008**, *41*, 1581–1593. [CrossRef]

55. Kwan, A.K.H.; Chen, J.J.; Fung, W.W.S. Effects of superplasticiser on rheology and cohesiveness of CSF cement paste. *Adv. Cem. Res.* **2012**, *24*, 125–137. [CrossRef]

56. Liu, J.; Wang, K.; Zhang, Q.; Han, F.; Sha, J.; Liu, J. Influence of superplasticizer dosage on the viscosity of cement paste with low water-binder ratio. *Constr. Build. Mater.* **2017**, *149*, 359–366. [CrossRef]

57. Papo, A.; Piani, L. Flow behaviour of fresh Portland cement pastes. *Part. Sci. Technol.* **2004**, *22*, 201–212. [CrossRef]

58. John, E.; Gettu, R. Effect of temperatura on fflow properties of superplasticized cement paste. *ACI Mater. J.* **2014**, *111*, 67–76.

59. Papo, A.; Piani, L.; Ceccon, L.; Novelli, V. Flow behavior of fresh very high strength Portland cement pastes. *Part. Sci. Technol.* **2010**, *28*, 74–85. [CrossRef]

60. Hanehara, S.; Yamada, K. Interaction between cement and chemical admixture from the point of cement hydration, absorption behaviour of admixture, and paste rheology. *Cem. Concr. Res.* **1999**, *29*, 1159–1165. [CrossRef]

61. Shu, X.; Zhao, H.; Wang, X.; Zhang, Q.; Yang, Y.; Ran, Q.; Liu, J. Effect of hydrophobic units of polycarboxylate superplasticizer on the flow behavior of cement paste. *J. Disp. Sci. Technol.* **2017**, *38*, 256–264. [CrossRef]

62. Ouyang, J.; Han, B.; Cao, Y.; Zhou, W.; Li, W.; Shah, S.P. The role and interaction of superplasticizer and emulsifier in fresh cement asphalt emulsion paste through rheology study. *Constr. Build. Mater.* **2016**, *125*, 643–653. [CrossRef]

63. Yuan, Q.; Liu, W.T.; Wang, C.; Deng, D.H.; Liu, Z.Q.; Long, G.C. Coupled effect of viscosity enhancing admixtures and superplasticizers on rheological behavior of cement paste. *J. Cent. South Univ.* **2017**, *24*, 2172–2179. [CrossRef]

64. Tan, H.; Zuo, F.; Ma, B.; Guo, Y.; Li, X.; Mei, J. Effect of competitive adsorption between sodium gluconate and polycarboxylate superplasticizer on rheology of cement paste. *Constr. Build. Mater.* **2017**, *144*, 338–346. [CrossRef]

65. Reales, O.A.M.; Jaramillo, Y.P.A.; Botero, J.C.O.; Delgado, C.A.; Quintero, J.H.; Filho, R.D.T. Influence of MWCNT/surfactant dispersions on the rheology of Portland cement pastes. *Cem. Concr. Res.* **2018**, *107*, 101–109. [CrossRef]

66. Wang, D.; Liu, Z.; Wu, Z.; Xiong, W.; Zuo, Y. Effect of viscosity modifying agents on the rheology properties of cement paste with polycarboxylate superplasticizer. In Proceedings of the Second International Symposium on Design, Performance and Use of Self-Consolidating Concrete, Beijing, China, 5–7 June 2009.

67. Qian, Y.; De Schutter, G. Enhancing thixotropy of fresh cement pastes with nanoclay in presence of polycarboxylate ether superplasticizer (PCE). *Cem. Concr. Res.* **2018**. [CrossRef]

68. Ma, B.; Peng, Y.; Tan, H.; Jian, S.; Zhi, Z.; Guo, Y.; Qi, H.; Zhang, T.; He, X. Effect of hydroxypropyl-methyl cellulose ether on rheology of cement paste plasticized by polycarboxylate superplasticizer. *Constr. Build. Mater.* **2018**, *160*, 341–350. [CrossRef]

69. Ma, B.; Peng, Y.; Tan, H.; Lv, Z.; Deng, X. Effect of polyacrylic acid on rheology of cement paste plasticized by polycarboxylate superplasticizer. *Materials* **2018**, *11*, 1081. [CrossRef]

70. Tan, H.; Guo, Y.; Zuo, F.; Jian, S.; Ma, B.; Zhi, Z. Effect of borax on rheology of calcium sulphoaluminate cement paste in the presence of polycarboxylate superplasticizer. *Constr. Build. Mater.* **2017**, *139*, 277–285. [CrossRef]

71. Rubio-Hernández, F.J.; Moreno-Lechado, S.; Velázquez-Navarro, J.F. Experimental study on the influence of two different additives onto the flow behaviour of a fresh cement paste. *Adv. Cem. Res.* **2011**, *23*, 255–263. [CrossRef]

72. Ng, S.; Justnes, H. Influence of lignosulfonate on the early age rheology and hydration characteristics of cement pastes. *J. Sustain. Cem.-Based Mater.* **2015**, *4*, 15–24. [CrossRef]

73. Bouras, R.; Kaci, A.; Chaouche, M. Influence of viscosity modifying admixtures on the rheological behavior of cement and mortar pastes. *Korea-Aust. Rheol. J.* **2012**, *24*, 35–44. [CrossRef]

74. Ouyang, J.; Tan, Y. Rheology of fresh cement asphalt emulsion pastes. *Constr. Build. Mater.* **2015**, *80*, 236–243. [CrossRef]

75. Ouyang, J.; Corr, D.J.; Shah, S.P.; Asce, M. Factors influencing the Rheology of fresh cement asphalt emulsion paste. *J. Mater. Civ. Eng.* **2016**, *28*, 1–9. [CrossRef]

76. Ouyang, J.; Tan, Y.; Corr, D.J.; Shah, S.P. Viscosity prediction of fresh cement asphalt emulsion pastes. *Mater. Struct.* **2017**, *50*, 59–69. [CrossRef]

77. Mehdipour, I.; Khayat, K.H. Effect of particle-size distribution and specific surface area of different binder systems on packing density and flow characteristics of cement paste. *Cem. Concr. Comp.* **2017**, *78*, 120–131. [CrossRef]

78. Páez-Flor, N.M.; Rubio-Hernández, F.J.; Velázquez-Navarro, J.F. Steady viscous flow of some commercial Andean volcanic Portland cement pastes. *Adv. Cem. Res.* **2017**, *29*, 438–449. [CrossRef]

79. Sébaïbi, Y.; Dheilly, R.M.; Quéneudec, M. A study of the viscosity of lime-cement paste: Influence of the physic-chemical characteristics of lime. *Constr. Build. Mater.* **2004**, *18*, 653–660. [CrossRef]

80. Favier, A.; Hot, J.; Habert, G.; Roussel, N.; De Lacaillerie, J.B.D. Flow properties of MK-based geopolymers pastes. A comparative study with standard Portland cement pastes. *Soft Matter* **2014**, *10*, 1134–1141. [CrossRef]

81. Janotka, I.; Puertas, F.; Palacios, M.; Kuliffayová, M.; Varga, C. Metakaolin sand-blended-cement pastes: Rheology, hydration process and mechanical properties. *Constr. Build. Mater.* **2010**, *24*, 791–802. [CrossRef]

82. Banfill, P.F.G.; Rodríguez, O.; de Rojas, M.I.S.; Frías, M. Effect of activation conditions of a kaolinite based waste on rheology of blended cement pastes. *Cem. Concr. Res.* **2009**, *39*, 843–848. [CrossRef]

83. Safi, B.; Benmounah, A.; Saidi, M. Rheology and zeta potential of cement pastes containing calcined slit and ground granulated blast-furnace slag. *Mater. Constr.* **2011**, *61*, 353–370. [CrossRef]

84. Rissanen, J.; Ohenoja, K.; Kinnunen, P.; Romagnoli, M.; Illikainen, M. Milling of peat-wood fly ash: Effect on water demand of mortar and rheology of cement paste. *Constr. Build. Mater.* **2018**, *180*, 143–153. [CrossRef]

85. Nachbaur, L.; Mutin, J.C.; Nonat, A.; Choplin, L. Dynamic mode rheology of cement and tricalcium silicate pastes from mixing to setting. *Cem. Concr. Res.* **2001**, *31*, 183–192. [CrossRef]

86. Páez-Flor, N.M.; Rubio-Hernández, F.J.; Velázquez-Navarro, J.F. Microstructure-at-rest evolution and steady viscous flow behavior of fresh natural pozzolanic cement pastes. *Constr. Build. Mater.* **2018**. [CrossRef]

87. Choi, M.; Park, K.; Oh, T. Viscoelastic properties of fresh cement paste to study the flow behavior. *Int. J. Concr. Struct. Mater.* **2016**, *10*, S65–S74. [CrossRef]

88. Struble, L.J.; Schultz, M.A. Using creep and recovery to study flow behavior of fresh cement paste. *Cem. Concr. Res.* **1993**, *23*, 1369–1379. [CrossRef]

89. Nehdi, M.; Martini, S.A. Estimating time and temperature dependent yield stress of cement paste using oscillatory rheology and generic algorithms. *Cem. Concr. Res.* **2009**, *39*, 1007–1016. [CrossRef]

90. Schultz, M.A.; Struble, L.J. Use of oscillatory shear to study flow behavior of fresh cement paste. *Cem. Concr. Res.* **1993**, *23*, 273–282. [CrossRef]

91. Bingham, E.C. *Fluidity and Plasticity*; McGraw-Hill Book Co. Inc.: New York, NY, USA, 1922; p. 440.

92. Feys, D.; Verhoeven, R.; De Schutter, G. Evaluation of time independent rheological models applicable to fresh self-compacting concrete. *Appl. Rheol.* **2007**, *17*, 1–10.

93. Herschel, W.H.; Bulkley, R. Measurement of consistency as applied to rubber-benzene solutions. *Am. Soc. Test. Proc.* **1926**, *26*, 621.

94. Robertson, R.E.; Stiff, H.A. An improved mathematic model for relating shear stress to shear rate in drilling fluids and cement slurries. *J. Soc. Pet. Eng.* **1976**, *16*, 31–36. [CrossRef]

95. Karam, G.N. Theoretical and empirical modeling of the rheology of fresh cement pastes. *Mater. Res. Soc. Symp. Proc.* **1993**, *289*, 167–172. [CrossRef]

96. Casson, N. A flow equation for pigment oil suspensions of the printing ink type. In *Rheology of Disperse Systems*; Mill, C.C., Ed.; Pergamon Press: London, UK, 1959; pp. 84–102.

97. Matsumoto, T.; Takashima, A.; Masuda, T.; Onogi, S. A modified Casson equation for dispersions. *Trans. Soc. Rheol.* **1970**, *14*, 617–620. [CrossRef]

98. Wessel, R.; Ball, R.C. Fractal aggregates and gels in shear flow. *Phys. Rev. A* **1992**, *46*, R3008. [CrossRef]

99. Michaels, A.S.; Bolger, J.C. The plastic flow behavior of flocculated kaolin suspensions. *Ind. Eng. Chem. Fundam.* **1962**, *1*, 153–162. [CrossRef]

100. Thomas, D.G. Transport characteristics of suspensions VII. Relation of hindered-settling floc characteristics to rheological parameters. *AIChE J.* **1963**, *9*, 310–316. [CrossRef]

101. Thomas, D.G. Turbulent disruption of flocs in small particle size suspensions. *AIChE J.* **1964**, *10*, 517–523. [CrossRef]

102. Chandler, H.W.; Macphee, D.E. A model for the flow of cement pastes. *Cem. Concr. Res.* **2003**, *33*, 265–270. [CrossRef]

103. Lapasin, R.; Papo, A.; Rajgelj, S. Flow behavior of fresh cement pastes. A comparison of different rheological instruments and techniques. *Cem. Concr. Res.* **1983**, *13*, 349–356. [CrossRef]
104. Hopkins, C.J.; Cabrera, J.G. The turning-tube viscometer: An instrument to measure the flow behaviour of cement-pfa pastes. *Mag. Concr. Res.* **1985**, *37*, 101–106. [CrossRef]
105. Min, B.H.; Erwin, L.; Jennings, H.M. Rheological behaviour of fresh cement paste as measured by squeeze flow. *J. Mater. Sci.* **1994**, *29*, 1374–1381. [CrossRef]
106. Phan, T.H.; Chaouche, M. Rheology and stability of self-compacting concrete cement pastes. *Appl. Rheol.* **2005**, *15*, 336–343.
107. Jarny, S.; Roussel, N.; Le Roy, R.; Coussot, P. Thixotropic behavior of fresh cement pastes from inclined plane flow measurements. *Appl. Rheol.* **2008**, *18*, 1–8.
108. Zhou, X.; Li, Z.; Fan, M.; Chen, H. Rheology of semi-solid fresh cement pastes and mortars in orifice extrusion. *Cem. Concr. Comp.* **2013**, *37*, 304–311. [CrossRef]
109. Kim, J.H.; Kwon, S.H.; Kawashima, S.; Yim, H.J. Rheology of cement paste under high pressure. *Cem. Concr. Comp.* **2017**, *77*, 60–67. [CrossRef]

![fluids logo] *fluids*

MDPI

Article

A Comprehensive Approach from Interfacial to Bulk Properties of Legume Protein-Stabilized Emulsions

Manuel Félix [1], Alberto Romero [2], Cecilio Carrera-Sanchez [3] and Antonio Guerrero [1,*]

[1] Departamento de Ingeniería Química, Escuela Politécnica Superior. Universidad de Sevilla, C.P.:41011 Sevilla, Spain; mfelix@us.es
[2] Departamento de Ingeniería Química, Facultad de Física. Universidad de Sevilla, C.P.:41012 Sevilla, Spain; alromero@us.es
[3] Departamento de Ingeniería Química, Facultad de Química. Universidad de Sevilla, C.P.:41012 Sevilla, Spain; cecilio@us.es
*   Correspondence: aguerrero@us.es; Tel.: +34-954-55-71-79

Received: 1 January 2019; Accepted: 31 March 2019; Published: 3 April 2019

Abstract: The correlation between interfacial properties and emulsion microstructure is a topic of special interest that has many industrial applications. This study deals with the comparison between the rheological properties of oil-water interfaces with adsorbed proteins from legumes (chickpea or faba bean) and the properties of the emulsions using them as the only emulsifier, both at microscopic (droplet size distribution) and macroscopic level (linear viscoelasticity). Two different pH values (2.5 and 7.5) were studied as a function of storage time. Interfaces were characterized by means of dilatational and interfacial shear rheology measurements. Subsequently, the microstructure of the final emulsions obtained was evaluated thorough droplet size distribution (DSD), light scattering and rheological measurements. Results obtained evidenced that pH value has a strong influence on interfacial properties and emulsion microstructure. The best interfacial results were obtained for the lower pH value using chickpea protein, which also corresponded to smaller droplet sizes, higher viscoelastic moduli, and higher emulsion stability. Thus, results put forward the relevance of the interfacial tension values, the adsorption kinetics, the viscoelastic properties of the interfacial film, and the electrostatic interactions among droplets, which depend on pH and the type of protein, on the microstructure, rheological properties, and stability of legume protein-stabilized emulsions.

Keywords: bulk rheology; droplet size distribution (DSD); dilatational rheology; emulsion stability; interfacial shear rheology

1. Introduction

A wide variety of commercial fluid products (e.g., for food or pharmaceutical applications) basically consist of colloidal disperse systems, such as emulsions and foams, whose primary characteristic is the formation and stabilization of a large interfacial area. Proteins are highly efficient for the stabilization of fluid–fluid interfaces as they tend to form two-dimensional (2D) microstructures at the interface, which are referred to as complex interfaces. Such complex interfacial microstructure leads to rheological complexity. In fact, the properties of the fluid-fluid interfaces (e.g., surface tension, surface dilatational, or surface shear moduli) may show a dominant effect on the overall dynamics of these systems. Hence, an understanding of the interfacial properties of adsorbed protein layers is considered to be essential for controlling the physio-chemical stability properties of such colloids [1,2]. However, the dynamic behavior of protein-adsorbed fluid-fluid interfaces can be analyzed by using different rheological approaches. Dilatational rheology has been related to the dynamics of emulsion formation [3,4] but involves changes in the surface area and surface concentration [5,6]. On the other hand, interfacial shear rheology is attracting increasing attention since it has been postulated to be more sensitive to the application of small deformations over which the surface area remains unaltered. In addition, the interpretation of these measurements is analogous to that of widely used techniques in bulk rheology [7,8]. However, a correct interpretation of interfacial films requires of both techniques (dilatational and interfacial shear rheology). Although several tools have been used to overcome the challenges of interfacial shear rheology, double wall ring (DWR) geometry has demonstrated many advantages, where the effect of adjoining subphases may be conveniently neglected in most of practical conditions [9,10].

The dynamics of model proteins (mainly coming from milk fractions), adsorbed at oil/water (O/W) interface, has been widely studied [9–11]. However, these proteins systems are not cost-competitive enough for the manufacturing of commercial emulsions. An attractive alternative for the formation and stabilization of emulsions has been the use of plant proteins, not only because they are more cost-competitive but also because they exhibit high nutritional quality and allow avoiding the use of animal sources. Among them: legume proteins, which constitute the world's fourth most important crop and are particularly interesting due to their excellent nutritional properties and composition as they are rich in carbohydrates, proteins, fibers, vitamins, and minerals [12,13]. In fact, the protein content in legumes is higher than in cereals (reaching up to 20 wt.%) and similar to that found in meat products [14].

In addition to the characterization of the emulsion at a nanoscale (interfacial properties), it is essential to relate the emulsion stability to its properties at a microscale (i.e., droplet size distribution, DSD), as well as to the macroscopic properties, such as the rheology of the continuous phase, both of which depend on the interactions among droplets [15] and can be used to predict some destabilization phenomena (e.g., creaming and flocculation) [16].

The aim of this work was to evaluate the links between the interfacial properties of legume protein-adsorbed O/W layers and the microscopic and macroscopic properties of emulsions stabilized by using these proteins, either from chickpea (CP) or faba bean (FB). Additionally, the influence of pH value was analyzed (pH 2.5 and 7.5). To achieve this objective, the interfacial characterization was carried out by means of dilatational and interfacial shear rheology measurements (using a DWR geometry). Moreover, the microstructure and stability of final emulsions were evaluated by means of droplet size distribution (DSD) and multiple light scattering (MLS) analysis whereas their bulk rheological properties were determined by using small amplitude oscillatory shear (SAOS) measurements.

2. Materials and Methods

2.1. Materials

Two protein systems were used in this study. The HerbaPro F65 (from faba bean) and the CD 300 (from chickpea). Both protein concentrates were supplied by Herba Ingredients (San José de la Rinconada, Seville, Spain). The former protein system was obtained by direct milling, followed by a dry densification process. The light effluent rich in proteins (56.4 ± 0.1 wt.%) was used in this work [17]. The later protein system was obtained by direct milling of chickpea, however it was concentrated by protein solubilization, followed by isoelectric precipitation (pH 4.0) [18]. The final protein content of this protein system (CP) was previously reported by the authors (65.2 ± 0.1 wt.%) [18]. On the other hand, the sunflower oil was purchased from a local producer (COREYSA, Seville, Spain) and all other chemical reagents were purchased from Sigma-Aldrich (St. Louis, MO, USA). The isoelectric point of each protein system was previously determined, being 3.5 for the FB protein system [17] and 4.0 for the CP protein system [18].

2.2. Interfacial Characterization

2.2.1. Pendant Droplet Measurements

Kinetics of protein adsorption was determined by pendant droplet measurements. Transient and steady-state interfacial dilatational measurements were carried out using the TRACKER pendant-droplet tensiometer (IT Concept, Nice, France). The shape of an axisymmetric droplet was analyzed through a charge-coupled device (CCD) camera coupled to a computer. Droplet profiles were processed according to the Laplace equation as was described by Castellani et al. [19]. Transient measurements were carried out at 0.62 rad/s and 10% strain amplitude. On the other hand, after reaching the pseudo-equilibrium state (i.e., after 10,800 s), the mechanical spectra were obtained by means of a frequency sweep test (from 0.048 to 0.62 rad/s). All the experiments were carried out, at least in triplicate, at the saturation protein concentration of the O/W interface, using an optical glass cuvette (8 mL), which contained the oil phase, at 20.0 ± 0.1 °C.

2.2.2. Interfacial Shear Rheological Properties

Protein adsorption kinetics was also determined by interfacial measurements. Transient and steady-state interfacial shear measurements were carried out using the Double wall-ring geometry, coupled to the high-sensitive DHR-3 rheometer (TA Instruments, New Castle, DE, USA). Interfacial shear characterization was carried out by means of interfacial small amplitude oscillatory shear measurements (i-SAOS) measurements [8]. Time sweep experiments were carried out during the protein adsorption (i.e., over 10,800 s) at 0.62 rad/s, obtaining the interfacial shear mechanical spectra. Subsequently, frequency sweep tests were performed after protein adsorption from 0.062 to 6.2 rad/s. Prior to frequency sweep tests, stress swept tests were performed to determinate the linear viscoelastic region (LVR). All these experiments were carried out at the saturation protein concentration of the O/W interface, at least in duplicate, where the experimental set-up was thermostated at 20.0 ± 0.1 °C by placing the double wall cup directly onto the Peltier bottom plate.

Moreover, the contribution of the subphase to i-SAOS results was assessed by calculating the Boussinesq number (Bo) (Equation (1)):

$$Bo = \frac{\eta_s}{a \times \eta_b} \tag{1}$$

where η_s and η_b represent the viscosities of the interface and the bulk, respectively and a is the characteristic length for the for geometry used (0.07 mm).

Bo was higher than 100 in all cases (Bo > 100), indicating that the response obtained is solely related to the interfacial contribution [8,20].

2.3. Emulsion Preparation

Emulsions were prepared following a two-stages method. In the first stage, protein dispersion (50/50) was adjusted at the selected pH value: either below the isoelectric point (IEP) (pH 2.5) or above the IEP (pH 7.5) [17,18]. Subsequently, high-oleic sunflower oil (Coreysa S.A. de C.V., Sevilla, Spain) was gradually blended with the aqueous protein dispersion. The protein concentration was selected according to the concentration of protein required for the saturation of the O/W interface: 2.5 wt.% for pH 2.5 and 4 wt.% for pH 7.5. Blends were subjected to high-shear mixing using the Ultraturrax T-50 (IKA, Staufen, Germany) over 2 min at 5000 rpm, obtaining pre-emulsions. The second stage consisted on passing once the pre-emulsions through the high-pressure valve homogenizer EmulsiFlex-C5 (Avestin, Mannheim, Germany) at 200 KPa. After that, the emulsions were ready for further characterization.

2.4. Emulsion Characterization

2.4.1. Droplet Size Distribution Measurements

DSD measurements were carried out by laser diffraction with the Mastersizer X (Malvern Instruments, Malvern, UK). To disrupt floccules, 1 wt.% of sodium dodecyl sulphate (SDS) was added to the water/emulsion dispersion, followed by a soft stirring [21]. The Sauter mean droplet diameter was calculated as follows:

$$D(3,2) = \frac{\sum n_i d_i^3}{\sum n_i d_i^2} \tag{2}$$

where n_i is the number of droplets which have d_i as diameter. DSD measurements were carried out the day after emulsion preparation and 28 days later.

2.4.2. Bulk Rheological Properties

Linear viscoelastic properties of legume protein-stabilized emulsions were determined by SAOS measurements, carried out using the AR-2000 rheometer (TA Instruments, New Castle, DE, USA). Prior to frequency sweep tests, stress sweep tests were performed at three different frequencies (0.62, 6.20, and 12.52 rad/s) to define the linear viscoelastic region (LVR). Subsequently, frequency sweep tests were carried out from 0.062 to 125 rad/s, at constant stress within the LVR. Serrated plates of 35 mm were used in these measurements to avoid slipping phenomena. SAOS measurements were carried out the day after emulsion preparation and 28 days later.

2.4.3. Backscattering Measurements

MLS measurements were carried out with a Turbiscan Lab Expert (L'Union, Toulouse, France). A light source on a glass tube, which contains the sample, was applied. The backscattering profile was obtained as a function of the tube length. The stability of these emulsions was analyzed during 28 days. Relative backscattering (ΔBS) as a function of time was defined as follows:

$$\Delta BS \ (\%) = (BS_0 - BS_t) \times 100 \tag{3}$$

where BS_0 and BS_t are the BS values obtained for the full profile at 50 mm tube length the day 0 and after time, t (1, 7, 14, 21, and 28 days), respectively.

2.5. Statistical Analysis

At least three replicates of each measurement were carried out. Measurement uncertainty was determined by means of standard deviation. Moreover, significant differences ($p < 0.05$) were analyzed by means of analysis of variance (ANOVA) tests (Excel statistical package). Different letters in tables indicate significant differences.

3. Results

3.1. Interfacial Characterization

3.1.1. Dilatational Measurements

Figure 1A shows the rheokinetics of the protein adsorption at O/W interface at two different pH values (2.5 and 7.5) for the two protein systems studied (FB and CP). As can be observed, the initial adsorption of the protein at the interface involves an increase in the apparent elastic modulus (E′$_s$), as well as a decrease in the apparent viscous modulus (E″$_s$) which corroborates that the interfacial film is progressively being formed in all cases [7,22,23].

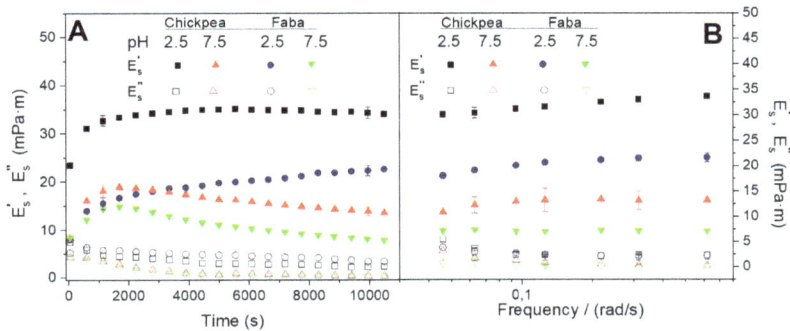

Figure 1. Evolution of the dilatational viscoelastic moduli (E′$_s$ and E″$_s$) of the oil/water O/W interface at two different pH values (2.5 and 7.5) for the two protein systems studied (faba bean (FB) and chickpea (CP)): (**A**) during protein adsorption; (**B**) after reaching the pseudo equilibrium state.

This initial behavior has been previously related with protein diffusion and penetration to the O/W interface. This first step is typically followed by protein opening and consequently unfolding, leading to a decrease in rate changes of the apparent viscoelastic moduli (E′$_s$ and E″$_s$), eventually showing a tendency to reach a constant value of E″$_s$ or E′$_s$. This is the behavior observed for both protein systems at pH 2.5, however at pH 7.5 the trend to a constant value is achieved after a shoulder in E′$_s$. This different behavior, regardless of the protein used, has been attributed to some protein rearrangements which might take place at this pH value [24], as well as to the formation of a protein multilayer [25]. In any case, previous studies indicated that the protein adsorption at low pH value (i.e., 2.5) took place faster since protein is denatured before reaching the O/W interface, leading to a faster protein unfolding at the interface.

This plot also reveals that the interfacial films developed are stronger for pH 2.5 than for pH 7.5, regardless of the protein system that studied FB or CP. Among them, CP interfacial films exhibit higher viscoelastic response. On the other hand, Figure 1B exhibits apparent dilatational viscoelastic moduli after 10,800 s protein adsorption at two different pH values (2.5 and 7.5) as a function of frequency value (from 0.048 to 0.62 rad/s). As can be observed, E′$_s$ is always above E″$_s$, corroborating that the formation of a gel-like interfacial film at the O/W interface takes place [9]. These gel-like results (low frequency dependence and high apparent viscoelastic moduli) have been previously obtained for other protein adsorbed at O/W interface [22,26,27]. Moreover, this plot also indicates that the pH value exerts a marked effect on both apparent viscoelastic moduli (E′$_s$ and E″$_s$). In this sense, the mechanical spectra obtained for the different systems studied also indicate that the best results (i.e., the strongest gel-like response) were obtained for CP-adsorbed films at pH 2.5. These results were related to unfolded proteins found at this pH value, allowing better interaction among different protein chains [28] and leading to stronger interfacial films at low pH value due to higher protein-protein interactions (related to hydrophobic forces) [29]. However, some precautions should be taken upon analysis of these results since it is recognized that the apparent dilatational moduli obtained in complex fluid-fluid interfaces may be affected by bending rigidity, as well as by the changes in the surface area which take place during dilation and contraction [30].

3.1.2. Interfacial Shear Measurements

Figure 2A shows the protein adsorption rheokinetics of the O/W interface at two different pH values (2.5 and 7.5) for the two protein systems studied (FB and CP) obtained by means of interfacial shear measurement.

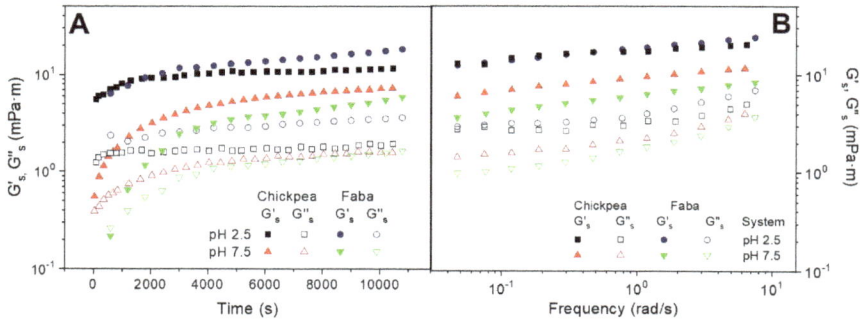

Figure 2. Evolution of the interfacial shear viscoelastic moduli (G'_s and G''_s) of the O/W interface at two different pH values (2.5 and 7.5) for the two protein systems studied (FB and CP): (**A**) during protein adsorption; (**B**) after reaching the pseudo equilibrium state.

The comparison between interfacial dilatational and shear measurements have been proposed as a key tool to understand the rheological behavior of complex interfaces. Thus, whereas the former may be affected by bending rigidity and changes in surface area, the latter is not affected by these phenomena. According to this plot, the evolution of G'_s and G''_s is similar to the behavior found from dilatational measurements (E'_s and E''_s), where G'_s experienced a fast initial increase followed by a tendency to reach a plateau zone. Interestingly, the interfacial shear measurements do not show the above-mentioned shoulder found at pH 7.5. Moreover, the G''_s modulus decreases until reaching a constant value. This response reflects the protein adsorption at the O/W interface which leads to the development of an interfacial film [31]. The interfacial films formed at low pH value exhibit higher elastic interfacial modulus than the interfaces stabilized at higher pH value, which is consistent with the results obtained from dilatational measurements.

In contrast, when comparing the two protein systems, the consistency between interfacial shear and dilatational measurements only remains at pH 7.5, at which CP displays higher values than FB. However, the strongest gel-like response found at pH 2.5 for CP over FB-adsorbed films, under dilatational measurement, is not observed by interfacial shear measurements.

On the other hand, Figure 1B shows the experimental mechanical spectra of the O/W interfaces obtained for CP and FB protein systems adsorbed at the O/W interface after reaching the pseudo-equilibrium state at two different pH values (2.5 and 7.5). These mechanical spectra confirm the development of the protein network, which is reflected by a 2D gel-like behavior. However, the G'_s values obtained by interfacial shear measurements are more frequency-dependent, suggesting that the E'_s values may be affected by Gibbs elasticity caused by changes in surface area, which takes place in dilatational measurements [7]. The above-mentioned inconsistency found between both techniques over protein adsorption at pH 2.5 remains for the mechanical spectra obtained after the pseudo-equilibrium. At pH 7.5, however, both techniques lead to highest mechanical spectra for CP-adsorbed films.

3.2. Emulsion Characterization

3.2.1. Droplet Size Distribution Measurements

Figure 3 shows the DSD profiles obtained for CP or FB-stabilized emulsions the same day of emulsion preparation (Figure 3A) and 28 days later (Figure 3B) at two different pH values (2.5 and 7.5).

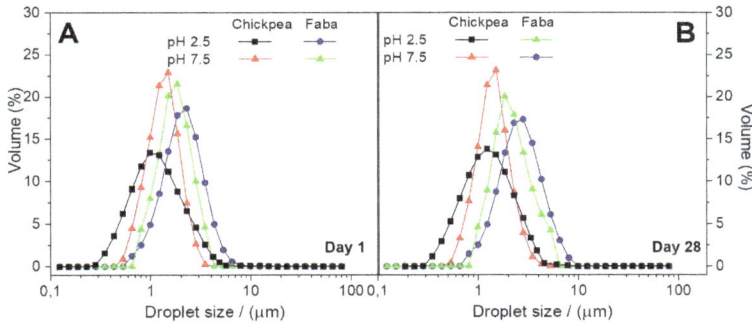

Figure 3. Droplet size distribution (DSD) profiles for FB and CP-stabilized emulsions at two pH values (2.5 and 7.5): (**A**) the same day of the emulsion preparation; (**B**) after 28 days emulsion storage.

In order to explain the results from DSD measurements, the equilibrium interfacial tension values for CP or FB-adsorbed O/W interfaces were previously measured by tensiometry (using a Whilhelmy plate fitted to a KSV-Sigma 701 tensiometer (KSV Instruments Ltd, Helsinki, Finland)). The values obtained for CP films at pH 2.5 and 7.5 were 5.0 ± 0.3 and 2.5 ± 0.5 mN/m, respectively, whereas FB-adsorbed films at pH 2.5 and 7.5 showed interfacial tension values of 9.3 ± 0.4 and 3.5 ± 0.3 mN/m, respectively.

According to Figure 3A the lower droplet sizes were obtained with CP protein at pH 2.5, despite the higher tension value shown at this pH as compared to pH 7.5. Similar results were found previously for other protein systems, being related with the ability of proteins to cover the overall surface droplet [32]. In fact, the adsorption kinetics is faster at low pH, as was shown in Figure 1A and Figure 2A, leading to faster droplet breakage. In addition, the faster development of the 2D network found at this pH also seems to contribute to protect the interface against recoalescence, which is particular important at the beginning of the emulsification process. Thus, higher interfacial viscoelastic response was related to smaller droplet sizes and stronger interfacial films for the same protein system [33]. In contrast, the high value obtained for the interfacial tension of FB at low pH seems to be dominant leading to a DSD profile with the highest droplet sizes. In any case, the lower interfacial tension values of CP-adsorbed films as well as its faster adsorption kinetics contribute to obtain lower droplet sizes for CP-stabilized emulsions. The lower values found for the loss tangent of CP-adsorbed layers (data not shown) also contributes to protect the O/W interface against coalescence.

Nevertheless, when comparing DSD profiles for emulsions measured at the same day of preparation (Figure 3A) with those measured 28 days after emulsion preparation (Figure 3B), it may be concluded that these emulsions are rather stable regardless of the pH or protein system used. The only exception is the FB-stabilized emulsion at low pH which shows a moderate evolution of the DSD profile towards higher sizes, which may be associated to some coalescence. According to previous studies on O/W emulsions stabilized by protein as the only emulsifier, an average size (D(3,2)) lower than 3 µm would lead to highly stable emulsions [34]. All the emulsions studied show D(3,2) values well below this value, excepting for the FB-stabilized emulsion prepared at pH 2.5 whose D(3,2) is close to this limit.

These excellent results of emulsion stability, obtained in absence of any stabilizer, are closely related to the gel-like viscoelastic response found for these O/W complex interfaces, since the development of strong interfacial films avoid droplet coalescence which eventually would lead to emulsion destabilization by creaming.

3.2.2. Bulk Rheology

Figure 4 shows the mechanical spectra obtained by means of frequency sweep tests for CP and FB-stabilized emulsions, just the same day after emulsion preparation at two different pH values (2.5

and 7.5). The frequency sweep tests showed that the values of G′ were always above G″ within the frequency interval studied, indicating that these emulsions have a gel-like behavior regardless of the protein and pH value studied. In agreement with the mechanical spectra obtained from interfacial films, the G′ and G″ values obtained from emulsions showed a moderate frequency-dependence, which is particularly remarkable for CP-stabilized emulsions at pH 2.5. In fact, this is the emulsion showing higher values for the viscoelastic functions, whose response is typical of a well-developed elastic network. On the other side, the lowest G′ values were obtained for the system stabilized with FB at pH 7.5, which confirm the weakness of the network formed [35,36]. As may be observed, CP-stabilized emulsions show higher viscoelastic responses at both pH values. This viscoelastic behavior is similar to the viscoelastic response found for full-fat emulsions stabilized by whey proteins [37]. This study indicated that the viscoelastic moduli of these emulsions exhibited a soft frequency dependence, whereas the elastic moduli reached a value of c.a. 2000 Pa.

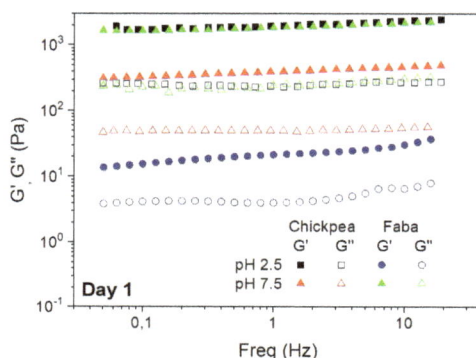

Figure 4. Mechanical spectra from frequency sweep tests obtained for FB and CP-stabilized emulsions at two different pH values (2.5 and 7.5) the same day of emulsion preparation.

SAOS results for emulsions are generally in agreement with DSD results shown in Figure 3, where the smaller droplet sizes correspond to the higher elastic moduli, and consequently to the most stable emulsions. Previous studies also found the same relationship between droplet sizes and viscoelastic properties, where the more stable systems correspond to the lower droplet sizes and higher viscoelastic moduli [38,39]. However, the viscoelastic properties of the emulsions are also strongly dependent on pH that may have a relevant role on the electrostatic interactions among droplets, thus conditioning the formation of the 3D elastic network.

3.2.3. Back Scattering Measurements

Figure 5 shows Backscattering ΔBS profiles and results for CP and FB-stabilized emulsions at two different pH values (2.5 and 7.5). Although one of the greatest advantages of these measurements is the prediction of emulsion destabilization in early stages, the profiles obtained for these emulsions are fairly constant after 28 days of ageing, especially for CP-based emulsions. Thus, ΔBS profiles were calculated in order to elucidate any slight change. These results indicate that only the emulsion stabilized with FB at the lowest pH value shows an apparent change over storage time. This poorer stability has been also observed by analyzing the evolution of DSD profiles. The rest of the emulsions show only very slight increases of droplet sizes over storage time and small slopes for ΔBS profiles (lower than 0.04), which reflect a good emulsion stability response against coalescence and creaming. This type of kinetic stabilization has been previously found for other emulsions systems, and it is able to provide them quite fairly long-term stability [40].

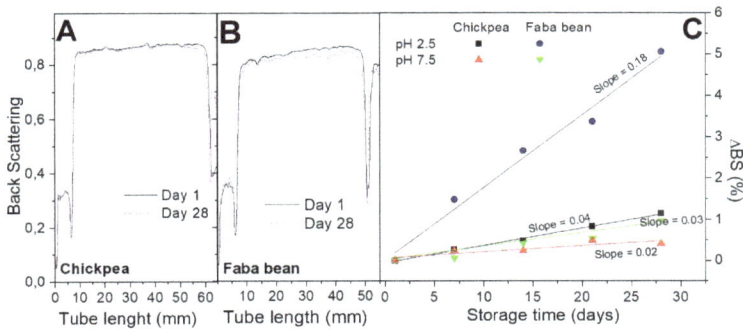

Figure 5. Results from multiple light scattering measurements as a function of storage time: (**A**) Backscattering (BS) profiles for CP-stabilized emulsions at pH 2.5; (**B**) BS profiles for FB-stabilized emulsions at pH 2.5; (**C**) ΔBS for FB and CP-stabilized emulsions at pH 2.5 and 7.5.

4. Conclusions

The results from the interfacial dilatational and shear viscoelastic measurements indicate that proteins are able to form a film with a gel-like behavior at the O/W interface, whose strength depends on the pH value and the type of protein used. Comparing the two techniques it can be inferred that, i-SAOS measurements have shown to be more sensitive to pH modifications and more useful to follow the evolution of viscoelastic properties over protein adsorption.

As for the macroscopic results obtained for legume protein-stabilized emulsions it may be concluded that DSD profiles, which is also strongly affected by pH and the type of protein, depend on the interfacial tension value and on the adsorption kinetics and on the viscoelastic properties of the interfacial film. SAOS and DSD results for legume protein-stabilized emulsions are generally in agreement, where the smaller droplet sizes correspond to the higher elastic moduli, and consequently to the most stable emulsions. However, electrostatic interactions among droplets also promote an enhancement of the 3D elastic network.

A comparison between the bulk and interfacial mechanical spectra of these systems reveals that both the pH of the continuous phase and the type of protein yield analogous effects on their corresponding bulk and interfacial viscoelastic moduli. However, it is also apparent that the changes induced on the mechanical spectra of the bulk are much more remarkable. There is a direct relationship between interfacial and bulk properties, which is more evident for the CP protein concentrate, particularly at pH 2.5. In any case, emulsions stabilized with FB at pH 2.5, at which the highest interfacial tension value was obtained, show the poorest DSD results and emulsion stability.

Moreover, results obtained from emulsion stability seem to agree with the interfacial rheology response of legume protein systems. This is particularly clear for the results obtained from i-SAOS measurements, since dilatational measurements seems to overestimate the solid character of the interface. Therefore, it may be concluded that emulsion stability cannot be solely predicted by interfacial rheology since interactions among droplets may also show a significant contribution on the bulk rheological response.

Author Contributions: Conceptualization, A.G. and C.C.-S.; Methodology, A.R.; Software, M.F.; Validation, A.R. and C.C.-S.; Formal Analysis, C.C.-S.; Investigation, M.F.; Resources, A.G.; Data Curation, A.R.; Writing-Original Draft Preparation, M.F.; Writing-Review and Editing, A.R.; Visualization, A.R.; Supervision, C.C.-S.; Project Administration, A.G.; Funding Acquisition, A.G.

Funding: Please add: This research was funded by University of Seville by a post-doc grant to Manuel Felix (call II.5–VPPIUS)

Acknowledgments: The authors acknowledge the Functional Characterization Service (CITIUS-Universidad de Sevilla) for providing full access to DHR-3 Rheometer.

Conflicts of Interest: The authors declare no conflict of interest.

References

1. Sagis, L.; Fischer, P. Nonlinear rheology of complex fluid–fluid interfaces. *Curr. Opin. Colloid Interface Sci.* **2014**, *19*, 520–529. [CrossRef]
2. Fuller, G.G.; Vermant, J. Complex Fluid-Fluid Interfaces: Rheology and Structure. *Annu. Rev. Chem. Biomol. Eng. Vol 3* **2012**, *3*, 519–543. [CrossRef]
3. Lucassen-Reynders, E.H.; Benjamins, J.; Fainerman, V.B. Dilational rheology of protein films adsorbed at fluid interfaces. *Curr. Opin. Colloid Interface Sci.* **2010**, *15*, 264–270. [CrossRef]
4. Romero, A.; Verwijlen, T.; Guerrero, A.; Vermant, J. Interfacial behaviour of crayfish protein isolate. *Food Hydrocoll.* **2013**, *30*, 470–476. [CrossRef]
5. Narsimhan, G. Characterization of interfacial rheology of protein-stabilized air-liquid interfaces. *Food Eng. Rev.* **2016**, *8*, 367–392. [CrossRef]
6. Danov, K.D.; Kralchevsky, P.A.; Radulova, G.M.; Basheva, E.S.; Stoyanov, S.D.; Pelan, E.G. Shear rheology of mixed protein adsorption layers vs their structure studied by surface force measurements. *Adv. Colloid Interface Sci.* **2015**, *222*, 148–161. [CrossRef] [PubMed]
7. Felix, M.; Romero, A.; Vermant, J.; Guerrero, A. Interfacial properties of highly soluble crayfish protein derivatives. *Colloid. Surface. A* **2016**, *499*, 10–17. [CrossRef]
8. Vandebril, S.; Franck, A.; Fuller, G.G.; Moldenaers, P.; Vermant, J. A double wall-ring geometry for interfacial shear rheometry. *Rheol. Acta* **2010**, *49*, 131–144. [CrossRef]
9. Perez, A.A.; Carrera, C.R.; Sanchez, C.C.; Santiago, L.G.; Patino, J.M.R. Interfacial dynamic properties of whey protein concentrate/polysaccharide mixtures at neutral pH. *Food Hydrocoll.* **2009**, *23*, 1253–1262. [CrossRef]
10. Dickinson, E. Interfacial, Emulsifying and Foaming Properties of Milk Proteins. In *Advanced Dairy Chemistry—1 Proteins: Part A/Part B*; Fox, P.F., McSweeney, P.L.H., Eds.; Springer US: Boston, MA, USA, 2003; pp. 1229–1260. ISBN 978-1-4419-8602-3.
11. Perez, A.A.; Sánchez, C.C.; Patino, J.M.R.; Rubiolo, A.C.; Santiago, L.G. Milk whey proteins and xanthan gum interactions in solution and at the air-water interface: A rheokinetic study. *Colloid. Surface. B* **2010**, *81*, 50–57. [CrossRef]
12. Tharanathan, R.N.; Mahadevamma, S. Grain legumes—a boon to human nutrition. *Trends Food Sci. Technol.* **2003**, *14*, 507–518. [CrossRef]
13. Du, S.; Jiang, H.; Yu, X.; Jane, J. Physicochemical and functional properties of whole legume flour. *LWT-Food Sci. Technol.* **2014**, *55*, 308–313. [CrossRef]
14. Almeida, G.; Silva, K.; Pissini, S.; Oliveira, A. Chemical composition, dietary fibre and resistant starch contents of raw and cooked pea, common bean, chickpea and lentil legumes. *Food Chem.* **2006**, *94*, 327–330. [CrossRef]
15. McClements, D.J. *Food Emulsions: Principles, Practice and Techniques*, 2nd ed.; CRC Press: Boca Raton, FL, USA, 2004.
16. Tadros, T. *Emulsion Formation and Stability*; Wiley: Hoboken, NJ, USA, 2013; ISBN 9783527647965.
17. Felix, M.; Lopez-Osorio, A.; Romero, A.; Guerrero, A. Faba bean protein flour obtained by densification: A sustainable method to develop protein concentrates with food applications. *LWT-Food Sci. Technol.* **2018**, *93*, 563–569. [CrossRef]
18. Felix, M.; Isurralde, N.; Romero, A.; Guerrero, A. Influence of pH value on microstructure of oil-in-water emulsions stabilized by chickpea protein flour. *Food Sci. Technol. Int.* **2018**, *27*, 555–563. [CrossRef] [PubMed]
19. Castellani, O.; Al-Assaf, S.; Axelos, M.; Phillips, G.O.; Anton, M. Hydrocolloids with emulsifying capacity. Part 2-Adsorption properties at the n-hexadecane-Water interface. *Food Hydrocoll.* **2010**, *24*, 121–130. [CrossRef]
20. Erni, P.; Fischer, P.; Windhab, E.J.; Kusnezov, V.; Stettin, H.; Lauger, J. Stress- and strain-controlled measurements of interfacial shear viscosity and viscoelasticity at liquid/liquid and gas/liquid interfaces. *Rev. Sci. Instrum.* **2003**, *74*, 4916–4924. [CrossRef]
21. Puppo, M.C.; Speroni, F.; Chapleau, N.; de Lamballerie, M.; Añón, M.C.; Anton, M. Effect of high-pressure treatment on emulsifying properties of soybean proteins. *Food Hydrocoll.* **2005**, *19*, 289–296. [CrossRef]

22. Felix, M.; Romero, A.; Guerrero, A. Viscoelastic properties, microstructure and stability of high-oleic O/W emulsions stabilised by crayfish protein concentrate and xanthan gum. *Food Hydrocoll.* **2017**, *64*, 9–17. [CrossRef]

23. Sánchez, C.C.; Patino, J.M.R. Interfacial, foaming and emulsifying characteristics of sodium caseinate as influenced by protein concentration in solution. *Food Hydrocoll.* **2005**, *19*, 407–416. [CrossRef]

24. Miller, R.; Fainerman, V.B.; Makievski, A.V.; Krägel, J.; Grigoriev, D.O.; Kazakov, V.N.; Sinyachenko, O. V Dynamics of protein and mixed protein/surfactant adsorption layers at the water/fluid interface. *Adv. Colloid Interface Sci.* **2000**, *86*, 39–82. [CrossRef]

25. Pérez, O.; Sánchez, C.; Pilosof, A.; Rodríguez, J.M. chang 2015. *J. Colloid Interface Sci.* **2009**, *336*, 485–496. [CrossRef]

26. Baldursdottir, S.G.; Fullerton, M.S.; Nielsen, S.H.; Jorgensen, L. Adsorption of proteins at the oil/water interface—Observation of protein adsorption by interfacial shear stress measurements. *Colloid. Surface. B* **2010**, *79*, 41–46. [CrossRef]

27. Schwenzfeier, A.; Lech, F.; Wierenga, P.A.; Eppink, M.H.M.; Gruppen, H. Foam properties of algae soluble protein isolate: Effect of pH and ionic strength. *Food Hydrocoll.* **2013**, *33*, 111–117. [CrossRef]

28. Tang, C.-H.; Shen, L. Dynamic adsorption and dilatational properties of BSA at oil/water interface: Role of conformational flexibility. *Food Hydrocoll.* **2015**, *43*, 388–399. [CrossRef]

29. Chang, C.; Tu, S.; Ghosh, S.; Nickerson, M.T. Effect of pH on the inter-relationships between the physicochemical, interfacial and emulsifying properties for pea, soy, lentil and canola protein isolates. *Food Res. Int.* **2015**, *77*(Part 3), 360–367. [CrossRef]

30. Sagis, L.M.C. Dynamic surface tension of complex fluid-fluid interfaces: A useful concept, or not? *Eur. Phys. Journal-Special Top.* **2013**, *222*, 39–46. [CrossRef]

31. Freer, E.M.; Yim, K.S.; Fuller, G.G.; Radke, C.J. Interfacial rheology of globular and flexible proteins at the hexadecane/water interface: Comparison of shear and dilatation deformation. *J. Phys. Chem. B* **2004**, *108*, 3835–3844. [CrossRef]

32. McClements, D. Protein-stabilized emulsions. *Curr. Opin. Colloid Interface Sci.* **2004**, *9*, 305–313. [CrossRef]

33. Dickinson, E. Caseins in emulsions: interfacial properties and interactions. *Int. Dairy J.* **1999**, *9*, 305–312. [CrossRef]

34. Romero, A.; Cordobés, F.; Puppo, M.C.; Guerrero, A.; Bengoechea, C. Rheology and droplet size distribution of emulsions stabilized by crayfish flour. *Food Hydrocoll.* **2008**, *22*, 1033–1043. [CrossRef]

35. Bengoechea, C.; Romero, A.; Cordobes, F.; Guerrero, A. Rheological and microstructural study of concentrated sunflower oil in water emulsions stabilized by food proteins. *Grasas Y Aceites* **2008**, *59*, 62–68. [CrossRef]

36. Bengoechea, C.; Cordobes, F.; Guerrero, A. Rheology and microstructure of gluten and soya-based O/W emulsions. *Rheol. Acta* **2006**, *46*, 13–21. [CrossRef]

37. Liu, H.; Xu, X.M.; Guo, S.D. Rheological, texture and sensory properties of low-fat mayonnaise with different fat mimetics. *LWT Food Sci. Technol.* **2007**, *40*, 946–954. [CrossRef]

38. Pal, R. Effect of droplet size on the rheology of emulsions. *AIChE J.* **1996**, *42*, 3181–3190. [CrossRef]

39. Huang, X.; Kakuda, Y.; Cui, W. Hydrocolloids in emulsions: Particle size distribution and interfacial activity. *Food Hydrocoll.* **2001**, *15*, 533–542. [CrossRef]

40. Dickinson, E. Milk protein interfacial layers and the relationship to emulsion stability and rheology. *Colloid. Surface. B* **2001**, *20*, 197–210. [CrossRef]

fluids

MDPI

Article

Piezo-Plunger Jetting Technology: An Experimental Study on Jetting Characteristics of Filled Epoxy Polymers

Alexander Kurz [1,2,*], Jörg Bauer [1] and Manfred Wagner [2]

[1] Fraunhofer Institute for Reliability and Microintegration (IZM), System Integration and Interconnection Technologies, Gustav-Meyer-Allee 25, D-13355 Berlin, Germany; Joerg.Bauer@izm.fraunhofer.de

[2] Berlin Institute of Technology (TU Berlin), Polymer Engineering and Physics, Ernst-Reuter-Platz 1, D-10587 Berlin, Germany; manfred.wagner@tu-berlin.de

* Correspondence: dr.alexander.kurz@gmx.de; Tel.: +49-30-314-24758; Fax: +49-30-314-21108

Received: 4 December 2018; Accepted: 22 January 2019; Published: 1 February 2019

Abstract: The droplet formation of Newtonian fluids and suspensions modified by spherical, non-colloidal particles has attracted much interest in practical and theoretical research. For the present study, a jetting technique was used which accelerates a geometrically defined plunger by a piezoelectric actuator. Changing rheological properties of materials and extending deformation rates towards nonlinear viscoelastic regimes created the requirement to extend dosage impulses towards larger magnitudes. To mimic the rheological characteristics of nonconductive adhesives we modified Newtonian epoxy resins by thixotropic additives and micro-scale glass spheres. Rheological analysis at steady shear and oscillatory shear ensured a differentiation between material and process-related factors. Evaluation of high-speed images allowed the investigation of drop dynamics and highlighted the dispense impulse reduction by material-specific dampening properties.

Keywords: complex fluids; drop formation; epoxy; jetting; polymers; polymer processing; prototyping; rheology

---

1. Introduction

The progressive evolution of micro-systems technology and its transition into consumer products requires continuous miniaturization of electronic components. The first step in the development of chip interconnection technology for advanced applications is the combination of electronic and optoelectronic devices for which suitable support structures are essential. For this strategic aim, jetting with piezo-controlled dosing valves is a technology with high potential due to its adaptability and flexibility. By only a few changes in dosing mechanics, this widely used technology can be customized from inkjet applications towards new and more complex materials [1,2]. General requirements for dispensing small quantities of fluids by jetting are high spatial resolution and short process times. To fulfill these requirements, it is necessary to improve the understanding of drop formation pinch-off and thread movement for fluids with complex rheological behavior [3]. As discussed in the following, fundamental differences exist in comparison to recent studies (as detailed below) on fluid application, drop formation, or dot deformation for plunger-based dispensing methods of paste-like fluids at short dosage times.

2. Background

Drop formation and drop constriction are intensively investigated topics of research. The wide interest in analyzing these phenomena in the context of jetting has resulted in an increasing number of scientific discourses. Major topics are, for example, studies on complex materials [4–7], on drop formation stages [8–11], or changes of the dispense mechanisms applied [11–15].

Dong, Carr, and Morris [16] focused their experimental research on drop-on-demand (DOD) formation. In combination with a piezoelectric inkjet print head and a 53-μm nozzle orifice, the authors used water and water-glycerin mixtures. A shockwave was directly coupled into the fluid leading to five stages of drop formation. Pinch-off from the nozzle orifice occurred during the stretching phase (primary break-up) whereas end-pinching evolved during the contraction phase (secondary break-up). With increasing surface tension and/or decreasing viscosity, the dispensed volumes of the primary drop and satellite drops increased. Additionally, the break-up depended critically on the fluid properties as well as on the type of shockwave applied.

Investigations by Rother, Richter, and Rehberg [17] focused on the pinch-off itself and revealed three categories of fluid flow. Drop formation measurements were executed by use of water-glycerin mixtures while the droplet release was induced only by gravitational and capillary forces. During the initial stage, drop shrinkage was observed in close proximity to the pinch point with the initiation of a self-similar motion. Two separate flow regimes became discernible by the transition from a viscosity-dominated Stokes regime towards an inertia-dominated Navier–Stokes regime which led to a slope change in the shrinkage velocity of the neck diameter. The transition points depended on the fluid viscosity. Due to self-similar motion, the classical linear stability analysis failed.

Henderson, Pritchard, and Smolka [18] analyzed the pinch-off of low-viscosity Newtonian fluids in a pendant drop setup with gravitationally induced drop detachment. Detailed observations of the fine structure of secondary breakups revealed wavelike instabilities which led to multiple breakups and a comprehensive number of satellite droplets. The primary pinch-off started with a monotonous necking near the drop and led to simultaneous breakups at different thread locations. Wavelike instabilities were not observed for the primary thread but were generated during the breakup stage.

Computational models give additional insight into drop formation characteristics. Xu and Basaran [19] focused their work on the detachment of incompressible Newtonian fluids from a 10-μm capillary. They used Galerkin/finite element method (FEM) methods to characterize the stripping of monodisperse droplets from a DOD-inkjet valve. The first of two stages started with an increasing fluid volume ejection directed to the nozzle orifice. Subsequently, a constant flow rate could be calculated. With increasing process time, the piezo actuator relaxed and the second phase commenced with reflux of the fluid. This stage was referred to as the meniscus retraction stage. Fluid inertia and disturbances led to drop shrinkage and primary drop breakup. For the starting time of flow inversion during jetting, the authors found a correlation with the Ohnesorge number. Additionally, the Weber number was used to characterize a positive axial velocity of the lower drop fragment and the meniscus retraction to the capillary orifice. Reflecting the importance of the Ohnesorge and Weber numbers, the authors developed a phase diagram for DOD drop formation.

The scope of the present research was to achieve a better understanding of mechanisms in plunger-based DOD drop formation for highly filled polymers. In contrast to studies on inkjet printing of Newtonian fluids or diluted polymer solutions [20,21] with equilibrium droplet constriction or pendant drop setups [17,22], a massive jet impulse is required for highly filled fluids. As a compromise between minimization of the drop size and microscale particle-based polymer modifications, we choose a nozzle orifice of 100 μm. The modifications of fluids investigated started with neat Newtonian resins and successively included high concentrations of thixotropic nanoscale additives and microscale glass spheres. The jetting process used is characterized by high-speed fluid deformation and short dispensing times. We could distinguish between dominating process conditions and optimal rheological properties for commercial adaption of the jetting process, e.g., in chip interconnection technologies.

3. Experimental

3.1. Materials

The selection of materials and their modification was motivated by the relevant application properties, see Table 1. Starting with Newtonian flow characteristics of different epoxy resins we increased the rheological complexity of the fluids by the addition of nanoscale and microscale inorganic fillers.

Table 1. Material composition overview.

Material Category	Matrix Resin	Microscale Filler (Glass Spheres)		Nanoscale Filler (Aerosil R805)	Fluid Characteristic	Sample Name
		wt.-%	Diameter	wt.-%		
I.	REPDTP	-	-	-	Newtonian	REPDTP
	REPDTP	-	-	5	Thixotropic	REPDTP-R
	R0161	-	-	-	Newtonian	R0161
	R0161	-	-	5	Thixotropic	R0161-R
II.	R0161	62	<24 μm	-	Newtonian-like	R0161-SiO
	R3001	Commercial system	<24 μm	-	Newtonian-like	R3001
III.	REPDTP	62	<24 μm	2	Paste-like	REPDTP-R-SiO
	R0161	62	<24 μm	2	Paste-like	R0161-R-SiO

To investigate the influence of nanoscale fillers on fluid cohesion by increasing elasticity and thixotropy, the Newtonian resins REPDTP[1] and R0161[2] (material category I) were modified by adding 5 wt.-% Aerosil R805[3]. The fluids of category II are equivalent to so-called Underfillers which are widely used in microsystems technology. They show Newtonian-like flow behavior but also include glass spheres[4] with diameters of around 20 μm (microscale fillers) at a typical concentration of 60–70 wt.-%. To avoid any kind of sedimentation only the high viscosity resin, R0161 was used as a matrix for the category II fluids which were compared with the commercial Underfiller R3001[5]. With fluids of category III of the material matrix, we aimed to investigate both epoxy resins REPDTP[6] and R0161[7] with the added combination of the nanoscale thixotropic agent Aerosil R805 and microscale glass spheres. To ensure a comparison between category II and category III fluids, we kept the glass sphere concentration at 62 wt.-%, but reduced the Aerosil R805 concentration to 2 wt.-% to guarantee optimal dispersion of the fillers.

[1] Ruetapox EPD TP (Trimethylolpropane triglycidyl ether, Martin Aerospace, Los Angeles, CA, USA)

[2] Ruetapox 0161 (Bisphenol F Diglycidyl ether)

[3] AEROSIL R805 (Business Line AEROSIL®, Evonik Degussa GmbH, Essen, Germany)

[4] SPHERIGLASS®Solid Glass Spheres: A Glass, Product Grade 5000 (Potters Industries Inc., Malvern, PA, USA)

[5] R3001iEX Nagase ChemteX Corporation (Osaka, Japan)

[6] Ruetapox EPD TP (Trimethylolpropane triglycidyl ether)

[7] Ruetapox 0161 (Bisphenol F Diglycidyl ether)

3.2. Jetting Apparatus

Considering the complex rheology of the test fluids, a plunger-based jetting technique had to be selected. The piezoelectric actuator ensured a well-defined material dosage. Small dimensional changes of the actuator are induced by electrical impulses and are translated into high plunger paths (needle lifts) with the assistance of a lever arm. As soon as the jetting sequence is triggered, the plunger moves upwards to a defined position that is given as a percentage of the overall plunger path. The jetting valve opens, and the test liquid is delivered into the fluidic module. To achieve high dosage impulses and ensure jetting of all viscous and paste-like fluids at the same comparable parameter set, the needle lift is set to its maximum height of 100%. Subsequently, the plunger is accelerated downwards to the nozzle (falling step). Due to the motion characteristic of the plunger, its kinetic energy is transferred to the test liquid and pushes the material through the nozzle orifice. Figure 1 shows a schematic of the needle lift as a function of time for a typical dispense sequence.

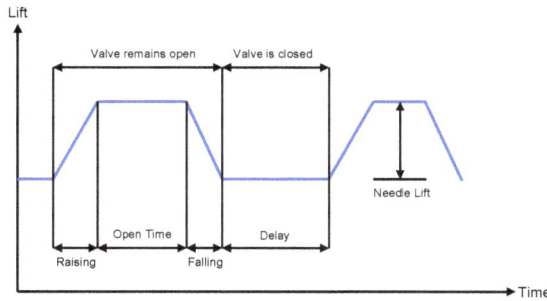

Figure 1. Schematic of dispense sequence and relevant process parameters.

The influence of individual motion parameters on the jetting process is obvious and, in combination with various nozzles and plunger geometries, a significant diversity of trigger impulses is achievable. Considering different materials with different relaxation times, zero-shear viscosities, and fluctuating networks, diverse dispense characteristics can be expected. When jetting particle-filled fluids, it is also important to avoid particle squeezing. Therefore, it is essential to use nozzles with a diameter of at least four to five times larger than the particle diameter. Thus, we choose a nozzle diameter of 100 µm for all experiments to avoid particle squeezing and nozzle blockage and, at the same time, to achieve the minimal drop diameters possible since the nozzle orifice and drop diameter are proportional to each other.

We used a ceramic plunger with a diameter of 1.5 mm for all of the fluid compositions examined, since it is superior with regard to the reduction of particle aggregation in comparison to metal plungers, and it provides the best fit to the inner nozzle curvature. By implementation of the high-speed camera Motion Pro-HS4 (Redlake MASD, LLC, Tucson, AZ, USA) and a zoom lens, drop formation and drop deformation at high dosage velocities were recorded. The magnification remained fixed and drop images were collected in a "free fall position" without a view of the target substrate (see Figure 2 for the camera placement). To achieve stable droplet geometries, the dispensing was performed in a so-called burst mode which is defined as jetting an infinite series of droplets at a set of fixed parameters. A delay time of 400 ms was programmed between drops while the image recording was triggered at every tenth drop.

With respect to high magnification and fast image sequences of 35,000 frames per second, a cold light source was used ensuring high light intensities without heating the test liquids. The test temperature was stabilized at 25 °C.

An ARES G2 rheometer (TA Instruments, New Castle, DE, USA) with plates of 25 mm in diameter in a plate-plate geometry and a gap of 500 µm was used to collect shear flow data. To correlate the rheological measurements with jetting experiments a test temperature of 25 °C was also used for the shear experiments. In order to investigate and correlate the complex rheology of the samples with a drop-on-demand formation, a wide range of shear flow conditions was applied. Shear rate sweeps were performed in the range of $0.001\ \mathrm{s}^{-1}$ to $1000\ \mathrm{s}^{-1}$, as well as strain sweeps with oscillatory shear at 6 rad/s from 0.001% to 100% shear deformation.

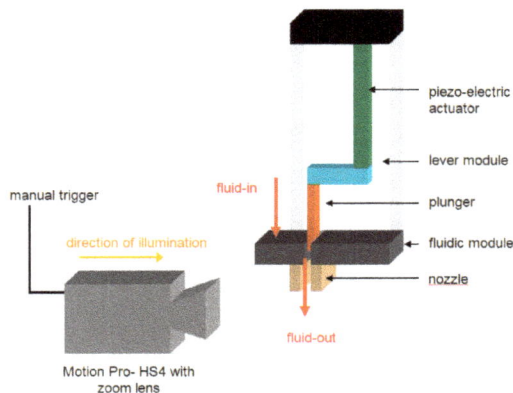

Figure 2. Experimental setup of jetting apparatus and high-speed image acquisition.

3.3. Material Deformation during Plunger-Based Jetting

Substantial dosage impulses had to be applied to utilize the valve system for dispensing high-viscosity materials with increasing shear rates in the nozzle orifice. In order to achieve a suitable correlation between the jetting sequence and fluid properties, laser vibrometer measurements of the path of the plunger (needle lift) were performed. In Figure 3 a plot of two dispense phases as a function of time is presented.

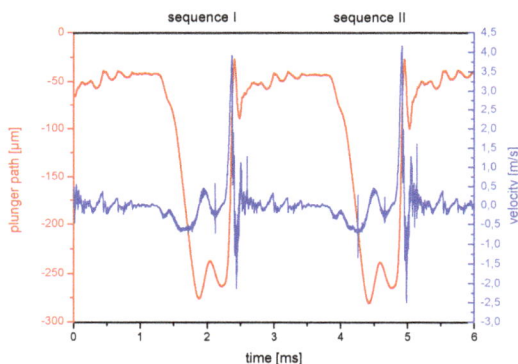

Figure 3. Laser vibrometer measurement of the linear plunger path as a function of time; parameters: Rt 0.5 ms, Ft 0.1 ms, Ot 0.5 ms, 100% needle lift.

For the jetting parameters indicated in Figure 3 a plunger velocity of 4.16 m/s and a needle lift of 234 μm were measured. The equation of Hagen and Poiseuille was used to calculate the maximum shear rate during the first jetting phase. Hereby, fluid incompressibility and a non-damped impulse transfer from the plunger to the dispensed fluid are valid assumptions. For a single-sided open valve geometry, a maximum shear rate of 3.8×10^4 s^{-1} was found, which led to solid-like fluid deformation.

4. Results

4.1. Rheological Characterization of Test Fluids

The change in shear flow behavior of the neat and modified epoxy resins is documented by shear rate sweeps, as shown in Figures 4 and 5. Newtonian flow behavior for the unmodified fluids is verified at the shear rate range investigated with viscosities of 150 mPa·s for REPDTP and 3700 mPa·s for R0161. For thixotropic modifications with Aerosil R805, a distinct correlation between shear rate and viscosity is observed. Due to the strong polarization effects of the Aerosil R805 nanoparticles, a temporary network of secondary forces is created in the polymer leading to a viscosity increase by several orders of magnitude. Increasing shear rates weaken the secondary forces and shear thinning results from a breakdown of the superstructure. By plotting shear stress as a function of shear rate, it can be observed that REPDTP-R is a plastic fluid with a yield stress and R0161-R is a pseudo-plastic fluid.

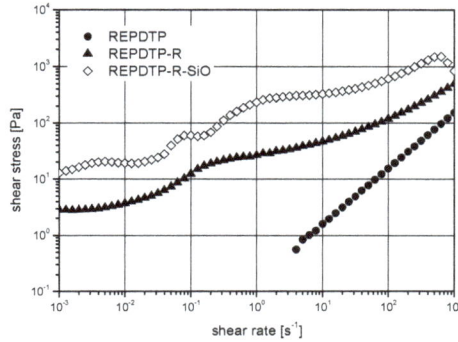

Figure 4. Shear rate sweeps of REPDTP and its modifications.

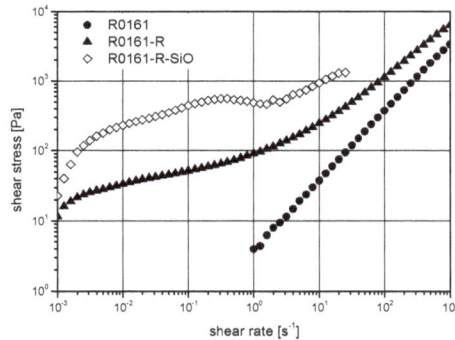

Figure 5. Shear rate sweeps of R0161 and its modifications.

The rheological data for REPDTP-R as a function of shear rate feature deviations from a monotonous trend. In contrast to other standard epoxy resins, a dilatancy plateau appears at a shear rate of 0.1 s^{-1}. Given the fact that Aerosil R805 is characterized by high nanoparticle polarity and additionally, by A high surface-to-volume ratio, the dipole interactions between the epoxy resin and polar additive lead to temporary network formation. The epoxy system REPDTP is a technical grade which is not fully epoxidized, and small quantities of polar groups remain in the resin. This leads to further network stabilization by long-range interactions of polar polymer groups with Aerosil R805. Up to shear rates of 0.1 s^{-1}, a moderate network deformation translates into nanoparticle orientation parallel to the flow direction. The viscosity decreases, and a viscosity plateau is reached at a shear rate of approximately 0.1 s^{-1}. However, due to its inherent nature, the superstructure is relatively weak and becomes gradually destroyed with increasing shear rates. The flow resistance is weakened in favor of a statistical distribution of the nanoparticles.

A combination of the resins REPDTP or R0161 with nanoscale fillers and glass spheres increases the shear viscosity substantially (materials of category III, Figures 4 and 5). The so-called dispersions show a pseudo-plastic characteristic. In comparison with REPDTP-R and R0161-R, the viscosity increases by one order of magnitude due to the addition of glass spheres. Furthermore, particle–particle interactions affect the low-shear regime around 0.002 s^{-1} for R0161-R-SiO and the superstructure plateau of REPDTP-R-SiO around 0.1 s^{-1} by an increase of flow resistance.

Figure 6 presents a shear rate sweep of fluids of material category II. As mentioned earlier, Underfillers consist of a low-viscosity polymer matrix and glass fillers of several micrometers in diameter. Incorporating a thixotropic effect in the property catalog of these materials is essential for achieving an optimized flow at low shear rates driven only by capillary forces. For system R0161-SiO, the steric impediment which is caused by particle–particle interactions can clearly be observed in the low-shear regime. Due to the lack of network formation, hard elastic effects dominate the flow behavior. Above 0.01 s^{-1}, a viscosity plateau is observed over a significant shear rate range. The second viscosity drop at high shear rates is caused by a breakdown of internal cohesion. Due to sensitivity limitations of the rheometer, the used shear rate sweeps of low-viscosity materials such as R 3001 are only reliable at shear stresses above 1 Pa. The observation of a constant shear stress slope for the test system is in accordance with the behavior of commercial Underfiller R 3001.

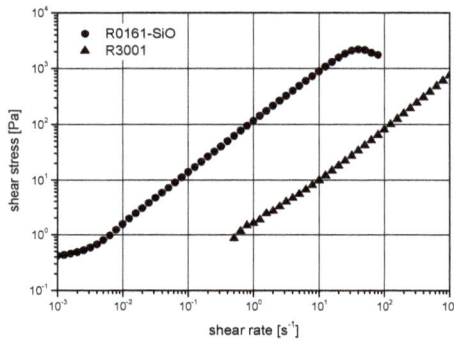

Figure 6. Shear rate sweeps of category II Underfillers.

To determine sol and gel states, as well as the sol-gel transition strain, sweep tests under oscillatory shear at a frequency of 6 rad/s were performed and are presented in Figure 7. Due to limited stress responses of low-viscosity fluids at small shear deformations, a critical value of the shear strain has to be reached in order to measure the storage modulus, G′, and loss modulus, G″, accurately. tan δ is referred to as the loss factor and is obtained as a ratio of the loss modulus G″ to storage modulus G′. It is important to note that a loss factor tan δ > 1 describes a gel state while a loss factor δ < 1 indicates a sol state of the viscoelastic fluid. According to the Winter–Chambon Spectrum [23], the so-called gel point is reached at a frequency-independent loss factor close to tan δ = 1. For the Newtonian systems REPDTP and R0161, a sol state at a constant modulus can be verified for the deformation range investigated. One can easily identify the viscosity difference between the two fluids by the different values of the loss moduli G″, see Figure 7A,B.

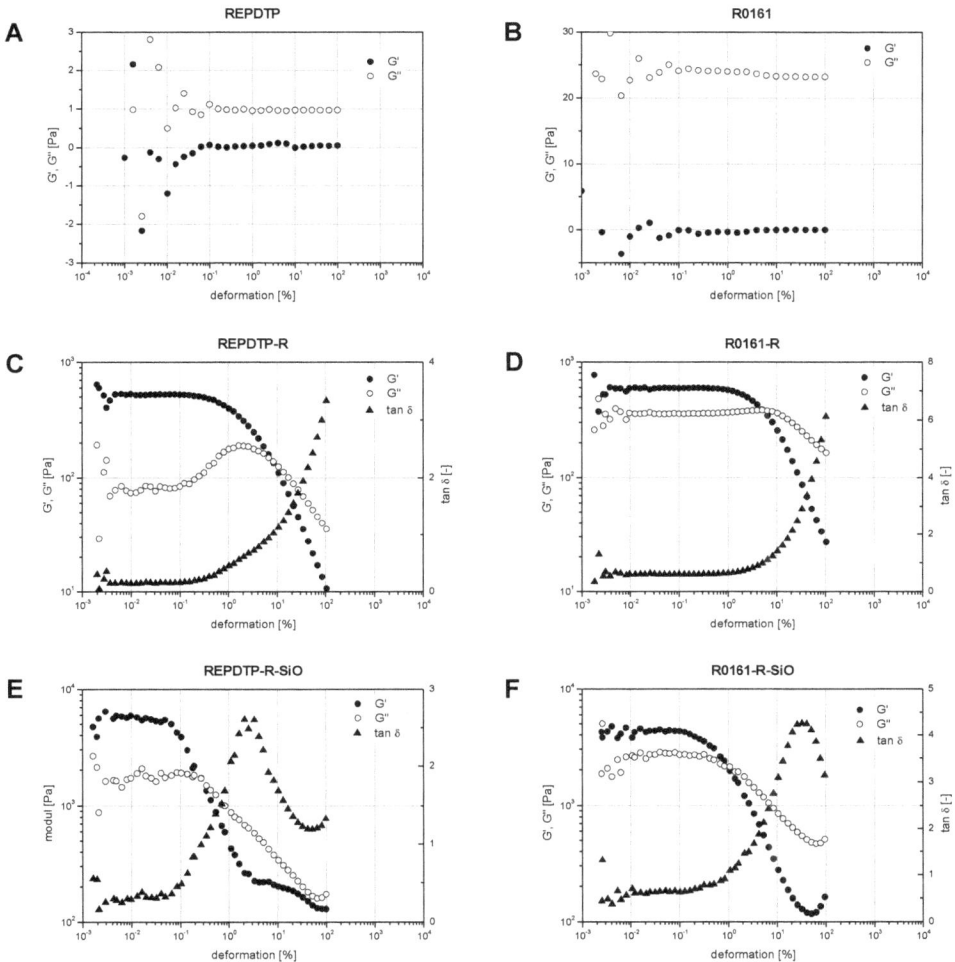

Figure 7. Strain sweep tests under oscillatory shear at 6 rad/s; material category I (**A–D**) and material category III (**E,F**).

The thixotropic modification, see Figure 7C,D, leads to a significant change in the material response. At low values of deformation, both materials show a linear viscoelastic response, i.e., G' and G'' remain constant and independent of the shear deformation. Since the values of G' are larger than the values of G', the materials are in the gel state under these conditions. When the non-linear viscoelastic regime is reached, a transition of the materials towards a sol state is observed. For both systems, the storage modulus G' decreases and the loss modulus G'' increases with increasing shear deformation. The characteristic slope of G'' is more pronounced for REPDTP-R compared to R0161-R and can be related to the superstructure effect discussed earlier. The resistance against structural deformations results from dipole interactions between polar groups of the resin and of the thixotropic additive which leads to an energy dissipation threshold for network deformation and network destruction. Given the fact that the resin R0161 is of higher purity than the technical grade REPDTP, the effects and consequences of the superstructure are negligible for R0161. Nevertheless, the storage moduli G' are comparable for REPDTP-R and R0161-R since the same quantities of Aerosil R805 are used and dynamic networks with equivalent levels of stability are formed. Differences in the underlying resin chemistry translate into individual G'' values and correspond with specific viscosities.

Both test systems of category III show an equivalent flow behavior in comparison with modified materials of category I, but with a Newtonian transition at smaller shear strain values, see Figure 7E,F. A key factor is the enhanced local shear strain in the vicinity of the glass spheres. Similar to the thixotropic modifications REPDTP-R and R0161-R, storage moduli of REPDTP-R-SiO and R0161-R-SiO are also comparable whereas the dissipative effect of glass spheres leads to a higher G'' level. For increasing shear strains, a peak in the loss tangent is found which is caused by hard elastic particle interactions between the glass spheres.

As discussed earlier, Underfillers show a viscosity plateau over a wide shear rate range. Due to a Newtonian flow, characteristic sol states are observed in strain sweeps for R0161-SiO and R3001 with a nearly constant loss modulus G'', see Figure 8. Although elastic softening is negligible for R0161-SiO, a local minimum in the storage modulus G' is seen with increasing deformation indicating the presence of hard elastic interactions of glass spheres. In agreement with viscosity measurements accomplished by strain rate sweeps, the loss modulus G'' of Underfiller R3001 is lower by several orders of magnitude compared to the model system R0161-SiO.

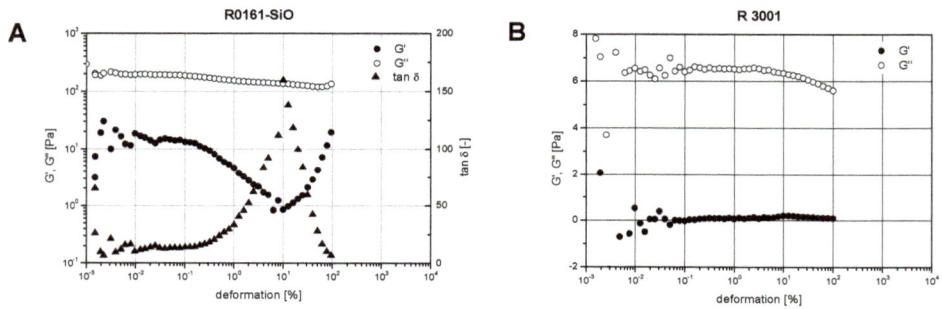

Figure 8. Strain sweep tests under oscillatory shear at 6 rad/s of material category II.

4.2. Plunger-Initiated Drop Formation

Figure 9 shows a schematic of the two jet stages as they evolve with time. For further analysis, we define a reference time, t_0, which corresponds to the time when the material exits the orifice. Jet phase I (from t_0 to t_1) is characterized by (a) moderate flow rates out of the nozzle, (b) by the plunger being in a raised position, and (c) an open valve status. Because of slow fluid accumulation at the nozzle orifice, a standard drop with spherical shape is formed. As soon as the opening interval reaches its pre-defined end-stage, the plunger is accelerated continuously downwards up to its maximum velocity. A major amount of fluid is subsequently pushed through the nozzle and the primary drop undergoes a geometrical change from a spherical to a primary thread (interval t_1 to t_2). When the plunger makes contact with the nozzle pan, vibrational states are injected into the primary thread and stage II of the jet process is reached. The primary thread elongates successively as the plunger impulse moves through the fluid and accelerates the tip.

To analyze the drop evolution, we defined two position parameters (x_1 and x_2) and followed their evolution in time. x_1 marks the beginning of the primary drop whereas x_2 refers to the primary drop end. Depending on the rheological characteristics of the fluid, the primary drop can transition into a primary thread. Hereby, the thread tip and thread end propagate at different velocities leading to primary thread elongation and thus to different slopes of the x_1 and x_2 curves as a function of time.

All geometrical changes are linked to the reference (trigger) time and the progressive time frame counting of the high-speed recording. Due to the field of view limitations, not all parameters are observable within one high-speed image until the thread breakup. Therefore, drop deformations which occur at the target substrate are not discussed in this study.

In contrast to other studies related to drop formation and drop deformation, we do not consider surface tension effects between the nozzle and fluid. Even for long dispense intervals, surface tension is insignificant compared to the dominance of the plunger impulse and thread contraction was not observed even for Newtonian systems.

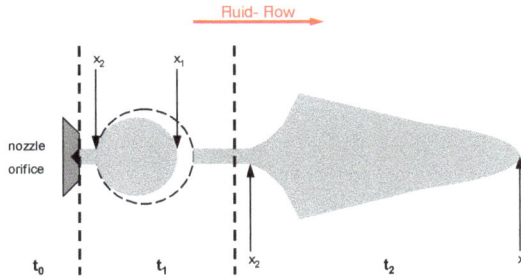

Figure 9. Schematic dispense progress over time.

4.2.1. Category I, Newtonian, and Thixotropic Materials

In this study, REPDTP is the system with the lowest structural stability and viscosity. Both phases of jet deformation are accessible and allow documentation of the drop development. Images 1 to 4 of Figure 10 show slow fluid accumulation at the nozzle orifice in phase I. The fluid exits the orifice at a low velocity until the opening time ends and the plunger acceleration starts. As soon as the plunger hits the nozzle pan, the impulse carries over to the epoxy resin (frame 5) and stretches the primary drop to a primary thread. The thread tip continues elongating while the filament end and the liquid tail move with lower velocity (images 6 to 9). The stretching of the filament leads to a reduction in the diameter and to multiple wavelike breakups (frame 12). The cohesion of this fluid is not strong enough to allow compact dispensing.

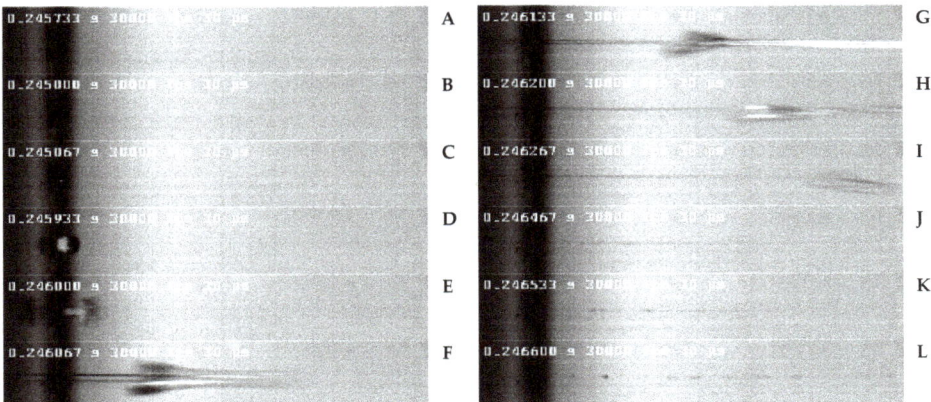

Figure 10. High-speed images of REPDTP drop formation with a time-interval of 67 μs between frames. Left from top to down: Frames 1 to 6. Right from top to down: Frames 7 to 12.

4.2.2. Category II, Underfillers

Similar to the neat resins REPDTP and R0161, the commercial Underfiller R3001 shows a classical start of drop formation in jet phase I, see frames 1 to 6 in Figure 11. As soon as the plunger comes into contact with the nozzle pan, a transition to jet phase II follows. Up to this transition point, the structural evolution with time is comparable to category I materials, except for different damping of the plunger impulse. Considering the Newtonian-like flow behavior and the low viscosity of R3001, its damping ability is impressive since the drop stays compact even after complete impulse transfer into the fluid (frame 7 to 12). During the filament-stretching phase, the evenly formed fluid tail shows a Rayleigh-type breakup [16,21,24] (frame 14).

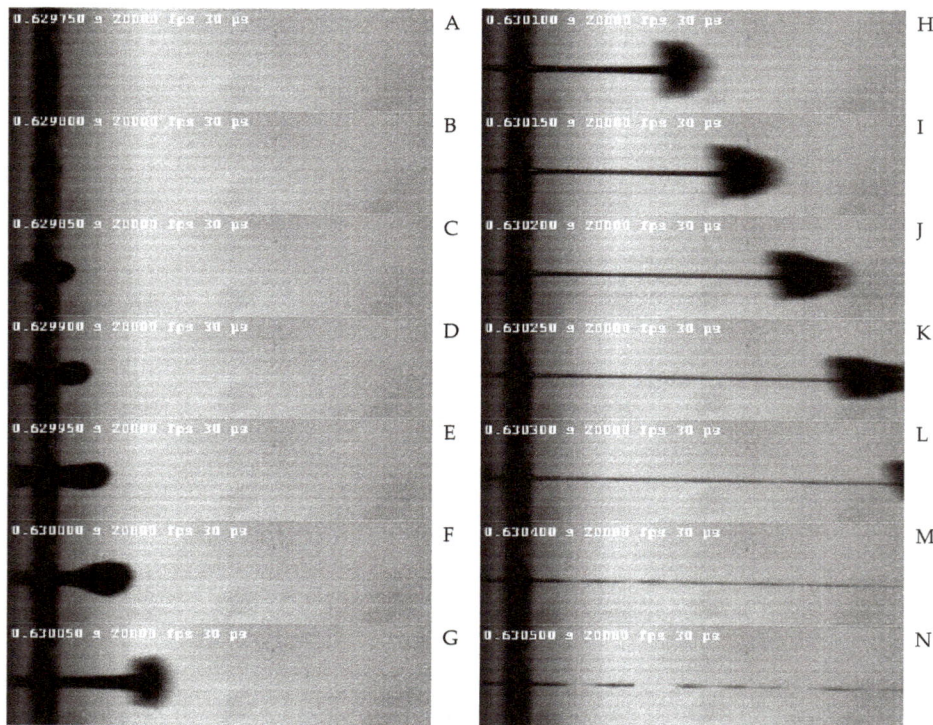

Figure 11. High-speed images of R3001 drop formation with a time-interval of 50 μs between frames. Left from top to down: Frames 1 to 7. Right from top to down: Frames 8 to 14.

4.2.3. Category III, Paste-Like Fluids

Paste-like systems of material category III show a shortened drop dynamic as evidenced by the dispense imagines of REPDTP-R-SiO in Figure 12. Jet phase I is completely overruled and the overall time period is reduced substantially to 0.07 ms in total. As soon as the plunger acceleration reaches a sufficiently large value, the impulse is transferred to the test liquid instantaneously. No damping characteristic can be detected, and a primary thread shoots out of the nozzle orifice. Additionally, no wavelike breakups are observed and the jet stream fractures at a few localized positions into ligands with diameters of approximately 60 μm.

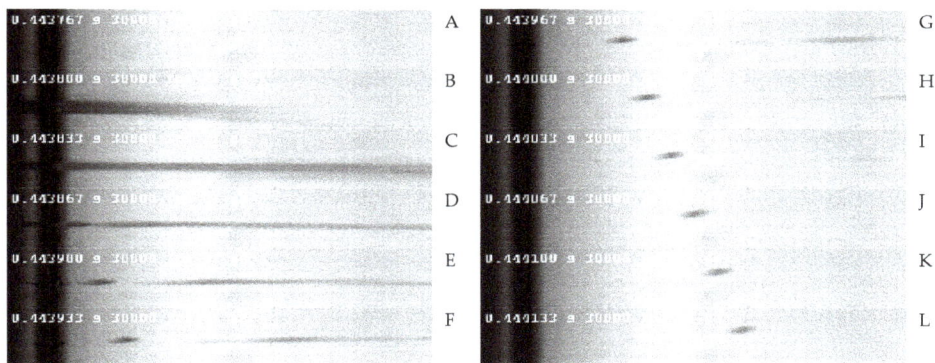

Figure 12. High-speed images of REPDTP-R-SiO drop formation with a time-interval of 33 μs between frames. Left from top to down: Frames 1 to 6. Right from top to down: Frames 7 to 12.

The time-dependent drop deformation stages in terms of positions x_1 and x_2 (beginning and end) of the primary drop/primary thread are presented in Figure 13. For the resins REPDTP and R0161, one can easily observe an abrupt change of slope at 0.35 ms which corresponds to an acceleration of the dispense velocity leading to a steep increase of x_1 and x_2 over time. Different slopes of x_1 and x_2 indicate different velocities of drop tip and drop end and thus a stretching of the thread. Samples REPDTP and REPDTP-R are characterized by a specific starting point for the steep increase of positions x_1 and x_2, as well as a continuous stretching and diameter reduction. We note that for sample REPDTP-R, a delayed transition to jet phase II takes place. During impulse transfer from the plunger to the fluid, the starting point of fluid stretching is shifted to later time frames compared with the unmodified REPDTP. This observation correlates with the rheological data and we conclude that the characteristic jet behavior of REPDTP-R is caused by enhanced viscoelasticity and thixotropy. Thixotropic modifications of REPDTP systematically preserve a low loss modulus G'', as shown in strain sweeps, see Figure 7. Comparing REPDTP-R and R0161-R, both systems have comparable network stability indicated by G', whereas R0161-R does not show a transition from jet phase I to jet phase II after impulse transfer from the plunger to the fluid.

In the case of Underfiller R3001, fluid stretching is delayed relative to impulse initiation as seen from a moderate transition of the slope between 0.2 ms and 0.5 ms in Figure 14. Furthermore, the drop tip and drop end detach from the nozzle with nearly identical velocities. Comparing the drop dynamics with the ones of REPDTP and REPDTP-R, the evenly dispersed glass spheres of R3001 create an additional way to dissipate energy. In contrast to fluids with thixotropic modification and strong network cohesion, particles in Underfillers interact with a non-elastic Newtonian matrix at low flow resistance. Therefore, an impulse injection into the primary drop leads to a viscous deformation. In addition, largely independent glass sphere realignment leads to additional dissipation of impulse energy and compact dispensing of R3001 until a Rayleigh-type breakup occurs.

An equivalent velocity and dispense profile is found for the primary drop of R0161. Once the neat epoxy polymer is modified with Aerosil R805, jet phase I is completely suppressed and only the second phase is observed, see R0161-R in Figure 14. Drop dynamics of REPDTP, R0161, and R3001 are ideal examples of plunger-based jetting and drop evolution as a function of the dispense time. With increasing rheological complexity and high dosage impulses, a strong dispense time decrease occurs for paste-like fluids. The plunger impulse is transferred to the material without any noticeable damping effects. The jetting characteristic of these materials which feature an elastic network in combination with solid particle interactions can be described as solid-like. To overcome the yield stress or the high zero-shear viscosity, all paste-like systems need a high plunger impulse and show identical drop dynamics.

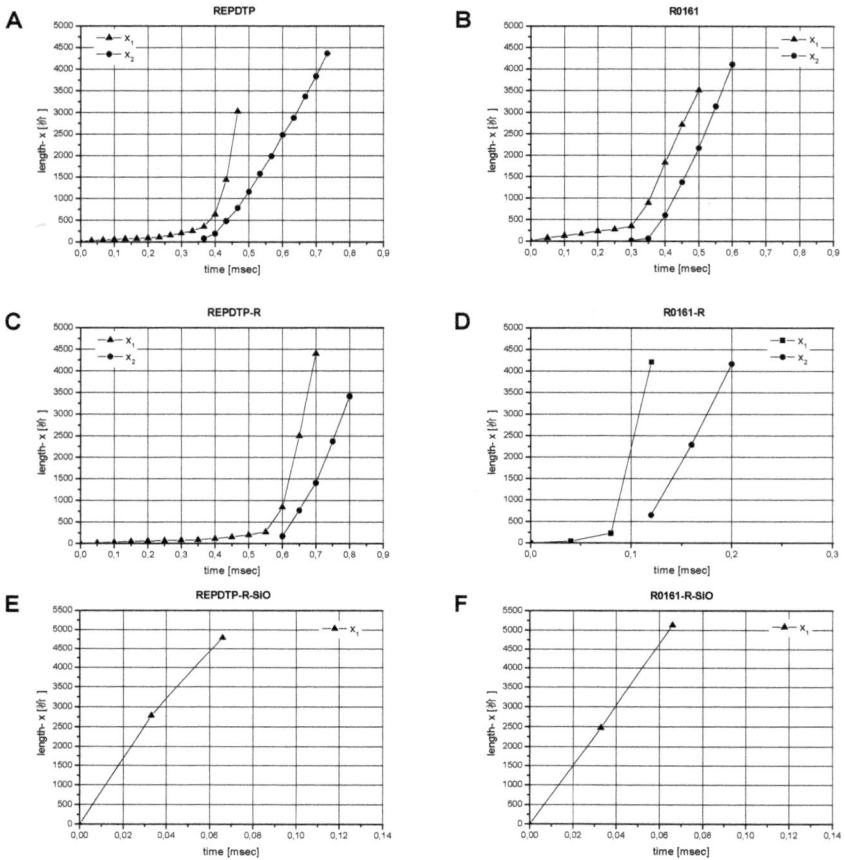

Figure 13. Drop dynamics (positions x1 and x2 of the beginning and end of the primary drop/primary thread) as a function of time for materials of category I (**A–D**) and materials of category III in plots (**E,F**).

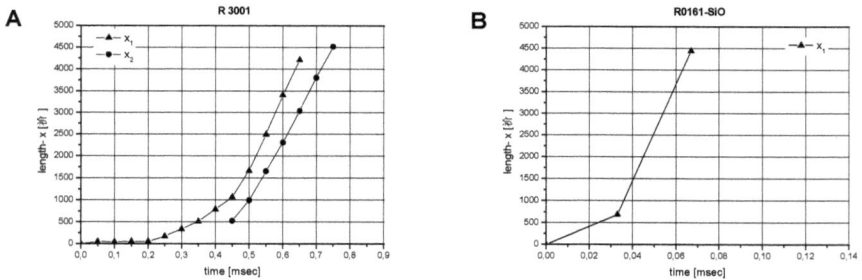

Figure 14. Drop dynamics as a function of time for materials of category II.

5. Discussion and Conclusions

The plunger-based jetting of complex polymer resins can be separated into two stages of material shaping: (a) classical drop formation with a stable structure at low deformation rates and (b) high-speed deformation with solid-like fluid dispensing. Here we presented a direct correlation between rheology and jet formation.

Test systems with low viscosity and limited structural stability offer a wide spectrum of accessible jet parameters. For systems of material category I, see Table 1, impulse damping and process time up-scaling are documented. The classical drop formation during jet stage I is characterized by various drop contours for identical jet parameters. Based on rheological data, the concept of a damping influence on fluid dispensing was developed. Without microscale particles, the ability of the fluid to reduce the jet impulse is limited, which leads to a continuous primary drop elongation. To verify the concept of a damping influence, the primary drop contour evolution was analyzed as a function of time. Here, a difference in slopes of drop tip and drop end positions reveals a significant damping effect. Changes in the slope of both position curves highlight a change in the relevant damping mechanism and thus a material-dependent influence of rheology.

A peculiarity is found for the commercial system, R3001, which allows for compact dispensing. In this case, only small interactions of the glass spheres with the polymer matrix exist as can be deduced from the sol character and the low viscosity of the fluid. Due to an adjustment of the physicochemical properties of R3001, the internal cohesion nevertheless maintains a compact drop-dispensing characteristic even though an elastic network could not be detected. However, neck thinning of R3001 leads to multiple Rayleigh-type breakups and multiple satellite droplets.

In the case of paste-like fluids, the impact of jet parameters and fluid properties on drop contour evolution becomes negligible. Equivalent drop shapes are obtained for various materials. Despite the fact that in the second jet stage fluid properties are more important than process conditions, a rheological study is generally advised in order to determine the level of dispensability. Glass spheres and thixotropy additives generate an enhanced network stiffness which is characterized by reduced glass sphere mobility within the dispersed Aerosil network. The average deformation range narrows down and shifts to instantaneous material acceleration and primary threads without drop contour. A brittle breakup at the nozzle orifice emerges.

In summary, for a variety of resins and resin modifications, the jetting process was investigated in detail by high-speed imaging. Based on rheological classifications of the materials chosen, the evolution of drop dynamics was studied in different stages of high-speed deformation. In general, the jetting process can be separated into two jet phases and a significant influence on the jet parameters was observed. With the increasing rheological complexity of the materials, jet phase I is suppressed due to dispense impulse reduction by material-specific dampening properties resulting in a stream-like material deposition. Plunger-based jetting was demonstrated to be a versatile technique which allowed for the dosage of all test fluids investigated, regardless of the large variation in the rheological characteristics.

Author Contributions: Conceptualization, A.K. and J.B.; Methodology, A.K.; Validation, J.B. and M.W.; Formal Analysis, A.K.; Investigation, A.K.; Resources, J.B.; Data Curation, A.K. and J.B.; Writing-Original Draft Preparation, A.K.; Writing-Review & Editing, M.W.; Visualization, A.K.; Supervision, J.B. and M.W.; Project Administration, A.K. and J.B.; Funding Acquisition, J.B.

Funding: This research received no external funding.

Acknowledgments: The authors thank VERMES micro dispensing for providing the experimental setup. Furthermore, we thank co-workers from IZM and TU Berlin as well as from the department of Micro Technologies of HTW Berlin for their assistance.

Conflicts of Interest: The authors declare no conflict of interest.

References

1. Rajiv, L.I.; Daryl, L.S. Experimental, Analysis of a Voice-Coil-Driven Jetting System for Micrograms Fluid Depositions in Electronics Assembly. *J. Microelectron. Electron. Packag.* **2017**, *14*, 108–121.

2. Mueller, M.; Franke, J. Feasability study of piezo jet printed silver ink structures for interconnection and condition monitoring of power electronics components. In Proceedings of the 2017 IEEE 19th Electronics Packaging Technology Conference (EPTC), Singapore, 6–9 December 2017; pp. 1–5.
3. Wijshoff, H. Drop dynamics in the inkjet printing process. *Elsevier Curr. Opin. Colloid Interface Sci.* **2018**, *36*, 20–27. [CrossRef]
4. Furbank, R.J.; Morris, J.F. An experimental study of particle effects on drop formation. *Phys. Fluids* **2004**, *16*, 1777–1790. [CrossRef]
5. Dravid, V.; Loke, P.B.; Corvalan, C.M.; Sojka, P.E. Drop Formation in Non-Newtonian Jets at low Reynolds Numbers. *J. Fluids Eng.* **2008**, *130*, 081504. [CrossRef]
6. Richards, J.R.; Beris, A.N.; Lenhoff, A.M. Drop formation in liquid-liquid systems before and after jetting. *Phys. Fluids* **1995**, *7*, 2617–2630. [CrossRef]
7. Soutrenon, M.; Billato, G.; Bircher, F. 3D printing of cellulose by solvent on binder jetting. *Print. Fabr.* **2018**, *4*, 166–169. [CrossRef]
8. Eggers, J. Theory of drop formation. *Phys. Fluids* **1995**, *5*, 941–953. [CrossRef]
9. Badie, R.; De Lange, D.F. Mechanism of drop constriction in a drop-on-demand inkjet system. *Proc. R. Soc. Math. Phys. Eng. Sci.* **1997**, *453*, 2573–2581. [CrossRef]
10. Craster, R.V.; Matar, O.K.; Papageorgiou, D.T. Pinchoff and satellite formation in compound viscous threads. *Phys. Fluids* **2003**, *15*, 3409–3428. [CrossRef]
11. Papageorgiou, D.T. On the breakup of viscous liquid threads. *Phys. Fluids* **1995**, *7*, 1529–1544. [CrossRef]
12. Chen, A.U.; Basaran, O.A. A new method of significantly reducing drop radius without reducing nozzle radius in a drop-on-demand drop production. *Phys. Fluids* **2002**, *14*, L1–L4. [CrossRef]
13. Thoroddsen, S.T.; Etoh, T.G.; Takehara, K. Microjetting from wave focussing on oscillating drops. *Phys. Fluids* **2007**, *19*, 052101. [CrossRef]
14. Meacham, J.M.; Varady, M.J.; Degertekin, F.L.; Fedorov, A.G. Droplet formation and ejection from a micromachined ultrasonic droplet generator: Visualization and scaling. *Phys. Fluids* **2005**, *17*, 100605. [CrossRef]
15. Shin, D.-Y.; Grassia, P.; Derby, B. Numerical and experimental comparisons of mass transport rate in a piezoelectric drop-on-demand inkjet print head. *Int. J. Mech. Sci.* **2004**, *46*, 181–199. [CrossRef]
16. Dong, H.; Carr, W.W.; Morris, J.F. An experimental study of drop-on-demand drop formation. *Phys. Fluids* **2006**, *18*, 072102. [CrossRef]
17. Rothert, A.; Richter, R.; Rehberg, I. Formation of a drop: Viscosity dependence of three flow regimes. *New J. Phys.* **2003**, *5*, 59. [CrossRef]
18. Henderson, D.M.; Pritchard, W.G.; Smolka, L.B. On the pinch-off of a pendant drop of viscous fluid. *Phys. Fluids* **1997**, *11*, 3188–3200. [CrossRef]
19. Xu, Q.; Basaran, O.A. Computational analysis of drop-on-demand drop formation. *Phys. Fluids* **2007**, *19*, 102111. [CrossRef]
20. Taur, A.; Doshi, P.; Yeoh, H.K. Dripping dynamics of Newtonian liquids from a tilted nozzle. *Eur. J. Mech. B/Fluids* **2015**, *51*, 8–15. [CrossRef]
21. Hoath, S.D.; Harlen, O.G.; Hutchings, I.M. Jetting behavior of polymer solutions in drop-on-demand inkjet printing. *J. Rheol.* **2012**, *56*, 1109–1127. [CrossRef]
22. Nazari, A.; Derakhshi, A.Z.; Nazari, A.; Firoozabadi, B. Drop formation from a capillary tube: Comparison of different bulk fluid on Newtonian drops and formation of Newtonian and non-Newtonian drops in air using image processing. *Int. J. Heat Mass Transf.* **2018**, *124*, 912–919. [CrossRef]
23. Winter, H.H.; Chambon, F. Analysis of linear viscoelasticity of a cross-linking polymer at the gel point. *J. Rheol.* **1986**, *32*, 367–382. [CrossRef]
24. Rayleigh, J.W.S. On the instability of a cylinder of viscous liquid under capillary force. *Philos. Mag.* **1892**, *34*, 145–154. [CrossRef]

![fluids logo] *fluids*

MDPI

Article

Relevance of Rheology on the Properties of PP/MWCNT Nanocomposites Elaborated with Different Irradiation/Mixing Protocols

Mercedes Fernandez [1],*, Arrate Huegun [2] and Antxon Santamaria [1]

[1] Polymer Science and Technology Department and Polymer Institute POLYMAT,
    University of the Basque Country (UPV/EHU), Joxe Mari Korta Building, Av. Tolosa 72,
    20018 Donostia-San Sebastian, Spain; antxon.santamaria@ehu.eus
[2] CIDETEC Research Center, Polymers and Composites, Po Miramón 191,
    20014 Donostia-San Sebastian, Spain; ahuegun@cidetec.es
* Correspondence: mercedes.fernandez@ehu.eus; Tel.: +34-943-018-185

Received: 3 December 2018; Accepted: 4 January 2019; Published: 9 January 2019

Abstract: Linear and nonlinear rheological features and electrical conductivity of two nanocomposite systems based on polypropylene/multiwall carbon nanotubes (PP/MWCNT) are investigated. The nanocomposites were irradiated with an electron beam following two different procedures. Protocol A, where the nanocomposite mixture is irradiated, and Protocol B where only the PP matrix is irradiated before mixing with MWCNT. The same irradiation dose adjusted to bring about long chain branching (LCB) but not crosslinking, is used in both types of nanocomposites. The modification of the polymer matrix viscosity caused by irradiation determines the MWCNT dispersion and therefore the rheological and percolation thresholds. Elongational flow results reveal that strain hardening, typical of irradiated PPs, is observed for the nanocomposites irradiated, but not for the nanocomposites prepared with the irradiated PP. The hypothesis of a shear flow modification that aligns the branches into the backbone, eliminating the strain hardening is considered.

Keywords: nanocomposites; LCB polypropylene; rheology; oscillatory flows; elongational flow

---

1. Introduction

One of the advantages of long chain branched PP is that it can be processed using techniques, like extrusion-blowing, blow molding, foaming extrusion and thermoforming, not affordable for linear PP. It is known that these industrial processing methods require the polymer melt showing a strain-hardening behavior in elongational flow, i.e., a rapid increase of the uniaxial extensional viscosity beyond a critical strain.

The inclusion of long chain branches (LCB) in a PP structure can be done by three different methods: Radical chemical modification, radical grafting with different monomers (for instance styrene and vinylsiloxane) and using electron-beam irradiation (EB) [1].

The modification of the molecular structure of linear isotactic polypropylene (PP) by electron-beam irradiation has been a subject of interest during the last few decades [1–8]. Adequate doses of irradiation lead to LCB, and chain scission (molecular weight reduction), still avoiding crosslinking. The primary effect of irradiation on thermoplastic polymers is to produce free radicals after the scission of bonds. These built radicals lead to further changes in molecular structure through chemical reactions. The main molecular effects are chain scission, chain branching, and crosslinking. Usually, all these reactions are in competition and take place simultaneously. Which reaction predominates depends on chemical structure and morphology of the polymer, as well as on the irradiation conditions (applied dose, dose rate and type of radiation) and annealing time and temperature. The specific conditions that favour the formation of LCB in polypropylene by EB irradiation and without additives, were determined by Rätzsch et al. [1]. According to these authors, the material should be irradiated under inert atmosphere and at a temperature below 100 °C, i.e., in solid state. Subsequent to the irradiation process, the polypropylene should be submitted to a thermal treatment for two reasons: First, the life time of macro-radicals is prolonged in order to raise the possibilities for branching, and second, the residual macro-radicals can be eliminated.

Very few papers refer to the effect of irradiation on polymer nanocomposites [9–14]. With the aim of improving the range of industrial applications, we have recently irradiated PP/MWCNT (Multiwall Carbon Nanotubes) nanocomposites, obtaining an electrical conductive polypropylene with strain-hardening behavior [12,14]. To our knowledge, elaboration of PP/MWCNT nanocomposites using irradiated PPs as a polymer matrix has not been reported in the literature, so far. The type, length, density and distribution of branches are known to dramatically affect polymer flow behavior and processing characteristics. Understanding the role of LCB in relation to the properties of these materials is of great academic and industrial interest, but, in general, the effect of the molecular architecture of the matrix has deserved little attention on the characterization of polymer nanocomposites. Espinoza-González et al. [15] reported that a patented method of extrusion assisted by a low-frequency and high-power ultrasound method was able to induce different states of dispersion, as well as chemical modifications promoted by branching reactions. The presence of branched structures affects the quality of the dispersion of CNTs. Vega et al. [16] demonstrated that high molecular weights or considerable molecular weight polydispersities can play a determining role in melt viscoelastic properties of polyethylene/carbon nanotube nanocomposites, which have important implications on determining rheological percolation thresholds, due to the viscoelastic screening effect of the CNTs.

The characterization of the structural modification produced by the electron-beam irradiation process in PP/MWCNT nanocomposites is a challenging task. Size Exclusion Chromatography (SEC) combined with MALLS (Multi-Angle Light Scattering) is a very suitable tool to ascertain the sparse long chain branching level in polyolefins [17]. However, this technique cannot be employed in the case of polymer nanocomposites, because proper solutions to be injected in SEC equipment cannot be prepared. Therefore, rheological methods are used to face this issue. Our research focuses on the influence of both, carbon nanotubes and the irradiation process on the rheological and electrical properties. The rheological characterization of nanocomposites is based on the study of the dynamic viscoelasticity and uniaxial elongational flow.

Two PP/MWCNT nanocomposites are compared: One is a PP/MWCNT nanocomposite that has been submitted to irradiation (Protocol A); the other is a PP/MWCNT nanocomposite obtained by mixing MWCNT with an irradiated PP (Protocol B). The same irradiation dose, adjusted to bring about LCB, but not crosslinking, was used in both types of nanocomposites. Furthermore, the mixing conditions (temperature, mixing time and applying torque) were the same in both elaboration protocols. The paper is organized focusing on the following issues: (a) Similarity and differences found between both nanocomposites; (b) effect of irradiation on polymer matrix viscosity and, subsequently, influence on MWCNT dispersion; (c) influence of long chain branching developed by irradiation: strain hardening in elongational flow; (d) capacity of the mixing process to annihilate LCB and, alternatively, to promote shear modification (alignment of long branches to the backbone); (e) explaining why only one of the nanocomposites displays strain hardening.

2. Materials and Methods

The basic material used in this study was a commercial isotactic polypropylene Moplen EP340K (M_w = 440,000, M_w/M_n = 4.4) provided by Basell polyolefins company (London, UK). We refer to this polymer as "neat PP". Non-functionalized Multiwall Carbon Nanotubes (MWCNT) supplied by Cheap Tubes Inc (Cambridgeport, VT, USA) were selected to be dispersed in neat PP matrix. These nanotubes have specified diameters of D = 30–50 nm, lengths of L = 10–20 μm and purity greater than 95%.

Before the melt-mixing process, polymer powder was prepared from pellets using a Mill Retsch® ZM 200 (Haan, Germany) and both polymer powder and MWCNT were properly dried. PP was dried at T = 80 °C in a vacuum oven during 2 h and MWCNT were dried at T = 100 °C during 2 h. The polymer and MWCNT were stirred to obtain a homogeneous mixture. A blend of Irganox 1010 and Irgafos 168 (BASF, Basel, Switzerland) was added to prevent the degradation of the irradiated polypropylene. The powder mixture was blended in a Haake Mini-Lab twin-screw extruder (Thermo Electron Coorp., Hamburg, Germany). The mixing was processed for 10 min at T = 180 °C and 100 rpm using a counter-rotating screw configuration.

Irradiated pure PP and MWCNT/PP nanocomposites were obtained by electron-beam irradiation. The process was performed under inert atmosphere at room temperature with an EB by IONMED (Cuenca, Spain). The acceleration energy was 10 MeV and the irradiation doses were 20, 50 and 80 kGy. After irradiation, samples were heated to react with the residual radicals.

The nanocomposites were prepared following two different protocols. Protocol A consisted of irradiating the prepared nanocomposites, whereas in protocol B neat PP powder was irradiated before the mixing process and, after, nanocomposites were prepared using irradiated neat PP as the matrix. To facilitate the comprehension to the reader, the specimens were named with the following sequential nomenclature: Protocol A: PP/MWCNT, load and irradiation dose and Protocol B: PP, irradiation dose/MWCNT, load.

The mixtures were compression molded using a hydraulic press at 180 °C for 5 min; the last 2 min at a maximum pressure of 100 bars. Then, the sheets were allowed to cool at room temperature outside the press. The thickness of the sheets was adjusted depending on the experiment.

The dynamic viscoelastic functions, storage or elastic modulus, G', loss or viscous modulus, G'' and complex viscosity, η^*, were determined in an ARG2 (TA Instrument, New Castle, DE, USA) rheometer with parallel-plate fixture (25 mm diameter). Dynamic frequency sweep experiments were carried out in a frequency range of 0.0628–628 rad/s at a temperature of T = 190 °C. Actually, small amplitude oscillatory shear (SAOS) measurements were carried out, ensuring that the strain amplitude was within the linear viscoelastic range. For the analysis of the non-linear viscoelasticity, large amplitude oscillatory shear (LAOS) measurements were performed using an ARES rheometer (Rheometric Scientific) to study the effect of the strain amplitude on G' and G'' at a constant frequency of 6.28 rad/s. Parallel plates of 12 mm diameter were employed. For data acquisition a 16-bit ADC card was used (National Instruments, Austin, TX, USA) and Fourier Transformation analysis was carried out employing LabView Software Program developed and kindly supplied by Prof. M. Wilhelm [18].

Extensional measurements were performed at a temperature of T = 180 °C using the Extensional Viscosity Fixture (EVF) of the ARES rheometer. The extensional viscosity was determined as a function of time at different extensional rates ranging from 0.01 to 1 s^{-1}.

Molecular parameters of non-irradiated and irradiated polypropylene were evaluated using high temperature size exclusion chromatography (SEC) coupled with an infrared detector (IR) as the primary detector and multiangle laser light scattering (MALLS). This was completed with an on-line viscometer (VI) to determine the intrinsic viscosity, which allowed for evaluation of the LCB degree. The solvent was trichlorobenzene (TCB) stabilized with 300 ppm butylated hydroxytoluene (BHT). Mass recovery was always higher than 95%, and not a single impediment was found to dissolve all the samples, which leads us to assume that irradiation did not cause crosslinking. Data were acquired and processed by Polymer Char Company (Valencia, Spain) following the procedure described elsewhere [12].

Electrical conductivity was measured at room temperature by means of the ARES Rheometer Dielectric Analysis option (DETA) with an Agilent E4980A Bridge. Dielectric data (ε'') were recorded for each sample as a function of frequency, and the conductivity was calculated using the equation:

$$\sigma\prime(\nu) = \varepsilon_0 2\pi\nu\varepsilon''(\nu)$$
$$\varepsilon_0 = 8.85 \times 10^{-12}\ \text{F·m}^{-1}$$

Transmission Electron Microscopy (TEM) images were obtained to study the dispersion state of the nanocomposites. Nanocomposites were trimmed using an ultramicrotome device at T = −60 °C (Leica EMFC6; Leica Microsystems, Vienna, Austria) equipped with a diamond knife. The ultrathin sections (100 nm) were placed on 300 mesh copper grids. The surfaces were observed by TEM (TECNAI G2 20 TWIN; FEI Company, Eindhoven, THe Netherlands), operating at an accelerating voltage of 200 KeV in a bright-field image mode. TEM results display a two dimensional cut of a 3D structure and, therefore, it is not possible to see through going pathways formed by agglomerates.

3. Results

3.1. Small Amplitude Oscillatory Shear (SAOS) Flow, Electrical Conductivity and TEM Results

The elastic or storage modulus G' as a function of frequency for the nanocomposites prepared, respectively, using Protocols A and B are presented in Figure 1a.

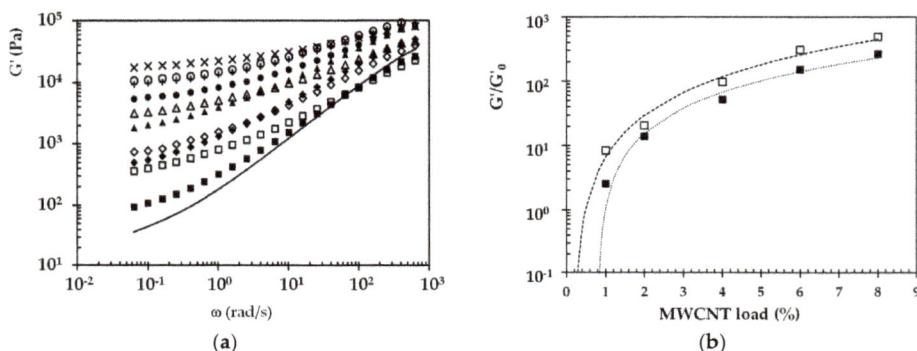

Figure 1. (a) Storage modulus as a function of frequency for both nanocomposites. Line corresponds to unloaded irradiated PP 80Kgy. Protocol A (filled symbols), PP/MWCNT 80 kGy: (■) 1%, (♦) 2%, (▲) 4%, (●) 6%, (+) 8%, Protocol B (empty symbols), PP 80 kGy/MWCNT: (□) 1%, (◇) 2%, (△) 4%, (○) 6%, (×) 8%; (b) percolation plot: G'/G'_0 versus MWCNT load for both nanocomposites. The values of the moduli are taken at a frequency of 0.0628 rad/s and G'_0 refers to the storage modulus of unloaded irradiated PP.

The results show that the elastic modulus tended to level off as the nanotube's concentration increased, giving a scale law $G \sim \omega^n$ with n values close to 0.5. Moreover, at these concentrations the elastic modulus overcame the viscous modulus, $G' > G''$ (not shown to avoid data overlapping). These results indicated that interactions between nanotubes and polymer chains where taking place, bringing about a percolation network that suppressed reptation relaxation in flow, characterized by frequency-dependent moduli, $G' \sim \omega^2$ and $G'' \sim \omega$, with $G'' > G$, at the terminal zone. Recently [19], the observation of a scaling of $G' \sim \omega^{0.5}$ at low frequencies was proposed as a criterion for the determination of the percolation threshold. In our case, an analysis of the data of Figure 1a at the light of the statistical percolation theory [13] was carried out, which allowed to evaluate the percolation threshold using the scaling law $G' = G'_0 (\varphi - \varphi_c)^t$, where φ_c is the percolation threshold and t is an adjustable parameter. The rheological percolation threshold was evaluated considering the values of G' taken at the lowest frequency (0.0628 rad/s) divided by the lowest value of the elastic modulus for neat PP, G' (0.0628 rad/s)/G_0 value. The experimental data fitted to the scaling law of the percolation theory are presented in Figure 1b: The obtained percolation thresholds were $\varphi_c = 1$ wt% for the nanocomposites of Protocol A and $\varphi_c = 0.3$ wt% for the nanocomposites of Protocol B. This difference, found for two nanocomposites which were prepared using different methods is discussed below.

Electrical conductivity results obtained at room temperature are displayed in Figure 2. Both nanocomposites showed electrical conductivities in the range of semiconductors for relatively low MWCNT concentrations, notwithstanding higher conductivities which were found for the nanocomposites prepared using protocol B. Actually, the electrical percolation threshold, which was determined using equation $X = X_0 (\varphi - \varphi_c)^t$, taking X as the electrical conductivity at a frequency of 20 Hz, was $\varphi_c = 3.5$ wt% for the nanocomposites of Protocol A and $\varphi_c = 2$ wt% for the nanocomposites of Protocol B. Therefore, in terms of industrial applications, mixing the nanocomposites after the PP matrix was irradiated (Protocol B) brought about more conductive nanocomposites than those obtained irradiating PP/MWCNT dispersions (Protocol A). This was certainly compatible with TEM results of Figure 3 which show a better MWCNT dispersion (smaller aggregates) for nanocomposites of Protocol B, as compared with Protocol A.

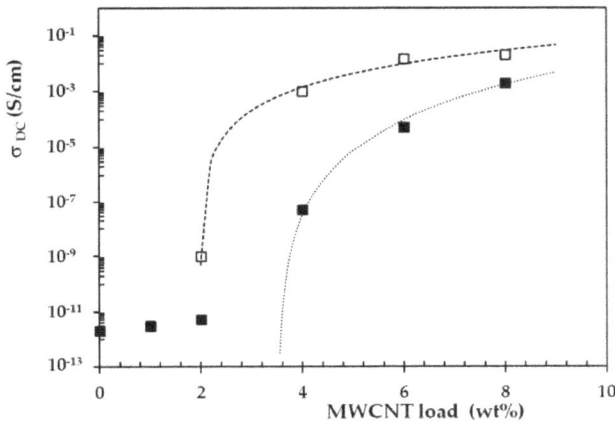

Figure 2. Electrical conductivity, σ, versus frequency for both nanocomposites elaborated using Protocol A (filled symbols) and Protocol B (empty symbols) respectively.

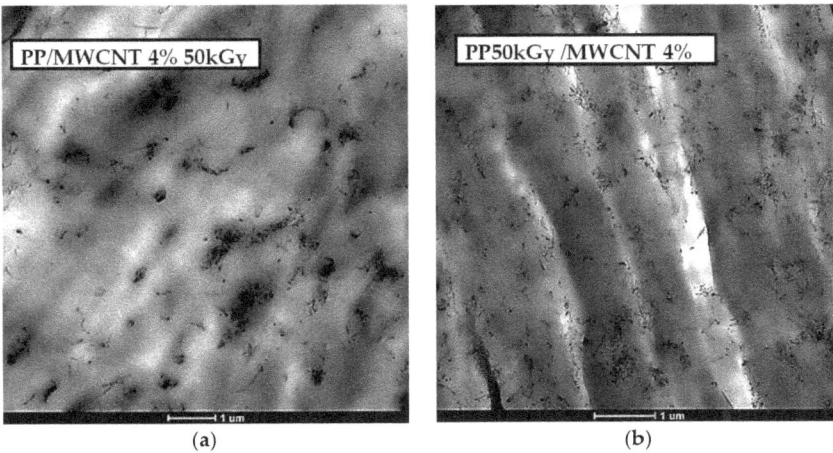

Figure 3. TEM microphotographs of 4% MWCNT nanocomposites. (**a**) Protocol A nanocomposite; (**b**) Protocol B nanocomposite. Better dispersion is observed for the nanocomposite obtained using Protocol B.

It is well known [20–23] that the rheological behaviour of the matrix during mixing plays an important role in the degree of dispersion of the nanoparticles. In our case, the same mixing conditions (see Experimental Part) were employed to prepare the nanocomposites using both Protocol A and Protocol B. However, the molecular characteristics of the matrices were different, since in Protocol A non-irradiated PP was used, whereas in Protocol B an irradiated PP was employed. The most relevant effect of irradiation is the constitution of LCBs that give rise to a strain-hardening effect observed in elongational flow measurements. However, irradiation also produces typically a molecular weight reduction that leads to a viscosity decrease. Eventually, depending on the dose and temperature conditions, irradiations also leads to crosslinking, which was not the case in our work. Besides the SEC-GPC analysis, which is shown in the last part of the paper, the viscosity results presented in Figure 4 reveal the double effect of long chain branching and molecular weight decrease. Irradiated PP displayed a lower viscosity, which constitutes a symptom of molecular weight reduction, in addition to a shear thinning effect at low frequencies (instead of a Newtonian response, like in non-irradiated PP) characteristic of long chain branching [2,5,24]. The observed lower viscosity of irradiated PP stood for the better dispersion noticed by TEM (Figure 3) and, consequently, for the lower rheological percolation threshold value (Figure 1), as well as the lower electrical percolation threshold value (Figure 2) of the nanocomposites manufactured with Protocol B. A dispersion improvement is also observed for polymers subjected to a high intensity ultrasound in the work of Espinoza-González et al. [15].

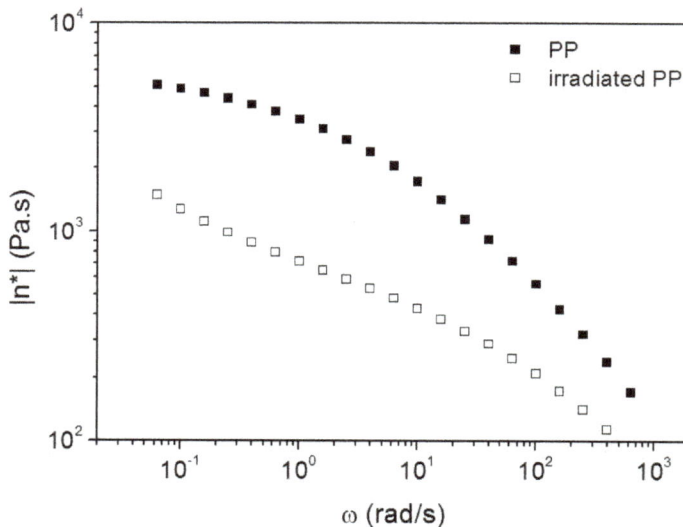

Figure 4. Comparison of the complex viscosity of pure PP before and after irradiation at 80 kGy. The irradiated sample shows shear thinning (concavity) behavior at low frequencies.

3.2. Elongational Flow and Large Amplitude Oscillatory Shear (LAOS) Flow: Differences between Both Types of Nanocomposites (Protocol A and Protocol B)

The results of linear elongational flow measurements of nanocomposites obtained respectively with Protocol A and Protocol B are displayed in Figure 5. The most remarkable outcome was the difference found at high times or elongational strains. Nanocomposites of Protocol A (Figure 5a) show a rapid increase of the elongational viscosity beyond a critical strain, which is referred to as strain hardening. Referring to polypropylene, this strain-hardening or dilatant behavior has been only reported for long chain branched samples obtained either by irradiation [2,5,14,25,26] or by peroxydicarbonate modification [27,28]. However, in the case of nanocomposites manufactured using Protocol B (Figure 5b), no strain hardening was observed (strain softening was observed instead), suggesting that the polypropylene matrix was not long chain branched. This is an intriguing result, because Protocol B implies mixing the nanotubes with a PP which was submitted to irradiation and indeed produced strain-hardening behavior. This can be noticed in Figure 5b, which includes the data of the irradiated PP matrix. The cause of this apparent vanishing of LCB is explained in the last part of the paper.

Basically, the method used to analyse large amplitude oscillatory shear (LAOS) data in this work is the Fourier Transform analysis that decomposes stress data in a time domain into a frequency-dependent spectrum, with the first harmonic as the excitation frequency ($\omega_1/2\pi$). For the case of a non-linear response, at large strain amplitudes the stress is no longer sinusoidal and further odd harmonics are obtained. The non-linearity can be quantified using the ratio $I_3/I_1 = I(n\omega_1)/I(\omega_1)$ with $I_n = I(n\omega_1)$ as the magnitude of the n-th harmonic and $I_1 = I(\omega_1)$ as the fundamental frequency. When the non-linear regime is reached, the third harmonic I_3 rises above the noise level and I_3/I_1 shows a quadratic scaling relationship as a function of strain amplitude [18,29].

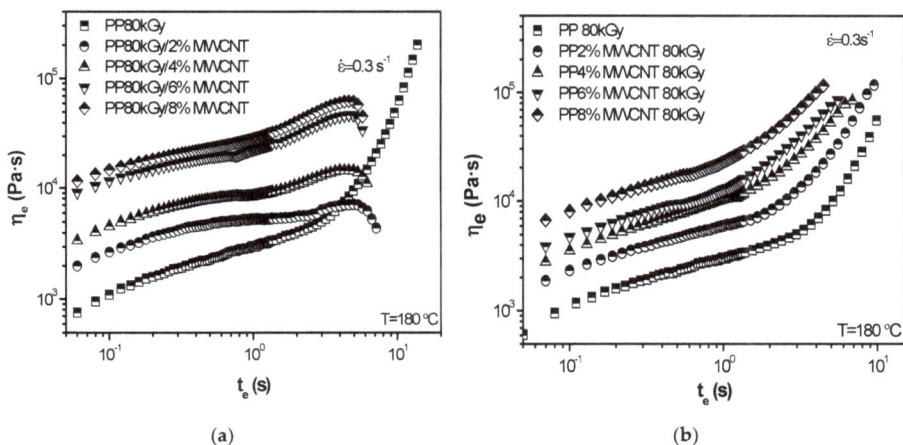

Figure 5. (a) Elongational viscosity vs. elongation time for the samples elaborated with Protocol A; (b) elongational viscosity vs. elongation time for the samples elaborated with Protocol B. Irradiated pure PP is also included for comparison purposes.

The capacity of LAOS to detect and evaluate long chain branching in irradiated MWCNT/PP nanocomposites was described in a paper of our group [12], where the effect of irradiation on LAOS results of pure PP and PP/MWCNT nanocomposites (equivalent to Protocol A of this paper) was investigated. SEC-GPC results came to demonstrate that irradiation caused long chain branching and LAOS data were correlated to the degree of LCB. An increase of the intensity of the third harmonic relative to the intensity of the response at the fundamental frequency, I_3/I_1, was observed, as LCB level increased, with respect to linear nonirradiated samples. The LAOS results of our nanocomposites (Protocol A and Protocol B) are presented in Figure 6. I_3/I_1 values at large amplitudes were higher for irradiated PP and Protocol A, than those of nonirradiated PP. A decrease of the intensity of the third harmonic at the higher strains is observed for the Protocol B nanocomposite. Concomitantly with elongational flow results, two groups of materials could be established: On the one hand, irradiated PP and nanocomposites of Protocol A which showed strain-hardening and relatively high values of I_3/I_1, and on the other hand, non-irradiated PP and nanocomposites of Protocol B which did not display any of these rheological features. The results of nanocomposites of Protocol B, which behave like a linear PP in both elongational flow and LAOS experiments, seemed to indicate that LCB disappeared during the extrusion accomplished to prepare the nanocomposite.

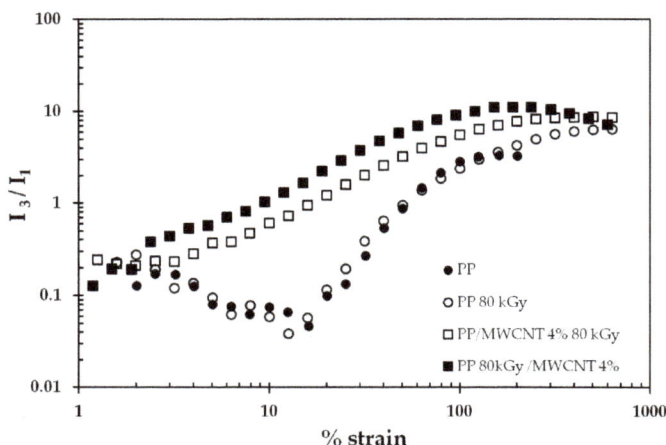

Figure 6. LAOS results (see text) for PP, irradiated PP (PP 80 kGy), Protocol A nanocomposite (PP/MWCNT 4% 80 kGy) and Protocol B nanocomposite (PP 80 kGy/MWCNT 4%). The values of I_3/I_1 at large amplitudes are higher for irradiated PP and nanocomposites of Protocol A, with respect to PP and Protocol B.

3.3. Shear Modification and Recovery: Alignment and Misalignment

An analysis of the PP matrix was required to investigate the molecular changes suffered by the irradiated PP during the extrusion carried out for mixing with MWCNTs to prepare the nanocomposites of Protocol B. A GPC-SEC analysis of the irradiated PP matrix before and after submitting it to mixing is shown in Figure 7. For comparison purposes, the results of PP irradiated at different irradiation doses are also presented in this figure and in Table 1. Long chain branches were detected in irradiated PP submitted to processing, although the level was considerably lower than that of the non-processed sample. As an exception, the LCB level of the PP irradiated at 80 kGy and submitted to extrusion was higher than that of the PP irradiated at 20 kGy (the lower irradiation dose) but not submitted to extrusion. Interestingly enough, the latter brought about strain-hardening behavior (not shown here), which was significant, because its LCB level was lower than that of the extruded PP (Figure 7 and Table 1). In consequence, although a considerable decrease of LCB level was observed in the extruded PP, this was not responsible for the complete vanishing of strain-hardening observed in Figure 5. Instead, the phenomenon of strain-hardening disappearance could be ascribed to the conformation change of the branched structure, as alignment of long branches to the backbone chains during processing took place. The aligned branches passed to behave like linear chains and, therefore, strain hardening was not observed. This reasoning is based on the concept of shear modification phenomenon, according to which the existence of long chain branches leads to rheological changes by processing involving straining and thermal history [30–32].

Table 1. Molecular parameters including LCB density evaluated using SEC Chromatography coupled with three detectors of non-irradiated, irradiated PP and processed irradiated PP.

Molecular Parameters	PP	PP 20 kGy	PP50 kGy	PP80 kGy	PP80 Kg (Processed)
Mw (g/mol)	298,000	287,000	237,000	168,000	191,060
Mn (g/mol)	33,000	41,000	40,000	34,000	26,339
Mz (g/mol)	755,000	1,350,000	1,360,000	900,000	706,371
Mw/Mn	9	7	6	5	7.4
LCBf/1000C	-	0.011	0.03	0.036	0.023

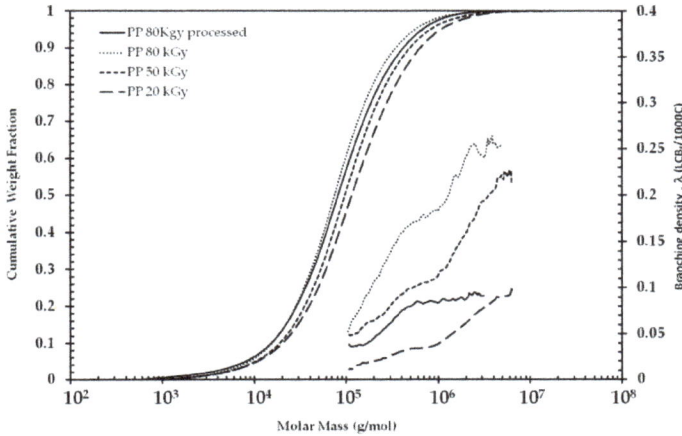

Figure 7. Branching density λ (LCB/1000C) and cumulative molar mass distribution for irradiated PP samples obtained by triple detection GPC-SEC.

To confirm this hypothesis, an analysis of the eventual recovery of the original conformation, when the sample was submitted to temperatures T > Tm in the quiescent state, had to be done. The elongational flow data of the extruded PP sample that was annealed at T = 160 °C during 8 h in a vacuum oven (presented in Figure 8) suggests the recovery of the conformation. This was manifested in the strain hardening shown by the annealed sample, which is in contrast to the sample not submitted to thermal treatment.

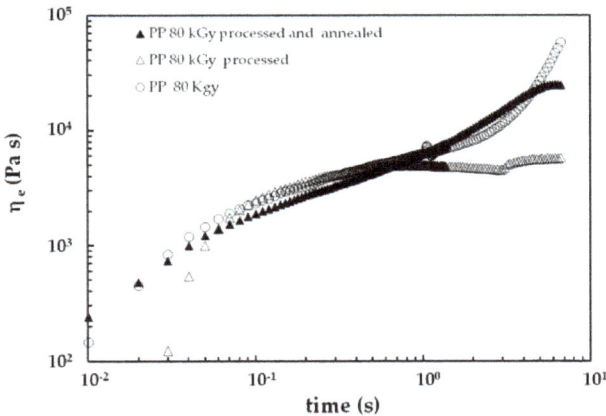

Figure 8. Uniaxial viscosity as a function of time obtained at T = 180 °C at 0.3 s^{-1} elongational rate. PP80 KGy is the Irradiated PP distinguished by the strain-hardening behaviour. PP80 kGy processed is the extruded irradiated PP distinguished by the absence of strain hardening. PP80 kGy processed and annealed shows the strain hardening behavior, which has been recovered during the annealing process at T = 160 °C in a vacuum oven during 8 h.

Considering this result, the question of an eventual repercussion of the recovery on LAOS results was posed. LAOS results of Figure 6 show a relative reduction of the value of I_3/I_1 at large deformations for nanocomposites of Protocol B with respect to Protocol A, denoting the effect of the extrusion process. At this point we wondered if a recovery of I_3/I_1 was possible, in parallel to the observed strain-hardening recovery. The positive answer to this question is given in Figure 9, which shows that annealing during 8 h at 160 °C in the extruded irradiated PP led to I_3/I_1 values close to those of non-extruded irradiated PP. The recovery similitude observed in two different experiments (elongational flow and LAOS) leads us to assume that annealing was able to produce a misalignment or return to the original chain conformation of the aligned long chain branches. Certainly, during extrusion for the preparation of nanocomposites by Protocol B, the LCB level of PP was reduced (as can be seen in Figure 7), but this reduction did not justify the strain hardening vanishing, because PP with a lower amount of LCB was able to bring about strain hardening. This leads us to conclude that although during processing two effects were produced, reduction of LCB level (irreversible) and chain alignment (reversible), only the last one was responsible for the elimination of strain hardening and the significant reduction of the value of I_3/I_1 at high deformations. The reversibility of the chain alignment process was confirmed by the results achieved with the PP submitted to extrusion in conditions similar to mixing and annealed during 8 h at T = 160 °C. Obviously, the thermal treatment could not be able to recover the eliminated LCB, but was capable of misaligning the chains leading them to the conformation previous to processing.

The analysis of the recovery was also carried out on the nanocomposites obtained with the irradiated matrix. Samples were annealed at T = 160 °C during 8 h in vacuum, but no recovery was observed. The viscoelastic response of the nanocomposites was clearly dependent on the MWCNT content. The conformational structure previous to processing could not be recovered because the viscous flow region was delayed to very long times, which prevented the misaligning capability of the long chains at the annealing conditions.

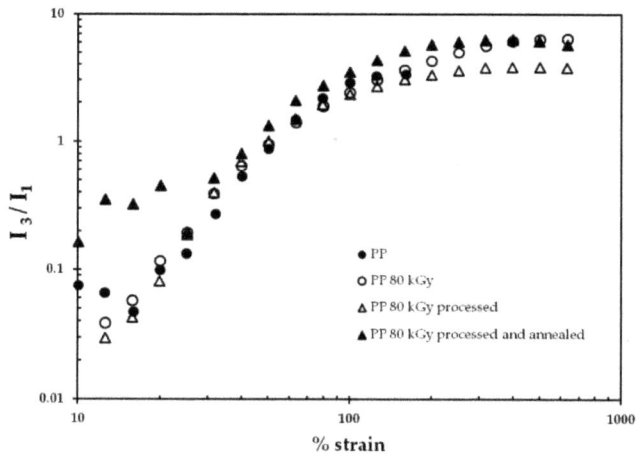

Figure 9. LAOS results (see text) for PP, irradiated PP, irradiated PP processed and irradiated PP processed and annealed during 8 h at T = 160 °C in vacuum.

4. Conclusions

Two protocols both involving an irradiation process were employed to prepare PP/MWCNT nanocomposites. Protocol A was achieved mixing first PP and MWCNTs in an extruder and then irradiating the obtained dispersion, and Protocol B was carried out irradiating first PP and then mixing it after with MWCNTs in an extruder. As noticed by TEM, Protocol B led to a better dispersion, bringing about lower rheological and electrical percolation thresholds. The reason for this outcome was the lower viscosity found in the irradiated PP matrix involved in Protocol B, which in turn was due to a molecular weight reduction.

Elongational flow results were striking at first sight. Strain hardening in elongational flow measurements, typical of irradiated PPs, was observed for the nanocomposite of Protocol A, but not for the nanocomposite of Protocol B. The practical consequences of this result were relevant, because strain hardening is a necessary condition to achieve certain industrial processes, such as extrusion-blowing, blow molding, foaming extrusion and thermoforming.

Moreover, LAOS results showed large I_3/I_1 values for irradiated PP and the nanocomposite of Protocol A, as compared with non-irradiated PP and the nanocomposite of Protocol B. Since it is accepted that strain hardening observed in irradiated PPs owes to LCB, and, on the other hand, larger I_3/I_1 values for long chain branched PPs have been reported, the initial conclusion was that no LCB was in the matrix of the nanocomposite of Protocol B. However, SEC GPC analysis of a polypropylene sample submitted to the same irradiation and mixing conditions as that carried out in Protocol B, revealed a level of LCB that should have been enough to provoke strain hardening. The hypothesis of a shear modification that aligned the branches into the backbone, eliminating the strain hardening, was considered. Recovery of the strain hardening and I_3/I_1 values by annealing the sample at T = 160 °C during 8 h to promote a misalignment of the branches, led us to demonstrate that although an irreversible destruction of LCBs was taking place during mixing, the actual reason for the observed rheological changes was shear modification.

Author Contributions: Conceptualization, A.S., A.H., and M.F.; Methodology, A.H., and M.F.; Software, A.H. and M.F.; Formal Analysis, A.S., A.H., and M.F.; Investigation, A.S., A.H., and M.F.; Data Curation, A.H. and M.F.; Writing—Original Draft Preparation, A.S.; Writing—Review & Editing, A.S. and M.F.; Supervision, A.S.; Project Administration, A.S.; Funding Acquisition, A.S., and A.H.

Funding: This research received no external funding.

Acknowledgments: The authors are grateful to the financial support from the Basque Government (GIC IT-586-13). A. Huegun would like to thank to the Spanish Government for the grant BES-2008-002469.

Conflicts of Interest: The authors declare no conflict of interest.

References

1. Rätzsch, M. Special PP's for a developing and future market. *J. Macromol. Sci. Pure Appl. Chem.* **1999**, *36*, 1587–1611. [CrossRef]

2. Pötschke, P.; Krause, B.; Stange, J.; Münstedt, H. Elongational viscosity and foaming behavior of PP modified by electron irradiation or nanotube addition. *Macromol. Symp.* **2007**, *254*, 400–408. [CrossRef]

3. Auhl, D.; Stadler, F.J.; Münstedt, H. Comparison of molecular structure and rheological properties of electron-beam- and gamma-irradiated polypropylene. *Macromolecules* **2012**, *45*, 2057–2065. [CrossRef]

4. Mishra, R.; Tripathy, S.P.; Dwivedi, K.K.; Khathing, D.T.; Ghosh, S.; Müller, M.; Fink, D. Electron induced modification in polypropylene. *Radiat. Meas.* **2001**, *33*, 845–850. [CrossRef]

5. Auhl, D.; Stange, J.; Münstedt, H.; Krause, B.; Voigt, D.; Lederer, A.; Lappan, U.; Lunkwitz, K. Long-chain branched polypropylenes by electron beam irradiation and their rheological properties. *Macromolecules* **2004**, *37*, 9465–9472. [CrossRef]

6. Krause, B.; Stephan, M.; Volkland, S.; Voigt, D.; Häußler, L.; Dorschner, H. Long-chain branching of polypropylene by electron-beam irradiation in the molten state. *J. Appl. Polym. Sci.* **2006**, *99*, 260–265. [CrossRef]

7. Nakamura, S.; Tokumitsu, K. Influence of electron beam irradiation on mechanical and thermal properties of polypropylene/polyamide blend. *AIP Conf. Proc.* **2014**, *1593*, 666–669. [CrossRef]

8. Hassan, F.; Entezam, M. Electron Beam Irradiation Method to Change Polypropylene Application: Rheology and Thermo-Mechanical Properties. *Polyolefins J.* **2018**, *6*, 53–61. [CrossRef]

9. Bee, S.T.; Sin, L.T.; Hoe, T.T.; Ratnam, C.T.; Bee, S.L.; Rahmat, A.R. Study of montmorillonite nanoparticles and electron beam irradiation interaction of ethylene vinyl acetate (EVA)/de-vulcanized waste rubber thermoplastic composites. *Nucl. Instrum. Methods Phys. Res. Sect. B* **2018**, *423*, 97–110. [CrossRef]

10. Lotfy, S.; Atta, A.; Abdeltwab, E. Comparative study of gamma and ion beam irradiation of polymeric nanocomposite on electrical conductivity. *J. Appl. Polym. Sci.* **2018**, *135*, 1–7. [CrossRef]

11. Martínez-Morlanes, M.J.; Castell, P.; Martínez-Nogués, V.; Martinez, M.T.; Alonso, P.J.; Puértolas, J.A. Effects of gamma-irradiation on UHMWPE/MWNT nanocomposites. *Compos. Sci. Technol.* **2011**, *71*, 282–288. [CrossRef]

12. Fernandez, M.; Huegun, A.; Muñoz, M.E.; Santamaria, A. Nonlinear oscillatory shear flow as a tool to characterize irradiated polypropylene/MWCNT nanocomposites. *Appl. Rheol.* **2015**, *25*, 1–12. [CrossRef]
13. Noll, A.; Burkhart, T. Morphological characterization and modelling of electrical conductivity of multi-walled carbon nanotube/poly(p-phenylene sulfide) nanocomposites obtained by twin screw extrusion. *Compos. Sci. Technol.* **2011**, *71*, 499–505. [CrossRef]
14. Huegun, A.; Fernández, M.; Muñoz, M.E.; Santamaría, A. Rheological properties and electrical conductivity of irradiated MWCNT/PP nanocomposites. *Compos. Sci. Technol.* **2012**. [CrossRef]
15. Espinoza-Gonzalez, C.; Avila-Orta, C.; Martinez-Colunga, G.; Lionetto, F.; Maffezzoli, A. A measure of CNTs dispersion in polymers with branched molecular architectures by UDMA. *IEEE Trans. Nanotechnol.* **2016**, *15*, 731–737. [CrossRef]
16. Vega, J.F.; da Silva, Y.; Vicente-Alique, V.; Núñez-Ramírez, R.; Trujillo, M.; Arnal, M.L.; Müller, A.J.; Dubois, P.; Martínez-Salazar, J. Influence of chain branching and molecular weight on melt rrheology and crystallization of polyehtylene/carbon nanotube nanocomposites. *Macromolecules* **2014**, *47*, 5668–5681. [CrossRef]
17. Cangussú, M.E.; de Azeredo, A.P.; Simanke, A.G.; Monrabal, B. Characterizing Long Chain Branching in Polypropylene. *Macromol. Symp.* **2018**, *377*, 1–9. [CrossRef]
18. Wilhelm, M. Fourier-transform rheology. *Macromol. Mater. Eng.* **2002**, *287*, 83–105. [CrossRef]
19. Hassanabadi, H.M.; Wilhelm, M.; Rodrigue, D. A rheological criterion to determine the percolation threshold in polymer nano-composites. *Rheol. Acta* **2014**, *53*, 869–882. [CrossRef]
20. McNally, T.; Potschke, P. *Polymer-Carbon Nanotube Composites: Preparation, properties and Applications*; Elsevier: New York, NY, USA, 2011.
21. Alig, I.; Pötschke, P.; Lellinger, D.; Skipa, T.; Pegel, S.; Kasaliwal, G.R.; Villmow, T. Establishment, morphology and properties of carbon nanotube networks in polymer melts. *Polymer* **2012**, *53*, 4–28. [CrossRef]
22. Du, F.; Scogna, R.C.; Zhou, W.; Brand, S.; Fischer, J.E.; Winey, K.I. Nanotube networks in polymer nanocomposites: Rheology and electrical conductivity. *Macromolecules* **2004**, *37*, 9048–9055. [CrossRef]
23. Kasaliwal, G.R.; Göldel, A.; Pötschke, P.; Heinrich, G. Influences of polymer matrix melt viscosity and molecular weight on MWCNT agglomerate dispersion. *Polymer* **2011**, *52*, 1027–1036. [CrossRef]
24. Rojo, E.; Muñoz, M.E.; Mateos, A.; Santamaría, A. Flow instabilities in linear and branched syndiotactic poly(propylene)s. *Macromol. Mater. Eng.* **2007**, *292*, 1210–1217. [CrossRef]
25. Schulze, D.; Mu, R. Rheological evidence of modifications of polypropylene by b-irradiation. *Rheol. Acta* **2003**, 251–258. [CrossRef]
26. Sugimoto, M.; Tanaka, T.; Masubuchi, Y.; Takimoto, J.I.; Koyama, K. Effect of chain structure on the melt rheology of modified polypropylene. *J. Appl. Polym. Sci.* **1999**, *73*, 1493–1500. [CrossRef]
27. Lagendijk, R.P.; Hogt, A.H.; Buijtenhuijs, A.; Gotsis, A.D. Peroxydicarbonate modification of polypropylene and extensional flow properties. *Polymer* **2001**, *42*, 10035–10043. [CrossRef]
28. Gotsis, A.D.; Zeevenhoven, B.L.F.; Hogt, A.H. The effect of long chain branching on the processability of polypropylene in thermoforming. *Polym. Eng. Sci.* **2004**, *44*, 973–982. [CrossRef]
29. Hyun, K.; Wilhelm, M.; Klein, C.O.; Cho, K.S.; Nam, J.G.; Ahn, K.H.; Lee, S.J.; Ewoldt, R.H.; McKinley, G.H. A review of nonlinear oscillatory shear tests: Analysis and application of large amplitude oscillatory shear (LAOS). *Prog. Polym. Sci.* **2011**, *36*, 1697–1753. [CrossRef]
30. Münstedt, H. The influence of various deformation histories on elongational properties of low density polyethylene. *Colloid Polym. Sci.* **1981**, *259*, 966–972. [CrossRef]
31. Yamaguchi, M.; Wagner, M.H. Impact of processing history on rheological properties for branched polypropylene. *Polymer* **2006**, *47*, 3629–3635. [CrossRef]
32. Luo, Y.; Xin, C.; Zheng, D.; Li, Z.; Zhu, W.; Wu, S.; Zheng, Q.; He, Y. Effect of processing history on the rheological properties, crystallization and foamability of branched polypropylene. *J. Polym. Res.* **2015**, *22*, 1–13. [CrossRef]

| 👁👁👁 *fluids* | | MDPI |

Article

A Comparison of the Effect of Temperature on the Rheological Properties of Diutan and Rhamsan Gum Aqueous Solutions

Mª Carmen García González, María del Socorro Cely García, José Muñoz García and Maria-Carmen Alfaro-Rodriguez *

Departamento de Ingeniería Química, Facultad de Química, Universidad de Sevilla, C/P. García González, 1, E41012 Sevilla, Spain; mcgarcia@us.es (M.C.G.G.); madelso40@hotmail.com (M.d.S.C.G.); jmunoz@us.es (J.M.G.)
* Correspondence: alfaro@us.es; Tel.: +34-954-557-180; Fax: +34-954-556-447

Received: 15 December 2018; Accepted: 19 January 2019; Published: 1 February 2019

Abstract: The rheological properties exhibited by gums make its use in applications interesting, such as foods, cosmetics, enhanced oil recovery, or constructions materials. Regardless of application field, the effect of temperature on these properties is of great importance, since these properties can be modified and cause the gum not to be useful for those conditions. Diutan and rhamsan gums are biopolymers, belonging to the sphingans, with similar structures which differ in the substituents of their side chains. It is known that both gums exhibit suitable viscoelastic properties and flow behavior when used as a stabilizer, gelling agent, or thickener. Both gums are widely used in food industry, personal care products, construction materials, oil operations, etc. For this reason, to know the effect of the temperature on their rheological properties is very helpful. For this purpose, small amplitude oscillatory shear measurements and flow curves, as a function of the temperature from 10 °C to 60 °C, have been performed, and the results obtained for both gums compared. The obtained results provide interesting information from an industrial point of view, since they reveal that the rheological properties remained almost unaltered in the temperature range assessed with diutan gum aqueous solutions, being slightly more viscous and viscoelastic than rhamsan gum solutions.

Keywords: diutan gum; rhamsan gum; rheology; viscoelasticity; flow properties; weak gel

1. Introduction

The rheological properties of macromolecular solutions are important in many industrial fields, such as the food industry, pharmaceutical and cosmetics, in the paper industry, in enhanced oil recovery (EOR), in water treatment, etc. Water-soluble polymers are used in these fields as thickeners, gelling agents, texture modifiers, stabilizers, etc. [1,2]. In the particular case of EOR, they are added to control mobility and reduce the permeability of the tank by increasing the viscosity of the injected water.

Hence, an important requirement for a potential polymer to be used in EOR is that the polymer-water solution has favorable rheological behavior. Gum diutan and rhamsan gum belong to that group of water-soluble polymers of high molecular weight. They are exopolysaccharides (EPS) belonging to the sphingans, as well as gellan gum or wellan gum [1,3,4]. The basic structure of all these polysaccharides is constituted by a tetrasaccharide repeat unit based on β-D-glucose, β-D-glucuronic acid, and α-L-rhamnose. Gum diutan, obtained by aerobic fermentation of Sphingomonas S.PATCC53159, has this basic structure, and each glucosidic residue that is next to a rhamnosil residue is substituted by two units of L-rhamnose. Rhamsan gum differs from gum diutan in the side chain substituents, which in this case is a disaccharide chain of β-D-glucopyranosyl or α-D-glucopyranosyl [5,6]. Both gums are non-gelling polysaccharides, unlike gellan gum, which belongs to the same family. It is precisely the difference in the side-chains responsible for it, namely the occurrence of different side-chains' influences on the behavior in aqueous media. Several studies reveal that native diutan and rhamsan gum exhibit a double helical conformation, which is maintained with the temperature as a consequence of their chains' strength [6,7].

Regarding applications, rhamsan gum has wider applications in the food industry compared to other sphingans, since it can tolerate high concentrations of sodium chloride and phosphates [8]. In general, they are used not only in the food industry but also in personal hygiene products, in construction materials, in oil operations, etc.

Nowadays, it is not only important to know the rheological properties of these gums in solution but also the modifications that these properties undergo when the temperature varies. This information is of great interest in numerous industrial processes and in the conditions of use and application of them. For example, in the food industry, sterilization, drying, extrusion, etc., are processes in which temperature is of vital importance. In this work, the influence of temperature on the viscoelastic properties and flow properties of aqueous solutions containing 1 wt % diutan gum or rhamsan gum were studied in detail, establishing a comparison between both gums. In order to reach this objective, stress and frequency sweep measurements and steady state tests were performed.

2. Materials and Methods

2.1. Materials

Diutan gum (CP Kelco 1A9722A) and rhamsan gum (OL9478A) used were cordially provided by CP-Kelco (San Diego, CA, USA). Both gums exhibit high molecular weight. Their values and the intrinsic viscosity values can be obtained from works published [6,7]. Sodium azide supplied by Panreac was utilized as preservative. Distilled water was used as solvent.

The sample composition is shown in Table 1.

Table 1. Sample composition.

Ingredients	wt %
Gum (Diutan or Rhamsan)	1
NaN$_3$	0.1
Water	98.9

2.2. Gum Aqueous Solutions' Preparation

Batches of 250 g were prepared. In order to obtain the sample, the gum amount necessary was added on the water containing sodium azide. The samples were homogenized by means of Ika-Visc MR-D1 (Ika, Staufen, Germany), at 1000 rpm for 3 h. The samples were kept refrigerated at 5 °C for some hours, and then they were stored at 25 °C.

2.3. Rheological Characterization.

The rheological tests were carried out by means of a controlled stress rheometer, Haake RS100 (Haake Thermo Scientific, Karlsruhe, Germany), using a serrated parallel plate geometry of 60 mm of diameter to avoid wall slip effects and using a measuring gap of 1 mm. To control the temperature (10 °C, 25 °C, 40 °C, and 60 °C), a Julabo circulator thermostat was utilized.

In order to prevent the mechanical history effects due to the sample placing in the sensor system, the samples were maintained at rest for 10 min before starting the measurements. To determine this equilibration time, time sweeps were performed.

All tests were carried out at least three times. Fresh sample were used in each measurement.

2.3.1. Stress Sweep Tests

To determine the linear viscoelastic region (LVR), stress sweep tests were carried out ranging from 1.0 Pa to 50 Pa, at 1 Hz.

2.3.2. Mechanical Spectra

The frequency sweep tests were performed from 0.01 to 10 Hz (0.0628 to 62.83 rad/s) at fixed shear stress within the linear viscoelastic region.

2.3.3. Steady State Tests

The flow curves were carried out in the 10–40 Pa shear stress range using a step-wise method to obtain the steady-state regime.

3. Results and Discussion

3.1. Influence of Temperature on the Linear Viscoelastic Region.

In order to study the viscoelastic behavior of gums studied, the linear viscoelastic regions were determined. Figure 1a,b show the stress sweep results of diutan gum and rhamsan gum, respectively, at the different studied temperatures (10 °C, 25 °C, 40 °C, and 60 °C). By increasing the stress, two regions can be clearly differentiated: (a) the so-called linear viscoelastic region, in which G' (storage or elastic modulus) and G" (loss or viscous modulus) remain practically constant, and (b) the non-linear viscoelastic region in which viscoelastic functions begin to decrease. It must be noted there was an increase of G" values before its decrease. This effect was observed in other systems and it was attributed to the reorganization of the gum solution structure before the output of the linear viscoelastic region [9,10].

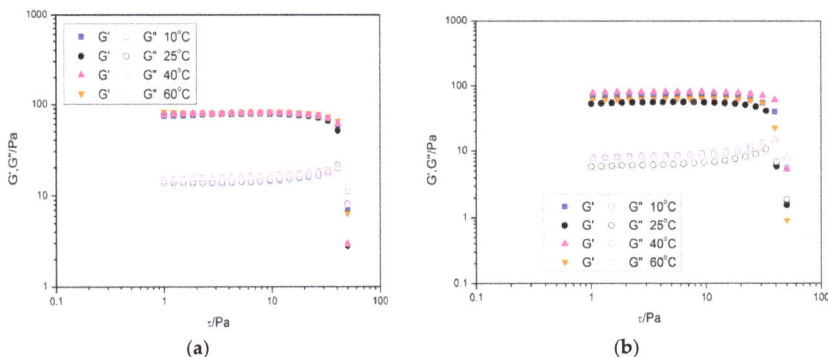

Figure 1. Influence of the temperature on the linear viscoelastic region of aqueous solutions containing (**a**) 1 wt % diutan gum and (**b**) 1 wt % rhamsan gum.

The shear stress from which the modules cease to be constant is called critical stress (τ_c), and it corresponds to stress that leads to the first non-linear changes in the structure. As can be observed in Figure 1, the G" values cease to be steady at smaller shear stress values than G'. For this reason, it has been considered that the onset of the non-linear viscoelastic region begins when G" ceases to be constant, particularly when the deviation of loss modulus becomes more than 5%. From Figure 1, it can be stated that there is no influence of the temperature on critical stress values of diutan and rhamsan gum aqueous solutions, being that these values are similar for both gums at 11.83 Pa and 11.68 Pa, respectively. Furthermore, it should be noted the broad extension of the linear viscoelastic zone, which indicated a wide zone where the sample is not damaged, namely, high stability to shear.

Regardless of temperature studied, G' is greater than G" within LVR, which indicates that diutan and rhamsan gum dispersions are more elastic than viscous. The complex module (G*) as a function of the shear stress is shown in Figure 2 in order to better compare the level of viscoelasticity of both polysaccharides. As can be observed, the values of G* were similar for both gums, however, they were slightly greater for the diutan gum solutions.

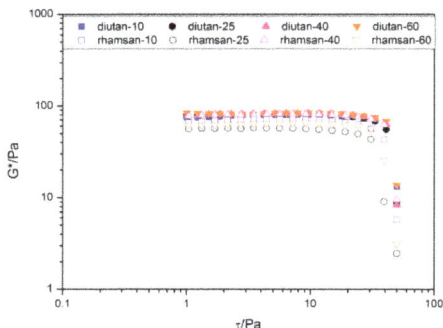

Figure 2. Comparison of the temperature effect on the linear viscoelastic region of the diutan gum aqueous solutions and rhamsan gum aqueous solutions.

3.2. Influence of Temperature on the Mechanical Spectra

Frequency sweep tests have been carried out at a fixed shear stress within the linear viscoelastic region, so that the sample structure is not damaged by the stress imposed during test.

Figure 3 shows the influence of the temperature on the mechanical spectra of diutan and rhamsan gums, respectively. Both gums (Figure 3) exhibited storage module values (G′) higher than those of loss module (G″) in the whole studied frequency range, with a small frequency dependence. According to bibliography [11,12], the shape of these mechanical spectra makes clarification of the diutan and rhamsan gum aqueous solutions as weak gels possible. In addition, no crossover point occurred. For this reason, it can be affirmed the solid character is higher than liquid character in aqueous solutions of these gums. This rheological behavior is characteristic of systems with a high degree of internal structuring, and coincides with that presented by other hydrocolloids, such as Sterculia apetala or striata [12,13]. Regarding temperature, Figure 3 illustrates the high thermal stability exhibited for both diutan and rhamsan gum.

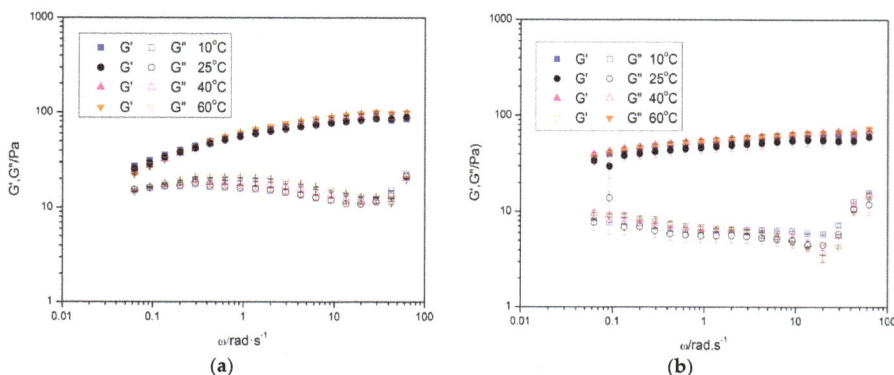

Figure 3. Influence of the temperature on the mechanical spectra of (**a**) 1 wt % diutan gum and (**b**) 1 wt % rhamsan gum.

In order to compare the results obtained for both gums, by way of example in Figure 4, the G′ and G″ values at 1 Hz against temperature for diutan and rhamsan gums are shown. As can be observed, the elastic module of diutan gum solutions was slightly higher than that of rhamsan gum solutions, while the viscous module was clearly greater. Additionally, it is important to note that there was no significant dependence on temperature in these results.

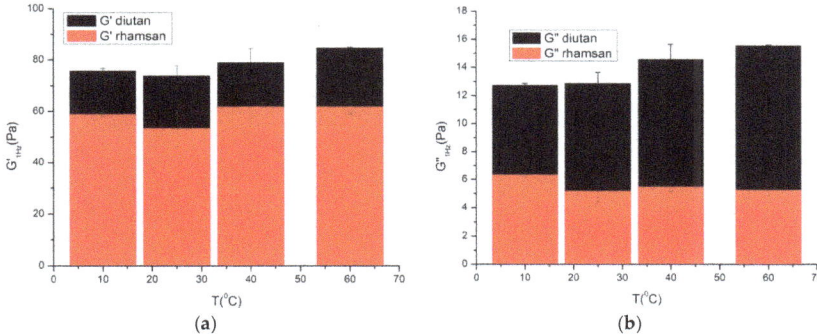

Figure 4. (**a**) Elastic module and (**b**) Viscous module, at 1 Hz, against temperature for 1 wt % diutan and rhamsan gum aqueous solutions. Additionally, standard deviation is plotted.

3.3. Steady State Measurements

The flow behavior of both gums is illustrated in Figure 5. These results pointed out the aqueous solution containing 1 wt % of gum exhibited a shear thinning behavior, which was distinguished by a fast decrease of the viscosity as shear rate increased. In addition, these samples exhibited a Newtonian region at low shear rate, so-called zero shear viscosity, η_0. This result was due to the fact that at low shear rate, polymeric molecules are disordered and partially aligned in the direction of the flow, resulting in a greater interaction between them, and therefore, a higher viscosity. By increasing the shear rate, the molecules were easily aligned, decreasing the physical interactions that occur between adjacent polymer chains. A marked shear thinning behavior is very interesting, since it facilitates pumping and imparts a finer consistency during swallowing if it is a food product [14]. The flow curves were fitted to the Ostwald-de-Waele Equation with a high degree of satisfaction.

$$\eta = \eta_1 \cdot \gamma^{n-1}$$

where η is the viscosity, η_1 is viscosity value at $1\ \mathrm{s}^{-1}$, γ is the shear rate, and n is the index flow.

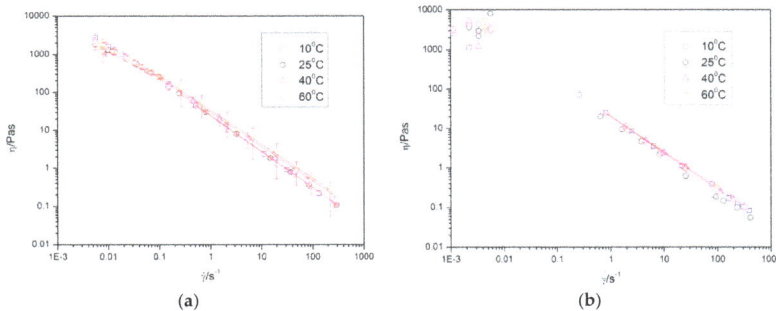

Figure 5. Influence of the temperature on the flow curves of aqueous solutions containing (**a**) 1 wt % diutan gum and (**b**) 1 wt % rhamsan gum.

The fitting parameters revealed a tendency to increase the viscosity at $1\ \mathrm{s}^{-1}$ for diutan gum solutions, while this parameter remained unaltered for rhamsan gum solutions. Additionally, this viscosity illustrated that aqueous solutions of diutan gum were more viscous than rhamsan gum solutions. In both gums, the flow index was independent of the temperature in the range of 10 °C to 60 °C, and its value revealed the occurrence of a marked non-Newtonian character, which was maintained in both cases when the temperature increased (Table 2).

Table 2. Fitting parameter to Ostwald–Waele model for aqueous solutions of diutan gum and rhamsan gum. SD is standard deviation of the mean (n = 3) for η_1 and n.

Gum	T (°C)	η_1 (Pa·s)	SDη_1	n	SDn	R^2
	10	23.67	0.22	0.05	0.01	0.999
	25	24.41	0.16	0.05	0.00	0.999
Diutan Gum	40	28.34	0.40	0.05	0.00	0.999
	60	37.55	1.25	0.05	0.01	0.995
	10	19.87	0.22	0.07	0.06	0.999
	25	14.89	0.37	0.09	0.03	0.998
Rhamsan Gum	40	20.32	0.01	0.08	0.00	1
	60	19.55	0.51	0.12	0.03	0.999

3.4. Cox-Merz Rule

The steady state properties have been compared with dynamic properties using the Cox-Merz rule. The apparent viscosity obtained from the flow curve, η, and the complex viscosity obtained from low amplitude oscillatory tests, $|\eta^*|$, have been plotted against the frequency and the shear rate. By way of example, the results are shown at a temperature of 60 °C, but similar results are obtained at all the temperatures studied.

For both diutan gum and rhamsan gum (Figure 6a,b), $|\eta^*|$ was slightly greater than η. Therefore, it can be deduced that the Cox-Merz rule is not met, even though the differences found between both viscosities are very small. This means that the application of shear did not produce a significant destruction of the structure.

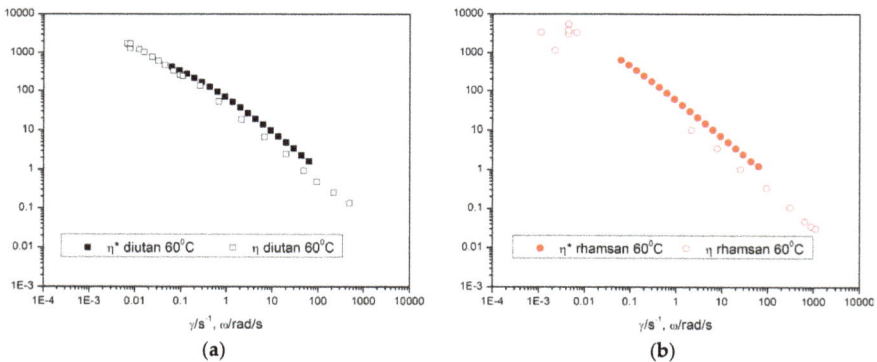

Figure 6. Apparent viscosity and complex viscosity versus shear rate and frequency of aqueous solutions containing (**a**) 1 wt % diutan gum and (**b**) 1 wt % rhamsan gum. Temperature 60 °C.

It is important to mention that these relationships are valid only for some types of materials or some ranges of conditions. It has been verified that it is satisfactorily fulfilled for polymers of homogeneous solutions and molten polymers, and that it failed in some other cases, such as linear and branched polyethylenes, block copolymers, and rigid molecules [15].

Other microbial polysaccharides, such as Aeromonas gum [16] and xanthan gum [17] and aqueous dispersions of exudates of gums [18], also demonstrated deviations from this rule. Clasen and Kulicke [19] proposed that deviations from this rule arise from the formation of associations in the relaxed state. In a shear test, the imposed shear stress destroys the intermolecular associations formed when the sample is relaxed. In contrast, in an oscillatory test, the gum solution is subjected to a sinusoidal shear stress within the linear viscoelastic range small enough so that these molecular associations are not destroyed. For this reason, the apparent viscosity calculated from a destructive test showed a lower value than the complex viscosity obtained from a non-destructive oscillatory test.

A comparison of results obtained for both gums does not provide additional information to that obtained in previous sections, namely the aqueous solutions of diutan gum are slightly more viscous and viscoelastic than those of rhamsan gum, and both solutions exhibit high thermal stability within the temperature range studied (10–60 °C).

4. Conclusions

The stress sweep measurements carried out to determine the linear viscoelastic region for both gums demonstrated the diutan and rhamsan gum aqueous solutions presented a high resistance to the shear, since its linear viscoelastic region was wide. In addition, this region remained practically unaltered, with temperature increase in the range of temperature studied (10–60 °C) being the critical stress similar for both gum solutions.

In both cases, the mechanical spectra showed a typical behavior of weak gel, whose structure remained thermally stable, although this stability was greater for the diutan gum than for the rhamsan gum. The flow curves exhibited a strongly non-Newtonian behavior, whose pseudoplasticity was maintained even at high temperatures. The viscosities at $1s^{-1}$ of the diutan gum solutions showed a slight tendency to grow with the temperature, while rhamsan gum solutions showed values of η_1 practically independent of temperature. On the other hand, diutan gum solutions presented higher values of viscosity.

Cox-Merz rule was not met. The viscous and elastic properties of diutan gum were higher than those of rhamsan gum, although both properties were very stable against temperature.

The rheological properties showed by both biopolymers and their high stability with temperature mean these gums have a great practical interest at the industrial level.

Author Contributions: Experimental data acquisition and calculations were obtained by M.d.S.C.G. Methodology was selected by J.M.G. Experimental design was carried out by M.-C.A.-R. Data analyses performed by M.-C.A.-R. and M.C.G.G. The manuscript was written by M.-C.A.-R. and M.C.G.G. with comments from J.M.G.

Funding: This research received no external funding.

Acknowledgments: The financial support received from the Spanish Ministerio de Economía y Competitividad (MINECO) and FEDER, UE is kindly acknowledged (project CTQ2015-70700-P).

Conflicts of Interest: The authors declare no conflict of interest.

References

1. Fialho, A.M.; Moreira, L.M.; Granja, A.T.; Popescu, A.O.; Hoffmann, K.; Sá-Correia, I. Occurrence, production, and applications of gellan: Current state and perspectives. *Appl. Microbiol. Biotechnol.* **2008**, *79*, 889–900. [CrossRef] [PubMed]
2. Bajaj, I.B.; Survase, S.A.; Saudagar, P.S.; Singhal, R.S. Gellan Gum: Fermentative Production, Downstream Processing and Applications. *Food Technol. Biotechnol.* **2007**, *45*, 341–354.
3. Schmid, J.; Sperl, N.; Sieber, V. A comparison of genes involved in sphingan biosynthesis brought up to date. *Appl. Microbiol. Biotechnol.* **2014**, *98*, 7719–7733. [CrossRef] [PubMed]
4. Kaur, V.; Bera, M.B.; Panesar, P.S.; Kumar, H.; Kennedy, J.F. Welan gum: Microbial production, characterization, and applications. *Int. J. Biol. Macromol.* **2014**, *65*, 454–461. [CrossRef] [PubMed]
5. Chowdhury, T.A.; Lindberg, B.; Lindquist, U.; Baird, J. Structural studies of an extracellular polysaccharide, S-657, elaborated by Xanthomonas ATCC 53159. *Carbohydr. Res.* **1987**, *164*, 117–122. [CrossRef]
6. Campana, S.; Ganter, J.; Milas, M.; Rinaudo, M. On the solution properties of bacterial polysaccharides of the gellan family. *Carbohydr. Res.* **1992**, *231*, 31–38. [CrossRef]

7. Xu, X.; Liu, W.; Zhang, L. Rheological behaviour of Aeromonas gum in aqueous solutions. *Food Hydrocolloids* **2006**, *20*, 723–729. [CrossRef]

8. Kang, K.S.; Pettitt, D.J. *In Industrial Gums: Polysaccharides and Their Derivatives*; Academic Press: San Diego, CA, USA, 1993; pp. 341–397.

9. Alemzadeh, T.; Mohammadifar, M.A.; Azizi, M.H.; Ghanati, K. Effect of two different species of Iranian gum tragacanth on the rheoligical properties of mayonnaise sauce. *J. Food Sci. Technol.* **2010**, *7*, 127–141.

10. Rincón, F.; Muñoz, J.; Ramírez, P.; Galán, H.; Alfaro, M.C. Physicochemical and rheological characterization of Prosopis juliflora seed gum aqueous dispersions. *Food Hydrocolloids* **2014**, *35*, 348–357. [CrossRef]

11. Steffe, J.F. *Rheological Methods in Food Process Engineering*, 2nd ed.; Freeman Press: East Lansing, MI, USA, 1996.

12. Ali, M.; Koocheky, A.; Razavi, S.M. Dynamic rheological properties of Lepidium perfoliatum seed gum: Effect of concentration, temperature and heating/cooling rate. *Food Hydrocolloids* **2014**, *35*, 583–589.

13. Pérez, L.M.; Ramírez, P.; Alfaro, M.C.; Rincón, F.; Muñoz, J. Surface properties and bulk rheology of Sterculia apetala gum exudate dispersions. *Food Hydrocolloids* **2013**, *32*, 440–446. [CrossRef]

14. Brito, A.C.F.; Sierakowski, M.R.; Reicher, F.; Feitosa, J.P.A.; de Paula, R.C.M. Dynamic rheological study of Sterculia striata and karaya poly-saccharides in aqueous solution. *Food Hydrocolloids* **2005**, *19*, 861–867. [CrossRef]

15. Vardhanabhuti, B.; Ikeda, S. Isolation and Characterization of Hydrocolloids from Monoi (*Cissampelos pareira*) Leaves. *Food Hydrocolloids* **2006**, *20*, 885–891. [CrossRef]

16. Morrison, F.A. *Understanding Rheology*; Oxford University Press: New York, NY, USA, 2001.

17. Rochefort, W.E.; Middleman, S. Rheology of xanthan gum: Salt, temperature, and strain effects in oscillatory and steady shear experiments. *J. Rheol.* **1987**, *31*, 337–369. [CrossRef]

18. Rincón, F.; Muñoz, J.; De Pinto, G.L.; Alfaro, M.C.; Calero, N. Rheological properties of Cedrela odorata gum exudate aqueous dispersions. *Food Hydrocolloids* **2009**, *23*, 1031–1037. [CrossRef]

19. Clasen, C.; Kulicke, W.M. A convenient way of interpreting steady shear rheo-optical data of semi-dilute polymer solution. *Rheol. Acta* **2001**, *40*, 74–85. [CrossRef]

Article

Influence of the Homogenization Pressure on the Rheology of Biopolymer-Stabilized Emulsions Formulated with Thyme Oil

Luis A. Trujillo-Cayado, Jenifer Santos *, Nuria Calero, Maria del Carmen Alfaro and José Muñoz

Departamento de Ingeniería Química, Facultad de Química, Universidad de Sevilla, C/P. García González, 1, 41012 Sevilla, Spain; ltrujillo@us.es (L.A.T.-C.); nuriacalero@us.es (N.C.); alfaro@us.es (M.d.C.A.); jmunoz@us.es (J.M.)
* Correspondence: jsantosgarcia@us.es; Tel.: +34-954-556-447

Received: 14 December 2018; Accepted: 14 February 2019; Published: 18 February 2019

Abstract: Different continuous phases formulated with ecofriendly ingredients such as AMIDET®N, an ecological surfactant, as well as welan and rhamsan gums were developed. An experimental design strategy was been in order to study the influence of the ratio of these two polysaccharides and the homogenization pressure applied in a microfluidizer on the critical shear stress for the continuous phases developed. A pure rhamsan gum solution was selected as the starting point for further study based on the production of thyme oil-in-water emulsions. The effect of the homogenization pressure on the physical stability, critical shear stress and droplet size distribution was analyzed for emulsions with optimized values of the rhamsan–welan ratio. These bioactive thyme oil-in-water emulgels could be considered as delivery systems with potential applications in the food industry.

Keywords: biopolymer; eco-friendly surfactant; microfluidization; rheology; thyme oil

---

1. Introduction

Emulsions are thermodynamically unstable and complex systems. They have several applications in many fields like pharmaceutics, the food industry, paints, agrochemistry and cosmetics due to their ability to act as drug delivery systems [1]. However, these fluids show some drawbacks because of their physical stability. Emulsions could present some destabilization mechanisms such as creaming, coalescence, flocculation and/or phase inversion. In order to extend the shelf life of emulsions, polysaccharides are added to the continuous phase. Aqueous solutions of polysaccharides possess some interesting rheological characteristics from the application point of view; i.e., viscoelastic properties and desirable flow behavior. In addition, there are several investigations that prove the essential role of polysaccharides to enhance the physical stability since they increase the continuous phase viscosity and hence reduce the movement of the droplets [2–4]. The most-used polysaccharides are xanthan and guar gum. However, there are some new ones that have recently attracted much attention, such as welan and rhamsan gums [5–7].

Both welan and rhamsan gums belong to the sphingans group. This group has a common linear tetrasaccharide backbone structure composed of glucose, glucuronic acid, rhamnose and mannose units. The difference between the gums of this group lies in the occurrence of distinct side groups. Welan gum has commercial application in the area of cement systems. It acts as a thickening, suspending, binding and emulsifying agent, as well as a stabilizer and viscosifier. On the other hand, rhamsan gum is considered a food-grade substance [8] and presents excellent suspension characteristics, even better than xanthan gum [4,9]. The role of rhamsan gum as a stabilizer in emulsions has been recently reported by Trujillo-Cayado et al. [5].

Microfluidization technology involves forcing a system pass through microchannels to a particular area by pressurizing compressed air up to about 150 MPa. It is an easy-to-use and effective method for the development of nanoemulsions [10–12]. However, this technique has been also used to modify functional properties of xanthan gum [13]. Xanthan gum can suffer ordered–disordered conformation transition using microfluidization. In addition, xanthan gum solutions present the occurrence of hydrogen bond and entanglements, which provoke the network formation [14]. Nevertheless, there are no studies about the influence of microfluidization on the properties of rhamsan or welan gums.

Lately, the use of essential oils in emulsions has increased in popularity due to their beneficial properties such as antimicrobial, antioxidant and anticancer activity. In addition, the natural characteristic of essential oils makes them very attractive to be studied as potential ingredients for emulsions [15–18]. Namely, thyme essential oil has been used to replace synthetic chemicals as food preservatives [19]. However, the major drawback of essential oils is their high tendency for oxidation and volatility.

AMIDET®N, an ecological surfactant mainly containing C18 unsaturated fatty acid, is derived from renewable European rapeseed oil. This emulsifier is very interesting to be used in ecological formulations because: (i) it has no aquatic toxicity, (ii) it is easily dispersible in aqueous solutions and (iii) it has good biodegradability. Furthermore, it has been recently used in ecological matrices for cosmetics applications [20].

In this research, an experimental design strategy was been in order to study the influence of biopolymer dispersions processed by a microfluidization technique on critical shear stress. The variables selected were the rhamsan–welan gum concentration ratio and the homogenization pressure. The influence of the rhamsan–welan ratio is interesting from the applied point of view since some mixtures of gums can present synergistic or antagonistic effects [14,21,22]. In addition, there are no reported studies about the influence of homogenization pressure for these gum solutions. Furthermore, the effect of the homogenization pressure on the physical stability, critical shear stress and droplet size distribution were analyzed for emulgels with optimized values of the rhamsan–welan ratio. These bioactive thyme oil-in-water emulgels could be considered as active ingredient delivery systems with potential applications in the food industry.

2. Materials and Methods

2.1. Materials

Industrial grade welan gum (K1A96) and rhamsan gum (K2C401) were used as supplied by CP Kelco Company (Atlanta, GA, USA). An ecological surfactant, AMIDET®N (INCI name: PEG-4 Rapeseedamide), was provided by KAO (Tokyo, Japan). Thyme oil (*Thymus vulgaris*) was purchased from Sigma–Aldrich (St. Louis, MI, USA). Water used in this study was Milli-Q water (Merk Millipore, Darmstadt, Germany).

2.2. Preparation of Samples

A scheme of the process followed for the preparation of samples is shown in Figure 1. The surfactant (2.5 wt %) was added to the biopolymers (0.4 wt %) dispersion batches of 200 g and mixed using an IKA Eurostar for 3 h at 700 rpm. In the first part of the study, the continuous phase was directly passed through a Microfluidizer M110P (Microfluidics, Westwood, MA, USA) at the corresponding homogenization pressure, taking into account the design of experiments. In the second part, the development of emulsions, samples were produced by adding the oil phase (20 wt %) to the continuous phase. Following this, they were homogenized for 120 s at 2000 rpm using an Ultraturrax T50 (IKA, Shanghai, China). The secondary homogenization was performed using a Microfluidizer M110P at different pressures for one pass. A total of 20 wt % thyme oil was chosen to obtain a non-concentrated emulsion, since essential oils are used in this type of emulsion. In addition, the concentrations of surfactant and polysaccharides were fixed, taking into account preliminary studies.

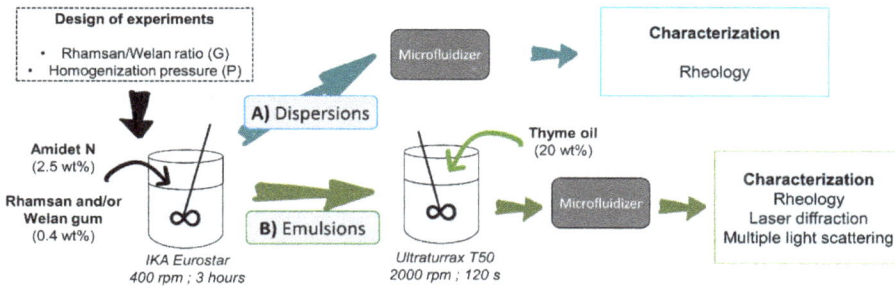

Figure 1. Scheme of the continuous phases and emulsion development.

2.3. Rheological Characterization

Oscillatory experiments were carried out in a controlled stress rheometer (Haake Mars II, Thermo Fisher Scientific, Waltham, MA, USA) equipped with serrated plate–plate (d = 60 mm) geometry. In order to determine the linear viscoelastic zone and critical shear stress, stress sweep tests (0.1–20 Pa) were conducted at a constant frequency (0.1 Hz). The temperature was fixed at 20 °C.

2.4. Droplet Size Distributions by Laser Diffraction Technique

A Malvern Mastersizer 2000 (Malvern, Worcestershire, UK) was used in order to measure the Droplet Size Distributions (DSD) for the thyme oil-in-water emulsions developed. The measurements were made in triplicate. The refraction indexes used were 1.495 and 1.33 for thyme oil and dispersed phase, respectively.

2.5. Physical Stability of Emulsions

In order to detect and quantify the destabilization mechanisms for the thyme oil-in-water emulsions developed, backscattering measurements (Turbiscan Lab Expert, Formulaction, Worthington, OH, USA) were carried out with aging time. These measurements were performed for at least 600 h at 25 °C. This technique, which is based in the multiple light scattering effect, has been used for different systems such as emulsions, suspensions and suspoemulsions [20,23,24]. Some authors have quantified the destabilization processes of emulsions by using the Turbiscan Stability Index (TSI) [20,25–27].

$$TSI = \sum_j \left| scan_{ref}(h_j) - scan_i(h_j) \right| \qquad (1)$$

where $scan_{ref}$ and $scan_i$ are the initial transmission value and the transmission value at a specific time, respectively and h_j is a specific height in the measuring cell.

2.6. Statistical Analysis

In this work, design of experiments (DoE) and response surface methodology (RSM) were used to study and optimize the biopolymers formulation and production conditions. Two independent variables, the rhamsan–welan gum ratio (X_1, G) and the homogenization pressure (X_2, P) at five levels (−1.414, −1, 0, 1 and 1.414), were carried out. The whole design, composed of 15 experimental runs performed in random order, is shown in Table 1. Each experiment was duplicated, and the average values were used in this paper. Experimental data were fitted to a quadratic model using Echip Software.

$$Y = \beta_0 + \beta_1 X_1 + \beta_2 X_2 + \beta_{11} X_1^2 + \beta_{22} X_2^2 + \beta_{12} X_1 X_2 \qquad (2)$$

where Y is the response variable, β_0 is a constant and β_i are coefficients. For model construction, terms with $p > 0.05$ were removed.

Table 1. Experimental design carried out for the continuous phases studied. X_1 = rhamsan–welan ratio. X_2 = homogenization pressure. (Rhamsan–welan ratio = 0 means 100 wt % of welan gum and 1 means 100 wt % of rhamsan gum).

Sample	X_1	X_2	Rhamsan–Welan Ratio	Pressure (bar)
1	−1	−1	0.3	100
2	1	−1	0.7	100
3	−1	1	0.3	600
4	1	1	0.7	600
5	0	0	0.5	350
6	0	0	0.5	350
7	0	0	0.5	350
8	−1.414	0	0	350
9	−1	0	0.3	350
10	1	0	0.7	350
11	1.414	0	1	350
12	0	−1.414	0.5	0
13	0	−1	0.5	100
14	0	1	0.5	600
15	0	1.414	0.5	700

3. Results and Discussion

3.1. Development of a Continuous Phase Containing Rhamsan and/or Welan Gums

The viscoelastic properties of the biopolymers used were determined via oscillatory shear tests. Oscillatory shear tests, and more specifically stress sweeps, can be used to analyze the structure of biopolymers. In a stress sweep, it is possible to distinguish two different zones, namely a linear viscoelastic region (LVR), in which the storage modulus (G') and the loss modulus (G'') are constant up to a critical shear stress (σ_c), and a non-linear one where G' and G'' start to diminish. In the LVR, which is defined as the stress and strain amplitudes ranges where G' as well as G'' remain independent of the applied stress and strain, the deformation in the material structure is reversible. Therefore, the amplitude of the linear viscoelastic range and the value of the critical stress can be utilized to describe polymer structure strength. Stress sweeps were performed as a function of the rhamsan–welan ratio (X_1) and the homogenization pressure (X_2) at 0.1 Hz. From Figures 2 and 3, it is possible to distinguish the two distinct zones, namely the LVR ($\sigma < \sigma_c$) and the non-linear zone ($\sigma > \sigma_c$). With increasing stress and after passing the critical shear stress, G' and G'' begin to decrease sharply due to the deformation of the gum samples. The effect of the rhamsan–welan gum ratio on the viscoelastic properties was studied, and the results are shown in Figure 2 and Table 2. The rhamsan–welan ratio was set at 0, 0.3, 0.5, 0.7 and 1 (where 0 means 100 wt % of welan gum and 1 means 100 wt % of rhamsan gum), while the homogenization pressure was fixed at 350 bar. In all tested samples as a function of the rhamsan–welan gum ratio (see Figure 2), the loss modulus was greater than the storage modulus in the linear viscoelastic range, as the viscous component dominates the elastic component. This fact may be due to the destruction of the three-dimensional structure of the biopolymer solution due to its introduction into the homogenizer at high pressures ($P = 350$ bar). In addition, it was observed that there was a tendency of higher values of the critical shear stress defining the LVR as the concentration of welan gum decreased and the rhamsan gum concentration increased. This fact can be verified by the comparison of the values of the critical shear stress of samples 5, 6, 7, 8, 9, 10 and 11 in Table 2 and in the inset of Figure 2. Therefore, it can be concluded that increasing the rhamsan gum concentration led to increase in the structural strength of samples and made it more rigid. The critical shear stress, storage modulus and loss modulus increased with the rhamsan gum concentration, which may be associated with the ability of this biopolymer to form stronger structures. This fact reveals that rhamsan gum solutions may possess a higher number of molecular aggregates through hydrogen bonds and polymer entanglement than welan gum solutions, similarly to xanthan gum solutions [14]. The use of welan and rhamsan gum in identical concentrations (samples 5, 6 and 7) did not lead to a positive synergistic effect on the stress sweeps, which was shown by other mixtures of gums [22]. This means that the resulting dispersion did not show more marked elastic properties. The influence of the homogenization pressure on the G', G" and σ_c is illustrated in Figure 3. To determine the effect of the homogenization pressure (P) on the viscoelastic properties, the rhamsan–welan gum ratio was 0.5 (50W/50R) while P was controlled at 0, 100, 350, 600 and 700 bar. It was observed that when the homogenization pressure increased, the critical shear stress values decreased (inset Figure 3), showing weaker structures. In addition, it is obviously shown in Figure 3 that the magnitudes of loss and storage moduli values at the LVR decreased with an increase in homogenization pressure, thus resulting in deteriorating structural strength of biopolymer solutions. Furthermore, the storage modulus was greater than the loss modulus in the linear viscoelastic range for homogenization pressures equal to or below 100 bar. The abovementioned entanglements could break via the application of the homogenization pressure by the microfluidization technique. This fact has been previously reported in xanthan gum solutions [13,14].

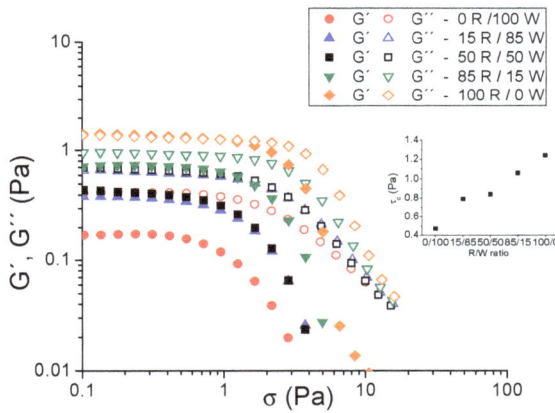

Figure 2. Stress sweeps for the continuous phases passed through a microfluidizer at 350 bar as a function of the rhamsan–welan ratio at 0.1 Hz.

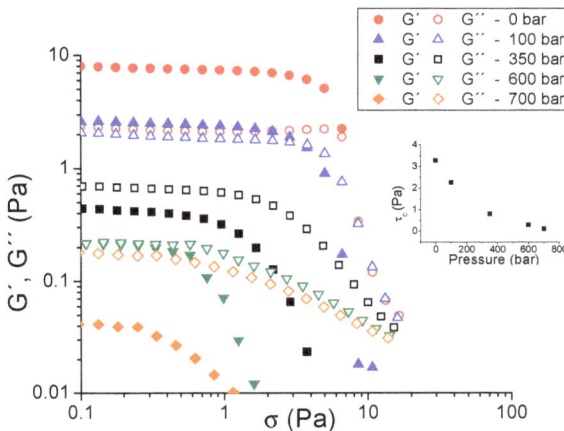

Figure 3. Stress sweeps for the continuous phase studied (Rhamsan-Welan ratio 0.5) as a function of the homogenization pressure applied in a microfluidizer at 0.1 Hz.

Table 2. Critical shear stress obtained from stress sweeps for the continuous phases studied.

Sample	Rhamsan–Welan Ratio	Pressure (bar)	τ_c (Pa)
1	0.3	100	1.45 ± 0.09
2	0.7	100	2.80 ± 0.14
3	0.3	600	0.13 ± 0.01
4	0.7	600	0.15 ± 0.01
5	0.5	350	0.83 ± 0.05
6	0.5	350	0.81 ± 0.06
7	0.5	350	0.84 ± 0.05
8	0	350	0.47 ± 0.02
9	0.3	350	0.79 ± 0.04
10	0.7	350	1.06 ± 0.08
11	1	350	1.24 ± 0.08
12	0.5	0	3.28 ± 0.19
13	0.5	100	2.27 ± 0.12
14	0.5	600	0.29 ± 0.02
15	0.5	700	0.11 ± 0.01

The design of experiment (DoE) and response surface methodology (RSM) allow the evaluation of multiple interactions between independent variables in experiments. The analysis of the interactions between the homogenization pressure (X_2) and gums ratio (X_1) and associated output response (critical shear stress values), instead of the conventional one-factor-at-a-time method, allows a better understanding of the influence of these input factors on the structure of biopolymer solutions. This can be observed in Table 2, where critical shear stress ranges from 0.13 ± 0.01 to 3.28 ± 0.19 Pa. Moreover, a quadratic model was performed to represent the critical stress as a function of the ratio of the gums and the homogenization pressure in the chosen ranges, which is written according to the following equation ($R^2 = 0.988$):

$$\sigma_C = 0.89 + 0.28 \cdot X_1 - 1.06 \cdot X_2 - 0.33 \cdot X_1 \cdot X_2 + 0.38 \cdot X_2^2. \tag{3}$$

Interestingly, the critical stress was sensitive to all studied variables. In fact, σ_c increased with the rhamsan gum concentration (lower rhamsan–welan ratio, X_1) as indicated by the comparison of systems 8, 9, 5/6/7, 10 and 11. This fact proves important in the selection of an adequate biopolymer as a viscosity modifier for emulsions since it also modifies the continuous phase viscosity. In contrast, the rise of the homogenization pressure (X_2) yielded the most significant effect as supported by their respective linear and quadratic coefficients values, since the critical stress decreased from 3.28 ± 0.19 to 0.11 ± 0.01 as the pressure increased from 0 to 700 bar. The contribution of the squared term coefficient of X_2 must be also taken into consideration. Finally, it an interaction between both factors, X_1 and X_2, was observed. This fact may indicate that the gums have different behaviors against the application of the homogenization pressure by means of the microfluidizer. Since the coefficient is negative, a higher concentration of welan gum (lower value of X_1) and an increase of homogenization pressures (higher value of X_2) reduce the critical stress. Thus, this biopolymer exhibits less resistance against the pressure applied, reducing to a greater extent the strength of its structure.

Figure 4 illustrates the three-dimensional response surface curve of critical shear stress (σ_c) for the rhamsan–welan gum concentration ratio and homogenization pressure. This figure is the representation of Equation (3), which relates the critical shear stress with the variables studied. An optimum formulation and homogenization pressure can be set for obtaining a biopolymer dispersion with enhanced critical shear stress. According to the surface response analysis, the maximum critical shear stress was obtained for $X_1 = 0$ (rhamsan–welan ratio = 0.5) and $X_2 = -1.41$ ($P = 0$ bar). However, taking into account Equation (3), the predicted and extrapolated maximum critical shear stress was obtained when $X_1 = 1.414$ (only rhamsan gum) and $X_2 = 0$ (0 bar). In order to produce emulsions with minimum droplet sizes and enhanced physical stability, the control of the formulation and processing conditions is required. As a result of these preliminary tests, it was suggested that the use of pure rhamsan gum solutions allows the production of continuous phases with higher critical shear stresses. For this reason, this ratio (100 rhamsan–0 welan) was fixed for further study.

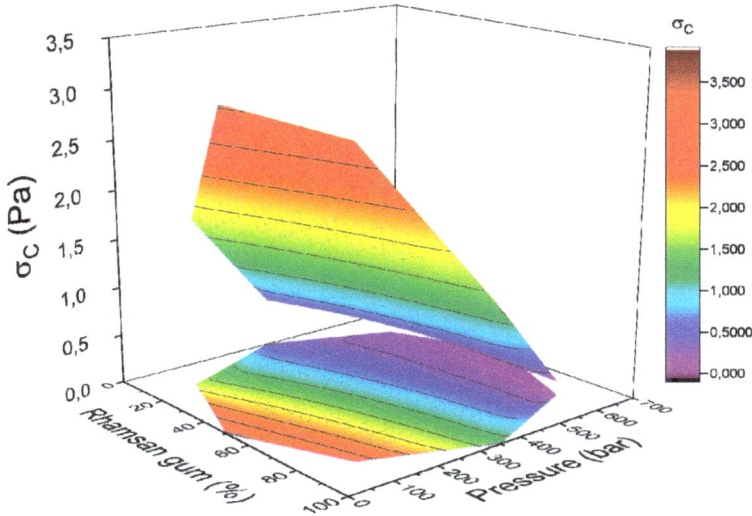

Figure 4. Three-dimensional response surface curve of critical shear stress (σ_c) for rhamsan gum (%) and homogenization pressure applied in a microfluidizer.

3.2. Development of Thyme Oil-in-Water Emulsions

Figure 5 shows the droplet size distribution (DSD) for the selected emulsion (0.4 wt % rhamsan gum) processed at different homogenization pressures in the microfluidizer. The pre-emulsion (emulsion not homogenized in the microfluidizer) presents a bimodal distribution with a second peak centered at above 50 μm. In contrast, emulsions developed using the microfluidizer exhibited DSDs with only one peak centered below 10 μm. There was a clear reduction of droplet size using the microfluidizer for these thyme oil-in-water emulsions. In addition, a more marked decrease of droplet size was related to higher homogenization pressures. This fact has been reported before by other authors for emulsions without gums [27–29]. However, this reduction of droplet size observed for emulsions containing gums had not been reported yet. It is important to highlight that the incorporation of rhamsan gum before microfluidization was not a drawback for the reduction of droplet size. Furthermore, there was no existence of recoalescence due to overprocessing. This seems to indicate that continuous phase development (surfactant and rhamsan gum) protects the interface from a possible recoalescence effect.

Stress sweeps at 1 Hz for the selected emulsion processed at different homogenization pressures in the microfluidizer are shown in Figure 6. Emulsions developed with no microfluidizer and at 350 bars exhibited similar behaviors. Both systems show a plateau zone of elastic and loss moduli (G′, G″), which is the linear viscoelastic range (LVR), followed by an abrupt decrease with stress. G′ is higher than G″ for the two abovementioned systems in the LVR. This is related to a gel behavior. However, the values of G′ and G″ for the emulsions developed at 350 bars are lower than those for the emulsion not processed in the microfluidizer. This fact is not in concordance with the DSD observed in Figure 5 since lower values of diameters are usually related to higher values of G′ and G″. In addition, the highest pressure applied (700 bar) to the selected emulsion provoked a decrease in both viscoelastic moduli, reaching similar values for G′ and G″. This emulsion presented the lowest mean diameter in Figure 5. Hence, the use of the microfluidizer in this emulsion not only reduced the droplet size of the emulsion but also provoked a loss of microstructure developed by the gums. This can be also observed in the length of the plateau zone, i.e., there was a reduction of the LVR with the increase in the homogenization pressure. These conformation changes using microfluidization have been previously reported in xanthan gum solutions [13].

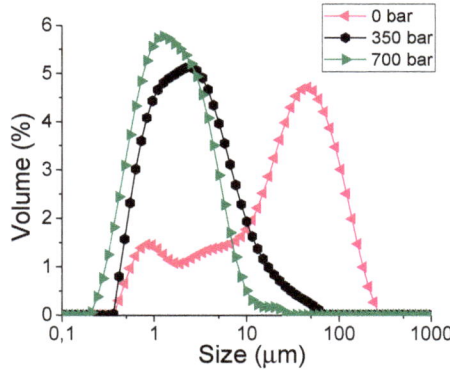

Figure 5. Droplet size distributions for thyme emulsions containing 0.4 wt % of rhamsan gum as a function of the homogenization pressure applied in the microfluidizer.

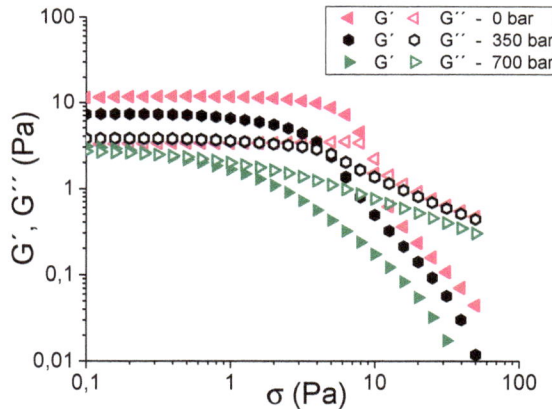

Figure 6. Stress sweeps for thyme emulsions studied containing 0.4 wt % of rhamsan gum as a function of the homogenization pressure applied in the microfluidizer.

Figure 7 shows the Turbiscan Stability Index (TSI) with aging time as a function of the homogenization pressure applied in the microfluidizer. It is important to mention that higher values of the TSI involve poorer physical stabilities [25,26]. All emulsions followed the same tendency in TSI values: a marked increase and subsequently a leveling off trend. However, the values were quite different. The emulsion processed at the highest homogenization pressure (700 bar) exhibited the lowest TSI values, proving its enhanced physical stability. Hence, although there is a loss of structuration using high homogenization pressures, the reduction of droplet size is the key parameter for the physical stability of these emulsions.

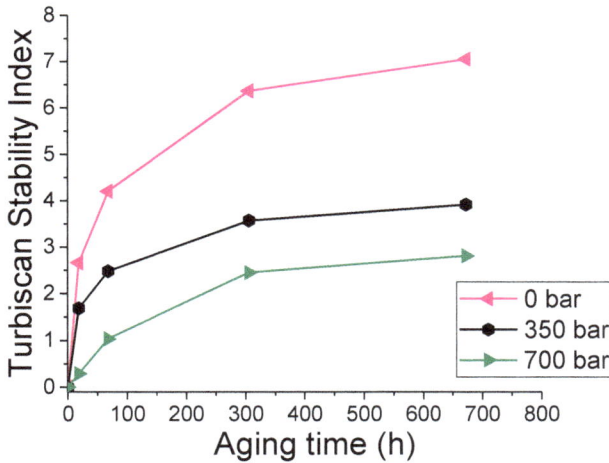

Figure 7. Turbiscan Stability Index parameter for thyme emulsions studied containing 0.4 wt % of rhamsan gum as a function of the homogenization pressure applied in the microfluidizer.

4. Conclusions

The influence of the gums' concentration ratio and homogenization pressure on the critical shear stress of dispersions containing a blend of welan and rhamsan gums was studied using design of experiments and surface response methodology. In order to rationally develop the formulation and processing conditions of biopolymer dispersions prepared with a microfluidizer, a quadratic model was required. An increase in the welan gum concentration and homogenization pressure provoked a decrease in critical shear stress values and weaker structures. An optimal combination of formulation and processing, with maximum critical shear stress, was obtained for $X_1 = 1.414$ and $X_2 = 0$. For this reason, the use of pure rhamsan gum and no pressure for the development of continuous phases was taken as a starting point for a further study of the influence of the homogenization pressure on the production of emulsions.

Thyme oil-in-water emulsions containing rhamsan gum were developed by using the microfluidization technique. The microfluidization technique has proved to be a powerful tool to reduce droplet size. However, this technique was also responsible for the decrease in the structuration grade of the studied emulsions. Hence, the break of gum structure was caused by the use of the microfluidizer, which was observed in the stress sweeps of the emulsions studied. This fact was pointed out not only by the lower viscoelastic functions but also by the smaller LVR. However, the reduction of droplet size led to the production of more stable emulsions, proved by the multiple light scattering technique. The emulsion processed at the highest homogenization pressure (700 bar) showed the greatest physical stability. This study can lay the foundation for the development of emulsions containing rhamsan gum via the microfluidization technique.

Author Contributions: Conceptualization, J.M. and M.d.C.A.; Methodology, L.A.T.-C.; Software, N.C.; Validation, N.C.; Formal Analysis, N.C.; Investigation, L.A.T.-C. and J.S.; Resources, J.M.; Data Curation, N.C.; Writing—Original Draft Preparation, J.S.; Writing—Review and Editing, L.A.T.-C. and J.S.; Visualization, L.A.T.-C.; Supervision, M.d.C.A.; Project Administration, J.M. and M.d.C.A.; Funding Acquisition, J.M. and M.d.C.A.

Acknowledgments: The financial support received (Project CTQ2015-70700-P) from the Spanish Ministerio de Economía y Competitividad and from the European Commission (FEDER Programme) is gratefully acknowledged.

Conflicts of Interest: The authors declare no conflict of interest.

References

1. Robins, M.M.; Wilde, P.J. Colloids and emulsions. *Science* **2003**, 1517–1524. [CrossRef]
2. Pérez-Mosqueda, L.M.; Ramírez, P.; Trujillo-Cayado, L.A.; Santos, J.; Muñoz, J. Development of eco-friendly submicron emulsions stabilized by a bio-derived gum. *Colloids Surf. B Biointerfaces* **2014**, *123*, 797–802. [CrossRef] [PubMed]
3. Qiu, C.; Zhao, M.; McClements, D.J. Improving the stability of wheat protein-stabilized emulsions: Effect of pectin and xanthan gum addition. *Food Hydrocoll.* **2015**, *43*, 377–387. [CrossRef]
4. Kang, K.S.; Pettitt, D.J. Xanthan, gellan, welan, and rhamsan. In *Industrial Gums*, 3rd ed.; Elsevier: Amsterdam, The Netherlands, 1993; pp. 341–397.
5. Trujillo-Cayado, L.A.; Alfaro, M.C.; Muñoz, J.; Raymundo, A.; Sousa, I. Development and rheological properties of ecological emulsions formulated with a biosolvent and two microbial polysaccharides. *Colloids Surf. B Biointerfaces* **2016**, *141*, 53–58. [CrossRef] [PubMed]
6. Trujillo-Cayado, L.A.; Alfaro, M.C.; Raymundo, A.; Sousa, I.; Muñoz, J. Rheological behavior of aqueous dispersions containing blends of rhamsan and welan polysaccharides with an eco-friendly surfactant. *Colloids Surf. B Biointerfaces* **2016**, *145*, 430–437. [CrossRef] [PubMed]
7. Xu, X.Y.; Dong, S.H.; Li, S.; Chen, X.Y.; Wu, D.; Xu, H. Statistical experimental design optimization of rhamsan gum production by *Sphingomonas* sp. CGMCC 6833. *J. Microbiol.* **2015**, *53*, 272–278. [CrossRef] [PubMed]
8. Hagiwara, A.; Imai, N.; Doi, Y.; Sano, M.; Tamano, S.; Omoto, T.; Asai, I.; Yasuhara, K.; Hayashi, S. Ninety-day oral toxicity study of rhamsan gum, a natural food thickener produced from Sphingomonas ATCC 31961, in Crl:CD(SD)IGS rats. *J. Toxicol. Sci.* **2010**, *35*, 493–501. [CrossRef] [PubMed]
9. Sanderson, G.R. Gellan gum. In *Food Gels*; Springer: New York, NY, USA, 1990; pp. 201–232.
10. Lee, L.; Norton, I.T. Comparing droplet breakup for a high-pressure valve homogeniser and a microfluidizer for the potential production of food-grade nanoemulsions. *J. Food Eng.* **2013**, *114*, 158–163. [CrossRef]
11. Jafari, S.M.; He, Y.; Bhandari, B. Optimization of nano-emulsions production by microfluidization. *Eur. Food Res. Technol.* **2007**, *225*, 733–741. [CrossRef]
12. Santos, J.; Trujillo-Cayado, L.A.; Calero, N.; Alfaro, M.C.; Muñoz, J. Development of eco-friendly emulsions produced by microfluidization technique. *J. Ind. Eng. Chem.* **2016**, *36*, 90–95. [CrossRef]
13. Lagoueyte, N.; Paquin, P. Effects of microfluidization on the functional properties of xanthan gum. *Food Hydrocoll.* **1998**, *12*, 365–371. [CrossRef]
14. Xu, L.; Xu, G.; Liu, T.; Chen, Y.; Gong, H. The comparison of rheological properties of aqueous welan gum and xanthan gum solutions. *Carbohydr. Polym.* **2013**, *92*, 516–522. [CrossRef] [PubMed]
15. Ruberto, G.; Baratta, M.T.; Deans, S.G.; Dorman, H.J.D. Antioxidant and antimicrobial activity of Foeniculum vulgare and *Crithmum maritimum* essential oils. *Planta Med.* **2000**, *66*, 687–693. [CrossRef] [PubMed]
16. Satou, T.; Hayakawa, M.; Goto, Y.; Masuo, Y.; Koike, K. Anxiolytic-like effects of essential oil from Thymus vulgaris was increased during stress. *Flavour Fragr. J.* **2018**, *33*, 191–195. [CrossRef]
17. Dobetsberger, C.; Buchbauer, G. Actions of essential oils on the central nervous system: An updated review. *Flavour Fragr. J.* **2011**, *26*, 300–316. [CrossRef]
18. Kwiatkowski, P.; Mnichowska-Polanowska, M.; Pruss, A.; Masiuk, H.; Dzięcioł, M.; Giedrys-Kalemba, S.; Sienkiewicz, M. The effect of fennel essential oil in combination with antibiotics on *Staphylococcus aureus* strains isolated from carriers. *Burns* **2017**, *43*, 1544–1551. [CrossRef] [PubMed]
19. Ryu, V.; McClements, D.J.; Corradini, M.G.; McLandsborough, L. Effect of ripening inhibitor type on formation, stability, and antimicrobial activity of thyme oil nanoemulsion. *Food Chem.* **2018**, *245*, 104–111. [CrossRef] [PubMed]
20. Santos, J.; Calero, N.; Trujillo-Cayado, L.A.; Muñoz, J. Development and characterisation of a continuous phase based on a fumed silica and a green surfactant with emulsion applications. *Colloids Surf. A Physicochem. Eng. Asp.* **2018**, *555*, 351–357. [CrossRef]
21. Zasypkin, D.V.; Braudo, E.E.; Tolstoguzov, V.B. Multicomponent biopolymer gels. *Food Hydrocoll.* **1997**, *11*, 159–170. [CrossRef]
22. Kennedy, J.F.; Woods, J.R.; Harding, S.E.; Hill, S.E.; Mitchell, J.R. Biopolymer Mixtures. *Bioseparation* **1998**, *7*, 64. [CrossRef]

23. Mengual, O.; Meunier, G.; Cayré, I.; Puech, K.; Snabre, P. TURBISCAN MA 2000: Multiple light scattering measurement for concentrated emulsion and suspension instability analysis. *Talanta* **1999**, *50*, 445–456. [CrossRef]

24. Santos, J.; Trujillo, L.A.; Calero, N.; Alfaro, M.C.; Munoz, J. Physical Characterization of a Commercial Suspoemulsion as a Reference for the Development of Suspoemulsions. *Chem. Eng. Technol.* **2013**, *36*, 1883–1890. [CrossRef]

25. Santos, J.; Calero, N.; Trujillo-Cayado, L.A.; Garcia, M.C.; Muñoz, J. Assessing differences between Ostwald ripening and coalescence by rheology, laser diffraction and multiple light scattering. *Colloids Surf. B Biointerfaces* **2017**, *159*, 405–411. [CrossRef] [PubMed]

26. Trujillo-Cayado, L.A.; Alfaro, M.C.; Muñoz, J. Effects of ethoxylated fatty acid alkanolamide concentration and processing on D-limonene emulsions. *Colloids Surf. A Physicochem. Eng. Asp.* **2018**, *536*, 198–203. [CrossRef]

27. Trujillo-Cayado, L.A.; Santos, J.; Ramírez, P.; Alfaro, M.C.; Muñoz, J. Strategy for the development and characterization of environmental friendly emulsions by microfluidization technique. *J. Clean. Prod.* **2018**, *178*, 723–730. [CrossRef]

28. Martin-Piñero, M.J.; Ramirez, P.; Muñoz, J.; Alfaro, M.C. Development of rosemary essential oil nanoemulsions using a wheat biomass-derived surfactant. *Colloids Surf. B Biointerfaces* **2019**, *173*, 486–492. [CrossRef] [PubMed]

29. Jo, Y.-J.; Kwon, Y.-J. Characterization of β-carotene nanoemulsions prepared by microfluidization technique. *Food Sci. Biotechnol.* **2014**, *23*, 107–113. [CrossRef]

fluids

MDPI

Article

Assessment of Piezoelectric Sensors for the Acquisition of Steady Melt Pressures in Polymer Extrusion

Sónia Costa, Paulo F. Teixeira *, José A. Covas * and Loic Hilliou *

Institute for Polymers and Composites/i3N, University of Minho, 4800-058 Guimarães, Portugal; sonia_25_21-09@hotmail.com
* Correspondence: p.teixeira@dep.uminho.pt (P.F.T.); jcovas@dep.uminho.pt (J.A.C.); loic@dep.uminho.pt (L.H.); Tel.: +351-253510320 (L.H.)

Received: 26 February 2019; Accepted: 1 April 2019; Published: 3 April 2019

Abstract: Piezoelectric sensors have made their way into polymer processing and rheometry applications, in particular when small pressure changes with very fast dynamics are to be measured. However, no validation of their use for steady shear rheometry is available in the literature. Here, a rheological slit die was designed and constructed to allow for the direct comparison of pressure data measured with conventional and piezoelectric transducers. The calibration of piezoelectric sensors is presented together with a methodology to correct the data from the inherent signal drift, which is shown to be temperature and pressure independent. Flow curves are measured for polymers showing different levels of viscoelasticity. Piezoelectric slit rheometry is validated and its advantage for the rheology of thermodegradable materials with viscosity below 100 Pa·s is highlighted.

Keywords: piezoelectric; pressure transducers; extrusion; rheology

---

1. Introduction

Accurate readings of steady melt pressure are important to monitor flow conditions during polymer extrusion and for rheometry experiments involving pressure flows. For example, it is well-known that the installation of flush-mounted pressure gauges along a slit die wall allows the assessment of shear viscosity up to relatively high shear stresses, with no need of the Bagley correction for non-viscometric flow effects at the edges of the die [1]. However, the conventional melt pressure transducers of the diaphragm type are bulky and have large front dimensions (typically a diameter of 7.8 mm and process connections with M18 (× 1.5) according to DIN 3852-1592, or 1/2-20 UNF), which may impact on the practical possibility of fixing them on small and/or curved flow channels. In the case of slits, consecutive transducers along the length maybe set apart more than desired and there may be little space for additional transducers. Indeed, if a microscope is also inserted in the slit, it is possible to image the structure of the material undergoing shear flow, thereby giving way to rheo-optical characterization [2]. Piezoelectric sensors have remarkable sensitivity, fast response and reduced size compared to conventional melt pressure sensors. They are often used in injection molding to measure instantaneous pressures along the production cycle. When mounted in customized miniaturized systems (such as capillary rheometry dies), and implementing oversampling techniques, piezoelectric sensors were used to establish relationships between rapid pressure fluctuations with small amplitudes and distortions found on the surface of the extruded materials [3–9]. As a result of these studies, slit dies equipped with piezoelectric transducers are now commercially available as accessories to capillary rheometers. Using these devices, it was possible to predict extrusion instabilities for a series of commercial polyethylenes [10]. A similar piezoelectric set-up was coupled to a laboratory screw extruder to demonstrate the feasibility of applying high sensitivity detection systems to practical polymer processing [6].

Unlike conventional diaphragm sensors, piezoelectric sensors exhibit a drift of the signal with time. The drift is inherent to both the charge amplification and transducer-to-amplifier cabling, which are needed to convert the transducer signal into volts. Therefore, it is necessary to adopt data treatment procedures to account for the drift of the piezoelectric signal and for the non-linearity of the transducer with pressure [11] or temperature [12]. Recently Kádár et al. [5] used piezoelectric sensors in a slit die attached to a capillary rheometer to estimate the first normal stress difference of polymer melts, N_1, via the so-called 'pressure hole effect' [13]. The method is very demanding for the pressure transducers, as small differences between the readings of two sensors are to be measured [14]. Kádár et al. [5] assumed the drift in the voltage V_{drift} to be a linear function of time t,

$$V_{drift}(t) = V_0 + st, \tag{1}$$

where V_0 is the value of the voltage at time 0 defining the start of the recording, and s is the slope that depends on the testing temperature and the pressure applied on the piezoelectric transducer. Eventually, a flow curve was obtained which compared reasonably well with small amplitude oscillatory shear data in a Cox-Merz representation, but a single slope was used to correct the signal drift recorded during the application of a ramp of steady shear rate steps, which correspondingly generated a ramp of pressures.

A direct comparison between the flow curves measured with piezoelectric sensors and conventional pressure transducers during steady shear has not yet been reported in the literature. Thus, the use of piezoelectric sensors for steady shear rheometry remains to be validated. This work proposes a methodology to correct the drift together with temperature and pressure effects. A modular slit die was designed and constructed in order to incorporate both piezoelectric sensors and conventional pressure transducers in a mirror-like arrangement along the channel, thus enabling the direct comparison between the flow curves acquired with the two types of transducers.

2. Experimental

2.1. Materials

An extrusion grade low density Polyethylene (ALCUDIA® LDPE 2221FG, from Repsol, Spain) with a density of 0.922 g/cm^3, a melt flow index of 2.1 g/10 min (190 °C/ 2.16 kg) and a processing temperature range between 150–180 °C was used for the assessment. This polymer has excellent thermal stability, thus minimizing the eventual influence of thermal degradation in the experiments. Two biodegradable polymers, a polyhydroxybutyrate (PHB P309 from Biomer®, Krailling, Germany) and a Polybutylene adipate terephthalate (PBAT, ecoflex® F Blend C1200, from BASF, Ludwigshafen, Germany) were also used to compare the operability windows of conventional and piezoelectric transducers.

2.2. Experimental Set-Up

A double slit rheometrical die for in-process characterization, recently developed by the authors [15], was modified and used in this study. It includes three modules: a central body (module 1) where the inlet circular channel from the extruder is progressively converted into a slit, which is then divided into perpendicular measurement (module 2) and extrusion (module 3) channels. The flow rate in the measuring slit is varied by a valve positioned at its entrance, whereas a second valve balances the flow rate in the extrusion slit to keep a constant pressure at the die inlet. Thus, rheometrical measurements can be performed while maintaining constant extrusion conditions (feed rate and screw speed). The initial design, construction, and rheological validation have been reported elsewhere [15]. For the present study, a new measurement channel (module 2) was developed. The new module comprises two halves bolted together. In one part, three conventional melt pressure transducers flush mounted can be inserted, while in the other part three piezoelectric transducers can be also flush mounted at the same axial position, see Figure 1. The measurement channel presents a cross section of 10 mm width by 0.8 mm height.

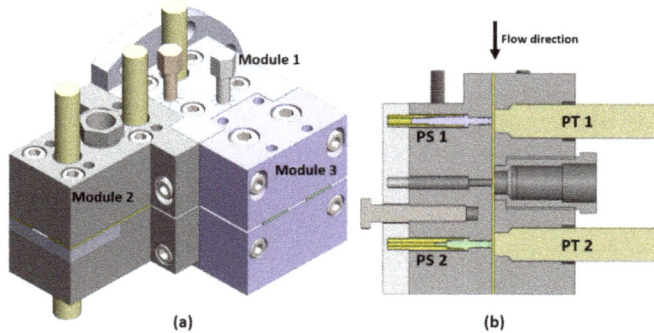

Figure 1. Schematic view of the double slit rheometrical die (**a**) and cross section view of the measurement channel (module 2) (**b**).

Two piezoelectric transducers from the Kistler Group, Winterthur, Switzerland (Kistler 6182B (PS 1) and a Kistler 6189A (PS 2), see main characteristics as provided by the manufacturers in Table 1) were used. PS 2 has the ability to perform simultaneously pressure and temperature measurements and was mounted closer to the die exit. In order to minimize the signal drift, as well as the influence of external electromagnetic noise, the sensors were separately shielded by means of a copper mesh covering the wires. Two conventional pressure transducers Dynisco PT422A (0-3000 PSI, Dynisco Inc., Franklin, MA, USA) with a sensitivity of ±0.5% and a front diameter of 7.8 mm were used (PT 1 mounted upstream and PT 2 mounted downstream). These transducers are sensitive to variations in the temperature in the range of ±0.005 MPa for ±1 °C. All sensors were flush mounted in the double slit die and adequate fixing was verified.

Table 1. Characteristics of the melt pressure transducers.

Characteristics	Kistler 6182 B	Kistler 6189A	PT422A
Front diameter	2.5 (mm)	2.5 (mm)	7.8 (mm)
Range	0–200 (MPa)	0–200 (MPa)	0–21 (MPa)
Sensibility, x	−2.5 (pC/bar)	−6.6 (pC/bar)	0.5%
Operating temperature range:			
Sensor, cable, connector box	0–00 (°C)	0–200 (°C)	
At the front of the sensor	< 450 (°C)	< 450 (°C)	0–400 (°C)

As a validation strategy, and in order to avoid the inherent fluctuations of pressure and flow rate in screw extruders, the double slit die was attached on the barrel of a Rosand RH10 capillary rheometer (Malvern Instruments, Malvern, UK), see Figure 2.

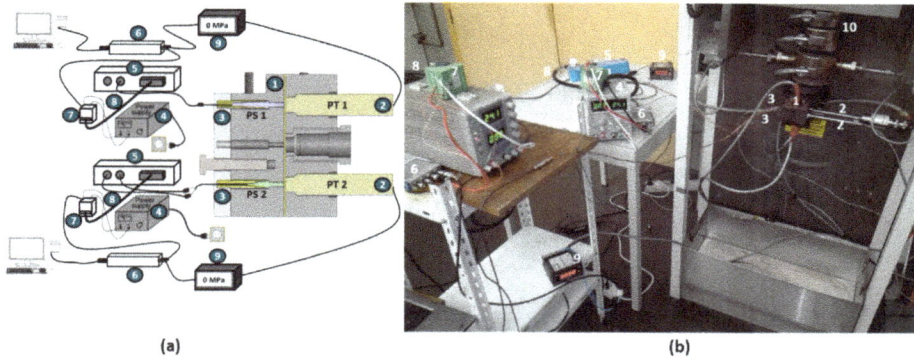

Figure 2. Schematic view (**a**) and photograph (**b**) of the experimental set-up. 1: double slit die. 2: conventional pressure transducers. 3: piezoelectric sensors. 4: power supply. 5: charge amplifier. 6: analog-to-digital converter (ADC) card. 7: terminal block for 25-Pin. 8:D-SUB Cable. 9: strain gage indicator. 10: capillary rheometer.

The double slit die is independently heated using two temperature controllers OMRON E5CSV (Omron Corporation, Tokyo, Japan). In order to acquire the signals from the piezoelectric sensors, the following equipment was used:

- Two charge amplifiers, a Kistler 5155A 2241 for PS 1 and a Kistler 5155A 22A1 (with separate channels to acquire pressure and temperature) for PS 2. Two amplification ranges are available which depend on the maximum charge delivered by the transducer, namely up to 20000 pC (which corresponds to 10 V in Range I) and up to 5000 pC (which corresponds to 10 V in Range II for larger amplification).
- Two power supplies (Matrix MPS-5LK-2, delivering a voltage between 18–30 V DC) to feed the amplifiers and command functions.
- Two analog-to-digital converter (ADC) boards (NI-9215 from National Instruments) driven by custom-written LabVIEW™ routines to digitize the amplifiers outputs and enhance the sensor sensitivity with on-the-fly oversampling techniques [16–18].
- Two DIN-Rail Mount Terminal Blocks for 25-Pin D-SUB Modules and two 25-Pin Shielded D-SUB cables (from National Instruments, Austin, TX, USA) in order to interface the power supply, the amplifier, and the ADC.

Each conventional pressure transducer was connected to a Dynisco 1390 strain gage indicator, with analog retransmission output accuracy span of ±0.2%. The strain gage indicator is also connected to a data acquisition system NI-9215 from National Instruments (see Figure 2) and driven by custom-written LabVIEW™ routines. The piezoelectric sensor PS 1 and the conventional pressure transducer PT 1 were connected to the same data acquisition system NI-9215 while PS 2 and PT 2 were connected to the second data acquisition system NI-9215, for time synchronization.

2.3. Experimental Procedure

2.3.1. Calibrations

Conventional transducers and piezoelectric transducers were coupled to a Terwin T1200 MkII hydraulic comparison test pump. Calibration curves were constructed by reporting the pressure returned by the transducers as a function of the pressure set by the test pump. For the piezoelectric transducers, the calibration was performed with the charge amplifiers set in the Range II. In this case, the voltage returned by the transducers are converted into pressure using the following equation:

$$PS\ 1 = PS\ 2 = (V_{cor} * 50)/x \tag{2}$$

where x is the sensibility given by the manufacturer (see Table 1) and V_{cor} is the output voltage $V(t)$ corrected from the drift, namely $V_{cor}(t) = V(t) - V_{drift}(t)$.

2.3.2. Pressure Measurements and Flow Curves

The whole set-up is switched on at least 30 min prior to any measurement to allow for a stabilized temperature in both capillary barrel and double slit die, and to warm up the electronics, therefore reducing the signal drift. The slit is then fed with melt (maintaining constant the piston velocity) for a period of 1 min and the material is left to relax until the pressure readings, PT 1 and PT 2, stabilize. Then, the zero balance of the conventional transducers is adjusted in order to define a zero pressure and to adjust the 80% span. The data acquisition by the ADC starts and the amplifier is switched on to record the piezoelectric drift $V_{drift}(t)$ over two minutes. This time is needed for measuring a consistent slope s. Afterwards, 6 successive incremental step increases of the piston velocity are performed in order to ramp up the corresponding shear rates. In each ramp the piston velocity varied from 10 mm/min to 60 mm/min, corresponding to apparent shear rates ranging from 27 s^{-1} to 165 s^{-1}. The shear rate was step increased only after steady state pressures were read from the graphic display provided under the LabVIEW™ environment. At the end of the ramp, the piston is retracted and the drift monitored for two minutes. The flow curves were constructed using the following analysis for slit rheometry [1], where the pressure drop dP = PS 1 − PS2 or dP = PT 1 − PT 2 and the volumetric flow rate, Q, are used to determine the wall shear stress σ and the shear rate $\dot{\gamma}$, respectively, with the following equations:

$$\sigma = \frac{H}{2(1 + H/W)} \frac{dP}{dx} \tag{3}$$

where W and H are the width and height of the flow channel respectively, and dx is the distance between transducers PS 1 and PS 2 or PT 1 and PT 2. The apparent shear rate $\dot{\gamma}_a$ is:

$$\dot{\gamma}_a = \frac{6Q}{WH^2} \tag{4}$$

where Q is obtained in an indirect way by measuring the weight of the extrudate. This methodology is preferred to the use of the piston speed for Q determination, as it can readily be extended to in-line rheometry during extrusion application. The true wall shear rate was calculated from:

$$\dot{\gamma} = \frac{\dot{\gamma}_a}{3} \left(2 + \frac{d \ln \dot{\gamma}_a}{d \ln \sigma} \right) \tag{5}$$

3. Results and Discussion

Figure 3 presents the time evolution of the voltage V acquired by transducer PS 1 mounted in an empty slit (no pressure applied), in order to assess the drift inherent to this measuring system. In Figure 3a, the effect of switching on the amplifier on the recorded voltage is evident: the signal jumps to a value V_0 before drifting with time. As expected [5], the signal presents a drift that follows Equation (1). Thus, the voltage can be corrected to give a flat signal $V_{cor}(t) = V(t) - V_{drift}(t)$, as shown in the inset to Figure 3a. Both jump to V_0 and drift are due to the closing of the electronic circuit (both amplification and pressure sensing), and as such should be independent of the pressure and temperature at the tip of the sensor [19].

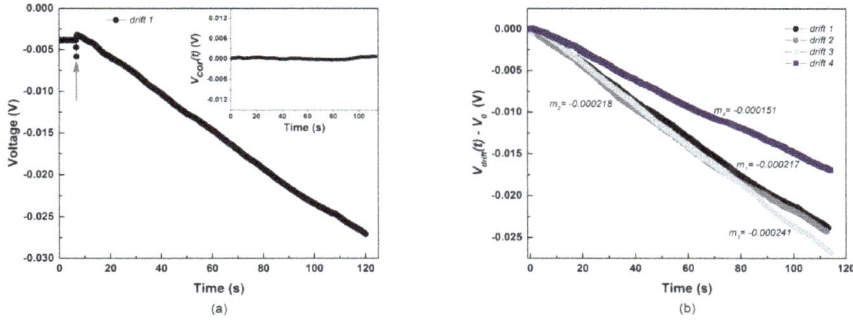

Figure 3. Drifts measurements with piezoelectric sensor PS 1 in an empty slit. (**a**) Time dependence of the voltage delivered by the charge amplifier. The arrow indicates the time at which the amplifier is switched on. The inset in (a) displays the corrected voltage $V_{cor}(t)$. (**b**) Reproducibility of the slope of the drift $V_{drift}(t) - V_0$.

Figure 3b shows the measurements of three consecutive independent drift measurements (drift 1, 2, and 3) and a drift measurement performed after 3 h (drift 4). The slope s does not change significantly in the three consecutive measurements, in fact there is almost an overlap of the data. Drift 4 deviates slightly from the previous measurements, but it reflects only a difference of 0.0019 MPa from the average of the consecutive drifts (using the sensibility of the manufacturer in the conversion). In order to check for any effect of pressure on $V_{drift}(t)$, the piezoelectric sensors were individually mounted on a calibrator and several pressures were successively applied. Figure 4 reports the time evolution of the output voltage $V(t)$ during the ramping up of pressure in the calibration pump. The inset in Figure 4 displays the pressure dependence of the slope s, recorded with sensor PS 1. The error bar associated to each data point results from the error computed from the fit of Equation (1) to $V(t)$ in each pressure step. The data show that in the applied pressure range, there is no significant effect of the pressure on the drift of the piezoelectric sensor.

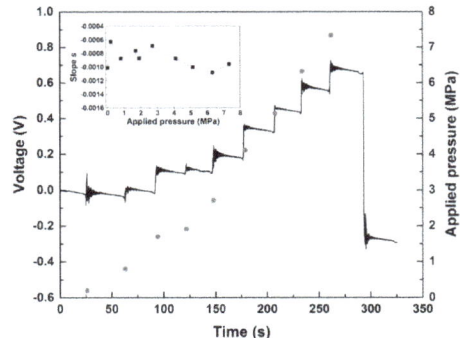

Figure 4. Time evolution of the output voltage $V(t)$ (solid line) of the PS 1 sensor during the ramping up of pressure (circles) in the calibration pump. The inset displays the pressure dependence of the slope s for the PS 1.

Figure 5 represents the calibration curves for PS 1 and PS 2. Each data point and error bar result respectively from the average and standard error of five measurements similar to the one displayed in Figure 4. The slopes returned by the linear fits to the data indicate that the response of the piezoelectric sensors is nicely linear for the range of pressures tested. In addition, the computed slopes do not significantly differ from the constants calculated with the sensibilities reported in the calibration certificate supplied by the manufacturer, namely $50/x = 20$ for PS 1 and $50/x = 7.58$ for PS 2 (see Table 1 for x). The intercepts to the origin for the linear fits are 0.28 ± 0.03 and 0.096 ± 0.001 for PS1 and PS2, respectively. These intercepts are smaller than those found for the conventional transducers (0.415 ± 0.015), which suggest a better sensitivity for piezoelectric sensors in the low-pressure range.

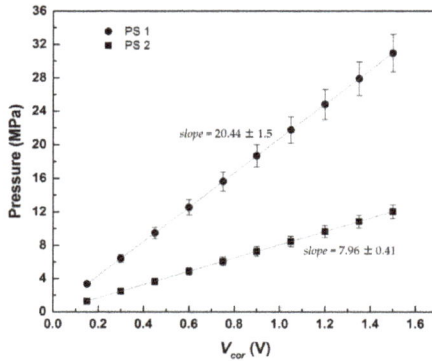

Figure 5. Calibration curves for PS 1 and PS 2, fitted with a linear function.

In order to estimate the eventual effect of the melt temperature on V_{drift}, experiments were performed with the piezoelectric transducers mounted on the heated double slit die, but with no material in the slit cavity. Figure 6 reports the temperature dependence of the slope s computed using Equation (1) from V_{drift} transients recorded with PS 2.

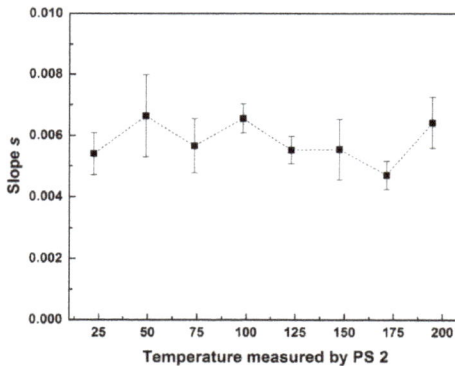

Figure 6. Average drift slope s as function of temperature.

Each point and error bar in the figure correspond to an average value and error computed from five measurements. For the temperature range tested in Figure 6, which corresponds to the range recommended by the manufacturer, no noteworthy variation of s with temperature is perceived. This result, together with the data displayed in Figure 4, confirms that V_{drift} is essentially due to the electronic imperfections of the whole piezoelectric acquisition system [11,12,19].

Figure 7 presents the time evolution of the outputs of the piezoelectric and conventional transducers recorded during the ramping up of piston velocities.

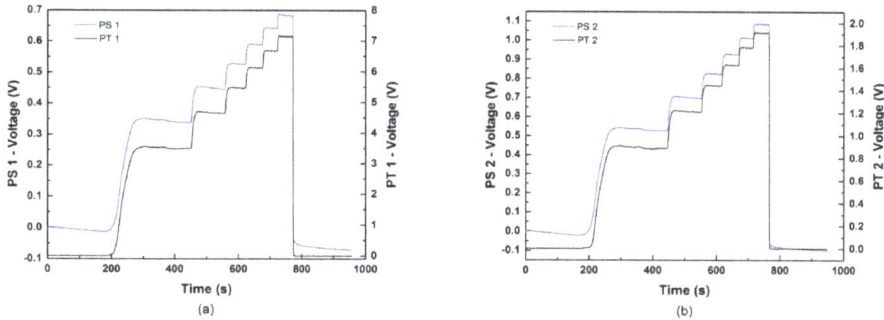

Figure 7. Time evolution of outputs (voltages) measured during the stepwise ramp in the piston velocity of the capillary rheometer fed with a low density polyethylene (LDPE) at 150 °C. (**a**)—Piezoelectric (PS 1) and conventional (PT 1) transducers located upstream. (**b**)—Piezoelectric (PS 2) and conventional (PT 2) transducers located downstream.

The double slit rheometer was fed with LDPE at 150 °C. The transients of face-mounted transducers show a satisfactory matching along the time scale. The piezoelectric signals show the expected drifts before the inception of melt flow, and after stopping and retracting the piston of the capillary rheometer. The piezoelectric data were corrected by computing $V_{cor}(t)$ using either $V_{drift}(t)$ measured before actuating the piston or $V_{drift}(t)$ measured at the end of the ramp. The steady state values of $V_{cor}(t)$ recorded for each piston velocity were then converted into pressure using the calibration curves, see Figure 5. The resulting pressures PS 1 and PS 2 are plotted in Figure 8 as a function of the piston velocities, together with the pressures PT 1 and PT 2 measured with the conventional transducers. Experiments performed at 180 °C are also reported in Figure 8. Each curve in Figure 8 is the result of the average of two ramps performed to check for data reproducibility. The corresponding error bars for each data point are smaller than the symbols used to represent the data. The error resulting from the reproducibility is larger (ranging from 0.02% to 3.4% for piezoelectric sensors and 0.02% to 1.2% for conventional transducers) than the error from the pressure reading (ranging from 0.13% to 0.3% for piezoelectric sensors and 0.16% to 0.5% for conventional transducers).

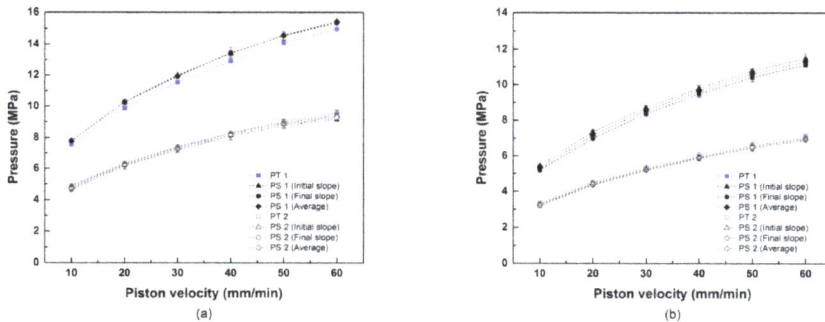

Figure 8. Comparison between the pressures obtained with the piezoelectric sensors and conventional transducers for each piston velocity. (**a**) LDPE at 150 °C. (**b**) LDPE at 180 °C. Initial slope and final slope refer to the values of *s* used in the fitting of V_{drift} before actuating the piston or at the end of the ramp in piston's velocity. Average indicates that drift correction is performed by computing the average of initial and final slopes *s*.

Data in Figure 8 indicate a moderate mismatch between the pressures returned by the two types of transducers, albeit their contour being virtually identical. Piezoelectric sensors measured somewhat larger values than the conventional sensors upstream, whereas the opposite occurs downstream. Generally, the method of drift correction (*s* fitted at the beginning or at the end of the ramp, or *s* computed from the average of these two values) does not impact significantly on the results, as the mismatch between conventional and piezoelectric measurements remains in the range of 0.4% to 4% (see Table 2). This conclusion is consistent with the results displayed in Figures 4 and 6 where the drift is shown to be independent of both pressure and temperature. Overall, the numbers reported in Table 2 compare well with the 2% scatter reported in the pressure measurements of a conventional transducer used for the capillary rheometry of a low-density polyethylene at 150 °C [20].

Table 2. Differences (in %) between the pressures measured with piezoelectric (PS) and conventional transducers (PT), and between the corresponding computed viscosities.

Temperature	Sensors	Initial Slope		Final Slope		Average Slope	
		Pressure	Viscosity	Pressure	Viscosity	Pressure	Viscosity
150 °C	PT 1 vs. PS 1	2.4–3.0%	6.7–9.3%	1.9–3.1%	9.1–11%	2.2–3.0%	8.0–10%
	PT 2 vs. PS 2	0.4–1.0%		1.6–3.6%		0.8–2.0%	
180 °C	PT 1 vs. PS 1	2.5–4.0%	6.7–12%	0.0–0.8%	5.0–7.1%	1.3–2.3%	5.6–9.2%
	PT 2 vs. PS 2	1.0–1.5%		2.3–3.9%		1.7–2.7%	

The data with the average slope method for drift correction displayed in Figure 8 were inserted in Equations (3)–(5) to compute the flow curves shown in Figure 9. The impact of the mismatch between the pressure readings of both type of transducers on the resulting flow curves is evident in Figure 9. The pressure mismatch produces a vertical shift between the flow curves, the differences between the measured viscosities being larger at smaller shear rates than at larger shear rates. Nonetheless, the discrepancy between the measured viscosities remains in the range 5.6% to 10% (see also Table 2), which can be assumed as acceptable if one considers the 10% experimental error usually reported for rotational rheometry [21] and recently confirmed with Newtonian viscosity standards [22] and with a low density polyethylene studied at 150 °C [14].

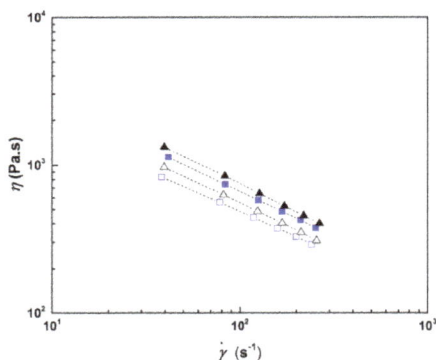

Figure 9. Flow curves measured with LDPE at 150 °C (solid symbols) and 180 °C (open symbols) using the piezoelectric (triangles) and the conventional (squares) pressure transducers.

The operability window of the piezoelectric transducers is now examined by testing materials showing different levels of viscoelasticity and comparing the resulting flow curves with those measured using conventional transducers. All drift corrections were performed using the average between the slopes recorded before and after the ramp in piston velocities. A PBAT grade designed for film blowing application was selected for exploring the upper viscoelastic part of the operability window, as this biodegradable material shows significant melt strength. The flow curves were measured at 190 °C and are presented in Figure 10. Overall, both sets of data, acquired either with conventional or piezoelectric transducers with an equally high level of precision (error bars are smaller than the symbols size), nicely overlap. This result validates the use of piezoelectric transducers for slit rheometry, at least for materials showing shear viscosities in the range 500–1000 Pa·s at shear rates between 20 s^{-1} and 200 s^{-1}. A bio-based and compostable PHB was selected for screening low viscosity materials. PHB is well known for its limited thermal stability and processability. As such, this is a challenging material for capillary rheometry, and flow curves of polyhydroxyalkanoates are scarcely found in the literature [23–25]. Indeed, true wall shear rates could not be computed from the data since steady state flow was not achieved for any piston velocity (see the inset in Figure 10). Thus, the apparent viscosity, η_a, is reported in Figure 10 as a function of the apparent shear rate, $\dot{\gamma}_a$, for the two flow curves measured at 190 °C. In spite of these difficulties, a monotonic shear thinning flow curve with satisfactory error bars is reported for the experiment carried out with the piezo transducers up to a shear rate of 150 s^{-1}. In contrast, larger error bars are obtained with the conventional transducers and the flow curve displays an unrealistic shear thickening at larger shear rates. Thus, this result confirms the good performance of piezoelectric transducers for detecting small pressure variations [3–6]. The slope of the PHB flow curve measured with the piezoelectric sensors is larger than −1, which suggests the occurrence of possible flow instabilities. However, the piezoelectric transient in the inset to Figure 10 does not show oscillations with frequency and amplitude signaling the presence flow instabilities as reported elsewhere [7]. Therefore, the steep slope in the PHB flow curve seems related to the fact that uncorrected apparent viscosities and shear rates are reported in the graph. In addition, the effect of thermal degradation on the shear viscosity further contributes to an apparent shear thinning.

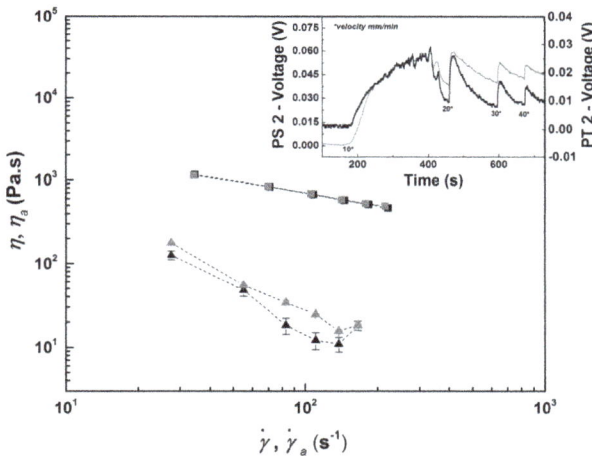

Figure 10. Flow curves measured with polybutylene adipate terephthalate (PBAT) at 180 °C (squares) and polyhydroxybutyrate (PHB) at 190 °C (triangles). Comparison between piezoelectric sensors (symbols and lines in grey) and conventional pressure transducers (symbols and lines in black). The inset represents the time evolution of outputs (voltages) measured with PHB by sensors located upstream during the stepwise ramp in the piston velocity of the capillary rheometer.

4. Conclusions

The use of piezoelectric transducers for steady melt pressure measurement has been assessed by directly comparing the flow curves measured with such transducers and with diaphragm transducers conventionally used in capillary rheometry and in polymer extrusion. A modular slit die was designed and constructed to allow for the direct comparison between the pressure data acquired by both types of transducers. Piezoelectric transducers were first calibrated and the drift inherent to the sensors electronics was analyzed as a function of both temperature and pressure. Various methodologies for the drift correction were proposed and the results confirmed that the drift does not depend on the temperature nor on the pressure. The pressure data acquired with piezoelectric transducers differ from the data returned by conventional transducers. The difference ranges from 0% to 4% for all shear rates and temperatures tested with a LDPE. Differences in pressure readings result in a maximum 10% difference in the shear viscosities measured with LDPE for all experimental conditions tested. This difference is acceptable given the 10% error usually reported for viscosities measured either with rotational rheometers or capillary rheometers. Accordingly, the results reported here validate the use of piezoelectric transducers for slit rheometry. A better agreement between the shear viscosities measured with the two types of transducers was achieved with a more elastic PBAT. In contrast to this, piezoelectric transducers outperformed conventional transducers for measuring the steady shear viscosity of a biodegradable PHB with values of the order of 20 Pa·s at shear rates of 100 s^{-1}.

Author Contributions: The conceptualization of the whole study was performed by P.F.T., J.A.C., and L.H. Experiments were performed by P.F.T. and S.C. P.F.T., J.A.C. and L.H. analyzed the data. P.F.T. drafted the paper and P.F.T., J.A.C., and L.H. wrote, reviewed and edited the paper.

Funding: This research was funded by National Funds through FCT—Portuguese Foundation for Science and Technology, Reference UID/CTM/50025/2013 and FEDER funds through the COMPETE 2020 Programme under the project number POCI-01-0145-FEDER 007688. L.H. acknowledges funding from the FCT Investigator Programme through grant IF/00606/2014.

Acknowledgments: The authors are grateful to Miguel Gomes and Jasper Cooremans from the Polymer Engineering Department of University of Minho for support in the set-up of experiments.

Conflicts of Interest: The authors declare no conflict of interest.

References

1. Macosko, C.W. *Principles, Measurements and Applications*; VCH: New York, NY, USA, 1994.
2. Teixeira, P.F.; Fernandes, S.N.; Canejo, J.; Godinho, M.H.; Covas, J.A.; Leal, C.; Hilliou, L. Rheo-optical characterization of liquid crystalline acetoxypropylcellulose melt undergoing large shear flow and relaxation after flow cessation. *Polymer* **2015**, *71*, 102–112. [CrossRef]
3. den Doelder, J.; Koopmans, R.; Dees, M.; Mangnus, M. Pressure oscillations and periodic extrudate distortions of long-chain branched polyolefins. *J. Rheol.* **2005**, *49*, 113–126. [CrossRef]
4. Filipe, S.; Becker, A.; Barroso, V.C.; Wilhelm, M. Evaluation of melt flow instabilities of high-density polyethylenes via an optimised method for detection and analysis of the pressure fluctuations in capillary rheometry. *Appl. Rheol.* **2009**, *19*, 23345.
5. Kádár, R.; Naue, I.F.; Wilhelm, M. First normal stress difference and in-situ spectral dynamics in a high sensitivity extrusion die for capillary rheometry via the 'hole effect'. *Polymer* **2016**, *104*, 193–203. [CrossRef]
6. Naue, I.F.; Kádár, R.; Wilhelm, M. A new high sensitivity system to detect instabilities during the extrusion of polymer melts. *Macromol. Mater. Eng.* **2015**, *300*, 1141–1152. [CrossRef]
7. Palza, H.; Filipe, S.; Naue, I.F.; Wilhelm, M. Correlation between polyethylene topology and melt flow instabilities by determining in-situ pressure fluctuations and applying advanced data analysis. *Polymer* **2010**, *51*, 522–534. [CrossRef]
8. Palza, H.; Naue, I.; Wilhelm, M.; Filipe, S.; Becker, A.; Sunder, J.; Goettfert, A. On-line detection of polymer melt flow instabilities in a capillary rheometer. *Kgk. Kautsch. GummiKunstst.* **2010**, *63*, 456–461.
9. Palza, H.; Naue, I.F.; Wilhelm, M. In situ pressure fluctuations of polymer melt flow instabilities: Experimental evidence about their origin and dynamics. *Macromol. Rapid Commun.* **2009**, *30*, 1799–1804. [CrossRef]
10. Cyriac, F.; Covas, J.A.; Hilliou, L.H.G.; Vittorias, I. Predicting extrusion instabilities of commercial polyethylene from non-linear rheology measurements. *Rheol. Acta* **2014**, *53*, 817–829. [CrossRef]
11. Hruska, C.K. Least-squares estimates of the nonlinear constants of piezoelectric crystals. *J. Appl. Phys.* **1992**, *72*, 2432–2439. [CrossRef]
12. Mullen, A.J. Temperature variation of the piezoelectric constant of quartz. *J. Appl. Phys.* **1969**, *40*, 1693–1696. [CrossRef]

13. Broadbent, J.M.; Kaye, A.; Lodge, A.S.; Vale, D.G. Possible systematic error in measurement of normal stress differences in polymer solutions in steady shear flow. *Nature* **1968**, *217*, 55–56. [CrossRef]

14. Teixeira, P.F.; Hilliou, L.; Covas, J.A.; Maia, J.M. Assessing the practical utility of the hole-pressure method for the in-line rheological characterization of polymer melts. *Rheol. Acta* **2013**, *52*, 661–672. [CrossRef]

15. Teixeira, P.F.; Ferrás, L.L.; Hilliou, L.; Covas, J.A. A new double-slit rheometrical die for in-process characterization and extrusion of thermo-mechanically sensitive polymer systems. *Polym. Test.* **2018**, *66*, 137–145. [CrossRef]

16. Wilhelm, M.; Reinheimer, P.; Ortseifer, M. High sensitivity fourier-transform rheology. *Rheol. Acta* **1999**, *38*, 349–356. [CrossRef]

17. Wilhelm, M. Fourier-transform rheology. *Macromol. Mater. Eng.* **2002**, *287*, 83–105. [CrossRef]

18. Hilliou, L.; van Dusschoten, D.; Wilhelm, M.; Burhin, H.; Rodger, E.R. Increasing the force torque transducer sensitivity of an rpa 2000 by a factor 5–10 via advanced data acquisition. *Rubber Chem. Technol.* **2004**, *77*, 192–200. [CrossRef]

19. Mack, O. New procedures to characterize drift and non-linear effects of piezoelectric force sensors. In Proceedings of the IMEKO TC3 Conference, Istanbul, Turkey, 17–21 September 2001.

20. Laun, H.M. Capillary rheometry for polymer melts revisited. *Rheol. Acta* **2004**, *43*, 509–528. [CrossRef]

21. Marquardt, W.; Nijmann, J. Experimental errors when using rotational rheometers. *Appl. Rheol.* **1993**, *3*, 120.

22. Laun, M.; Auhl, D.; Brummer, R.; Dijkstra, D.J.; Gabriel, K.; Mangnus, M.A.; Rullmann, M.; Zoetelief, W.; Handge, U.A. Guidelines for checking performance and verifying accuracy of rotational rheometers: viscosity measurements in steady and oscillatory shear (IUPAC Technical Report). *Pure Appl. Chem.* **2014**, *86*, 1945–1968. [CrossRef]

23. Yamaguchi, M.; Arakawa, K. Effect of thermal degradation on rheological properties for poly (3-hydroxybutyrate). *Eur. Polym. J.* **2006**, *42*, 1479–1486. [CrossRef]

24. Ramkumar, D.; Bhattacharya, M. Steady shear and dynamic properties of biodegradable polyesters. *Polym. Eng. Sci.* **1998**, *38*, 1426–1435. [CrossRef]

25. Cunha, M.; Fernandes, B.; Covas, J.A.; Vicente, A.A.; Hilliou, L. Film blowing of phbv blends and phbv-based multilayers for the production of biodegradable packages. *J. Appl. Polym. Sci.* **2016**, *133*. [CrossRef]

fluids

MDPI

Article

Milligram Size Rheology of Molten Polymers

Salvatore Costanzo [1], Rossana Pasquino [1], Jörg Läuger [2] and Nino Grizzuti [1,*]

[1] Department of Chemical, Materials and Industrial production engineering (DICMaPI), University of Naples Federico II, Piazzale V. Tecchio 80, 80125 Napoli, Italy; salvatore.costanzo@unina.it (S.C.); r.pasquino@unina.it (R.P.)

[2] Anton Paar Germany GmbH, Helmuth-Hirth-Strasse 6, D-73760 Ostfildern, Germany; joerg.laeuger@anton-paar.com

* Correspondence: nino.grizzuti@unina.it; Tel.: +39-081-768-22-85

Received: 11 January 2019; Accepted: 14 February 2019; Published: 18 February 2019

Abstract: During laboratory practice, it is often necessary to perform rheological measurements with small specimens, mainly due to the limited availability of the investigated systems. Such a restriction occurs, for example, because the laboratory synthesis of new materials is performed on small scales, or can concern biological samples that are notoriously difficult to be extracted from living organisms. A complete rheological characterization of a viscoelastic material involves both linear and nonlinear measurements. The latter are more challenging and generally require more mass, as flow instabilities often cause material losses during the experiments. In such situations, it is crucial to perform rheological tests carefully in order to avoid experimental artifacts caused by the use of small geometries. In this paper, we indicate the drawbacks of performing linear and nonlinear rheological measurements with very small amounts of samples, and by using a well-characterized linear polystyrene, we attempt to address the challenge of obtaining reliable measurements with sample masses of the order of a milligram, in both linear and nonlinear regimes. We demonstrate that, when suitable protocols and careful running conditions are chosen, linear viscoelastic mastercurves can be obtained with good accuracy and reproducibility, working with plates as small as 3 mm in diameter and sample thickness of less than 0.2 mm. This is equivalent to polymer masses of less than 2 mg. We show also that the nonlinear start-up shear fingerprint of polymer melts can be reliably obtained with samples as small as 10 mg.

Keywords: rheometry; polystyrene; linear viscoelasticity; start-up shear

1. Introduction

The laboratory production of new materials usually yields small quantities. This is the case, for example, of polymer synthesis, where specific architectures with controlled morphologies and narrow molecular weight distributions can be produced [1–3]. With very low amounts (order of hundreds of milligrams), the analysis and characterization of the chemical and physical properties of the material become, as such, difficult and sometimes even impossible to perform with the available quantities, if further purification is needed [4,5]. Moreover, the synthetic process of complex architectures can be time consuming as, for example, in the case of regular branched structures [6–9]. On the other hand, time is an important variable in research, and laboratory resources are limited; hence, the necessity of obtaining data in a short time does not allow to gather sample material via a series of synthetic processes [10].

One particular situation where both small samples and short measuring times are required is that of high throughput experimentation (HTE), particularly relevant in the field of polyolefin synthesis [11,12]. Here, a large number of polymer samples are synthesized under different conditions (pressure, temperature, feed composition, catalyst), simultaneously yielding several tens of different polymers, each one in the amount of a few milligrams.

Depending on the type of material, the amount of sample required for rheological characterization varies from several grams to a few milligrams. More specifically, for polymer melts, a linear frequency response with the canonical approach is obtained with less than 100 mg. For example, a standard 8mm plate-plate geometry requires roughly 25–50 mg [13]. However, as mentioned above, this quantity can be larger than the usual amounts obtained from laboratory-scale synthesis. For these reasons, there is an increasing need to find an accurate and controlled way to measure small samples. Recent commercial rheometers facilitate this process, as they are able to measure very low torques (few nanonewton meters) and resolve extremely low angular displacements (hundreds of nanorads). Nevertheless, when different variables are forced to the maximum of their sensitivity in the same measurement, and secondary effects also play a role, the viscoelastic response becomes arguable and a more critical approach to the measurement protocol and its elaboration should be developed [14]. Indeed, experimental techniques measure displacements and loads, and convert them in stress or deformation by assuming ideal conditions. Unfortunately, ideal conditions can be invalidated by various factors such as fluid and instrument inertia [15], underfilling and overfilling, sample volume, and surface tension [16,17], to name a few. These effects can concurrently contribute in the same rheological measurement, and may invalidate results and sometimes cause measured properties to incorrectly appear nonlinear or non-Newtonian. As an example, surface tension can generate torques in steady shear for Newtonian fluids (which could be mistaken as shear thinning) and the effect could grow with slight overfilling, which increases contact line rotational asymmetry [18].

A smaller amount of sample can be obviously employed by using smaller size geometries. This is a straightforward approach, for example, when working with parallel plates on rotational rheometers. Concerning parallel plates, the possibility of experimental errors, however, increases as the plate size decreases. For a given material in the same test conditions, the torque varies with the third power of the radius. Hence, reducing the radius by a factor of two means reducing the measured torque by eight times. In addition, if gap and strain are fixed, the applied angular displacement decreases linearly with the radius. Another issue arising with the use of small geometries is the difficulty of machining flat surfaces with small diameters, thus determining plate misalignment. On top of that, the squeeze flow of air between plates induces gap errors [19,20]; hence, the zero gap position must be carefully checked.

Nonlinear measurements are even more challenging. Nonetheless, there is a great interest to obtain information on the rheology of fluids in fast flow conditions, for example in fast shear, for processing applications. On the other hand, transient shear experiments are hindered by flow instabilities such as edge fracture. A possibility to perform reliable measurements in start-up shear is provided by the so-called cone-partitioned plate (CPP) geometry [21–24]. Such a geometry is formed by two parts, namely a bottom standard cone and a top plate that is split into an inner measuring plate and an outer non-measuring corona. The sample exceeds the area of the inner measuring plate so that, when shear fracture sets in at the edge of the sample, it requires a certain time to reach the inner part. During this time, reliable rheological measurements are still possible. However, even when working with small diameters of the inner plate, the overfilling to prevent edge fracture requires more sample compared with a standard cone and plate geometry with the same diameter.

In this study, we perform linear and nonlinear rheological measurements on a monodisperse polystyrene of average molecular weight 200,000 Da, with home-made tools characterized by small diameters and variable gap size. We try to minimize sample quantities and show that the linear response obtained with 8-mm parallel plates is reproducible, within experimental error, by using a plate of 3 mm. This means that, starting from the standard mass of 25–50 mg, one can obtain an accurate viscoelastic measurement with samples as small as 2 mg. In addition, we performed nonlinear start-up shear experiments with two cone-partitioned plate geometries, having an inner diameter of 6 mm and 4 mm, respectively. We demonstrate that reliable and reproducible nonlinear measurements are obtained with an amount of 10 mg.

2. Materials and Methods

For both linear and nonlinear tests, we used a linear polystyrene with a molecular weight of 200,000 Da, henceforth coded as PS200k. The polymer was purchased from Agilent Technologies (molecular weight of the highest peak, M_p = 202,100 Da; PDI = 1.03, as supplied by the producer; Batch No.: 0006314460). The samples were shaped to discs of different diameters by means of vacuum compression molding. Briefly, a proper amount of sample is weighed and inserted in a mold with a specific diameter (Figure 1a). The mold consists of a bottom plate (E), a cylinder (D) with holes of specific diameters, and corresponding pistons (A). The clearance between the pistons and the holes is approximately 50 μm. The cylinder with holes is fixed to the bottom plate by means of four screws. The sample fills in a specific hole and the corresponding piston is placed on top. The mold is then inserted in a small vacuum chamber (B,C) and put in a hot press.

The samples were shaped for 8 min at 150 °C. The normal force applied was approximately 200 kg. After 8 min, the normal force was released. The mold was then cooled down to room temperature and the samples were extracted. We shaped samples with diameter values of 8, 6, 5, 4, and 3 mm. The corresponding masses at 25 °C, nominal gap (ratio of the total volume to the cross-section), actual gap used for the measurements, and ratio of nominal to actual gap are reported in Table 1. To calculate the nominal gap, we evaluated the nominal volume as the product of the sample mass by the density of polystyrene at 150 °C ($\rho_{PS}(150\ °C) = 0.994$ g/cm^3).

Table 1. Diameter, mass, and gap of the samples. The values in parenthesis refer to nonlinear tests. The mass was weighed at 25 °C. The actual gap was measured at 150 °C. The uncertainty for mass is ±0.1 mg, and for the actual gap ±1 μm.

Diameter (mm)	Mass (mg)	Nominal Gap (mm)	Actual Gap (mm)	Gap Ratio (-)
8	25.0 (23.0)	0.500	0.408	1.23
6	15.4 (10.0)	0.548	0.490	1.12
5	8.9	0.456	0.376	1.21
4	4.5	0.360	0.280	1.29
3	1.9	0.270	0.189	1.43

Linear rheological measurements were performed on a Physica MCR702 (Anton Paar GmbH, Ostfildern, Germany) equipped with a Peltier plate and hood (H-PTD200) for temperature control. A shaft for disposable tools was used in order to mount homemade stainless-steel plates with different diameters (Figure 1b). The linear frequency response was measured at three different temperatures, namely 130 °C, 150 °C, and 170 °C. The applied strain was 6% at 170 °C, 5% at 150 °C, and 2% at 130 °C. The thermal expansion of the tools was taken into account while measuring at different temperatures.

The terminal regime was obtained by performing creep measurements at 170 °C (applied stress = 50 Pa) and converting the creep compliance data into dynamic moduli. The creep conversion was performed by using the NLreg software [25].

Nonlinear rheological tests were carried out on an ARES strain-controlled rheometer (TA instruments, New Castle, DE, USA) equipped with a convection oven for temperature control. We used two homemade cone-partitioned plate (CPP) geometries, one having an inner plate radius of 6 mm (referred to as CPP6) and another having an inner plate radius of 4 mm (referred to as CPP4). Details on the construction of the plates are provided elsewhere [21,26,27]. The bottom cone is a homemade cone with a truncation of 105 μm and an angle of 5.5°.

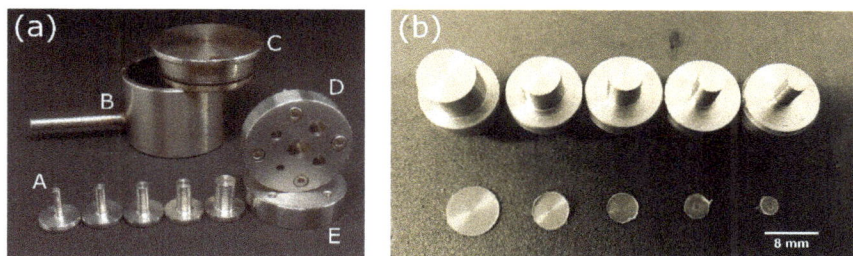

Figure 1. (**a**) Vacuum mold used to shape the samples: A, inner pistons (diameters in mm are indicated in the figure); B, vacuum chamber; C, sealing cap; D, mold; E, lower plate. (**b**) Parallel plates of different diameters used for linear measurements, along with samples obtained with vacuum compression molding technique.

3. Results and Discussion

Figure 2 shows the frequency response of PS200k at different temperatures. The data are taken with parallel plates having different diameters. For all samples, the following measuring protocol was applied. The sample was loaded at 150 °C; after 20 min, necessary to guarantee thermal equilibration and relaxation of the sample, a frequency sweep test in the range from 100 to 0.1 rad/s was performed (applied strain = 5%). Next, the temperature is increased to 170 °C; after 20 min, a frequency sweep test was performed (applied strain = 6%); at the end of the test a creep measurement was completed for approximately 30 min (applied stress = 50 Pa). The creep test at high temperature is used to complement the frequency measurements in the low frequency range. Subsequently, the temperature is lowered to 130 °C; after 40 min for thermal equilibration and sample relaxation, a frequency sweep test was performed (applied strain = 2%).

Figure 2 demonstrates good reproducibility of the tests with different diameters, within experimental error. The maximum deviation among the frequency sweep tests at 150 °C is of the order of 10%. At 170 °C, the deviation is even lower (7%). The maximum discrepancy among the tests is observed at 130 °C (Figure 2c). There are multiple reasons for this discrepancy. First, the extent to which the sample is pressed in between the plates is not the same for all the samples, as can be noted from Table 1. In fact, the ratio of the nominal gap (calculated as the ratio of the total volume of the sample to its cross-section) to the actual gap is not the same for all samples. During the loading, the sample needs to be pressed between the plates to allow for optimal adhesion of the polymer to the plate surfaces. Moreover, the volume between the plates must be completely filled by the polymer. It is quite difficult, however, to perfectly center the sample between the plates at high temperature. Therefore, it is sometimes necessary to squeeze the sample in between the plates in order to fill the measuring volume completely. In doing so, some parts of the sample exceed the area of the plate and wrap around the plate itself, causing an extra torque contribution, thus inducing an apparent increase in the viscoelastic moduli. Such an increase depends on the level of overloading and can shift the viscoelastic moduli up to 20% with respect to the true value. Another possible explanation for the vertical shift between the data is the gap setting. The zero gap was performed by software and manually checked by bringing the plates in contact with each other. However, it is difficult to machine completely flat plates, especially when dealing with small-diameter geometries. Small imperfections can lead to a few-microns-gap error. When working with standard gaps of approximately 1 mm, such an error is not relevant. However, it becomes important when working with small gaps, i.e., below 0.2 mm. Another possible source of the vertical shift has to do with the cooling of the sample from 170 °C to 130 °C. A rapid decrease in temperature can cause partial detachment of the sample from the plates or inhomogeneity of the edges due to the different thermal expansion coefficients of the polymer and measuring tools. The typical cooling rate of the Peltier element used here is approximately 20 °C/min in the temperature range explored. Moreover, the response of the thermal control unit is generally overdamped. This causes a temperature overshoot with respect to the set point during heating, or an undershoot during cooling. Therefore, the sample experiences a larger temperature difference with respect to the set points.

We point out that the shift of the curves at a fixed temperature in Figure 2 is mainly vertical, even though a minor horizontal shift can be also detected from phase angle data. It is worth mentioning the slight deviation of the measurements in the high frequency part when measuring with 4-mm and 3-mm parallel plates (green and orange curves). This discrepancy is evident at 150 °C and 170 °C, where the elastic and viscous moduli tend to "bend" at high frequencies. We attribute such a deviation to the torsional compliance of the instrument, which affects the values of the moduli mainly in the high frequency range [28]. We note that the instrument compliance is the total compliance resulting from the motor, measuring system, and Peltier element; therefore, it is specific for each combination of rheometer and plate [29]. If the compliance of the measuring instrument is known, then the measured viscoelastic moduli can be corrected accordingly in order to obtain the true values. The expression for such a correction with parallel plates has been demonstrated in previous works [28–30]. Since the error here is small, we did not attempt to correct the compliance effect of the measuring tools.

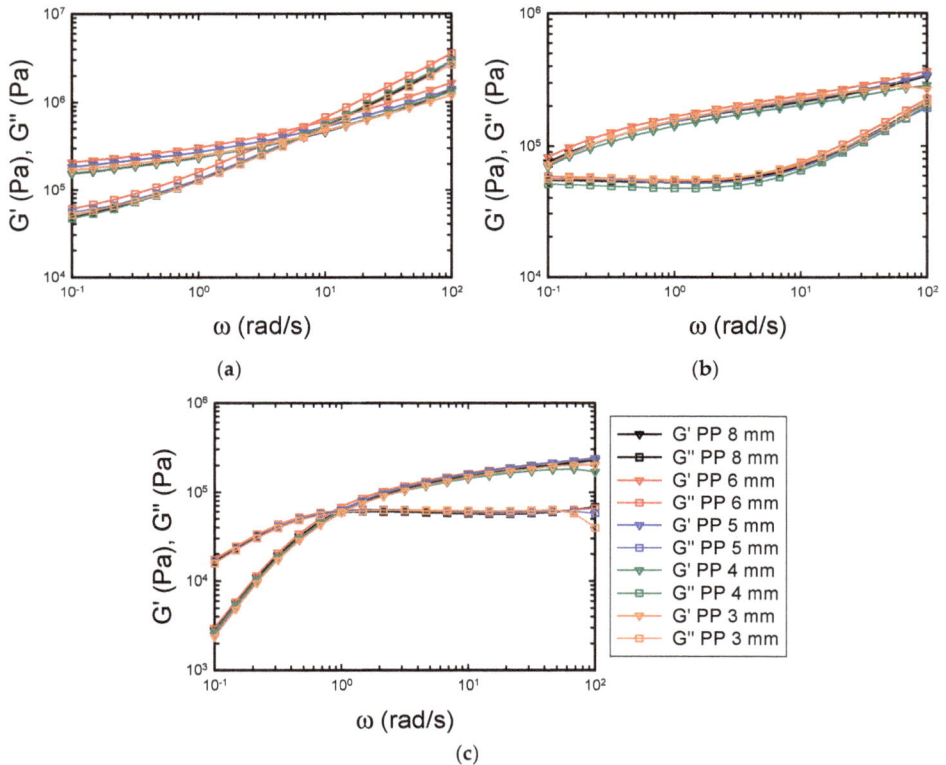

Figure 2. Dynamic frequency sweep tests on PS200k performed with different plate diameters as reported in the legend: (**a**) T = 130 °C, (**b**) T = 150 °C, (**c**) T = 170 °C. The color legend applies to the three panels.

The data in Figure 2 were used to build rheological mastercurves according to the time–temperature superposition (TTS) principle [31,32]. Figure 3 shows the mastercurves built from the dynamic frequency sweep tests in Figure 2. Since only three temperatures were available, it was difficult to obtain a good empirical estimate of the horizontal shift factors. Therefore, the values of the William-Landel-Ferry (WLF) fit constants of polystyrene were taken from literature [31,33,34]. In order to build the mastercurves in Figure 3, we calculated the horizontal shift factors a_T according to the WLF fit equation, $\log(a_T) = -c_1 \left(T - T_{ref} \right) / \left(c_2 + T - T_{ref} \right)$, at the reference temperature of 150 °C, using $c_1 = 6.9842$ °C and $c_2 = 102.08$ °C. The calculation yields the following values of the shift factors: $a_T(130\,°C) = 50.327$, and $a_T(170\,°C) = 0.0717$. The vertical shift factors were also calculated theoretically, according to temperature-density compensation. In fact, the vertical shift factors are given by $b_T = (\rho_0 T_0) / (\rho T)$, where ρ_0 is the density at the reference temperature T_0 (in K), and ρ is the density at the temperature T. The density of polystyrene (in kg/m^3) at the different temperatures was evaluated according to $\rho(T) = 1250 - 0.605T$ [35], with the temperature expressed in K.

The mastercurves are reported in Figure 3 with the same color legend as in Figure 2. The agreement between the different datasets is good: the curves measured with the different plates virtually overlap with each other over the whole frequency range. The maximum deviation between points of the different mastercurves at the same frequency is approximately 12%; therefore, it can be considered within experimental error.

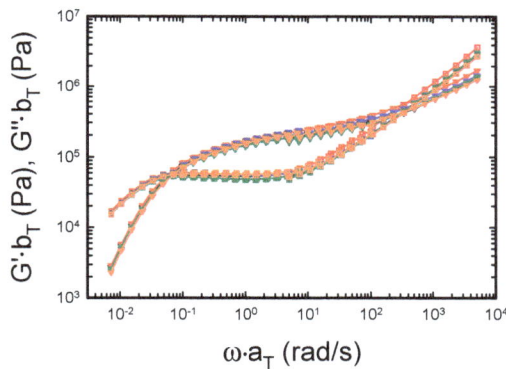

Figure 3. Mastercurves from data in Figure 2. The reference temperature is 150 °C. The color legend is the same as in Figure 2.

The deviation between measurements with different diameters can be better highlighted by plotting the relevant variables read by the rheometer during dynamic measurements, namely torque, and phase shift between torque and strain waves. Figure 4a shows the mastercurves obtained normalizing the torque at different diameters.

In the linear regime, the torque signal is proportional to the applied strain and to the radius raised to the third power. Therefore, by dividing the torque signal by the strain and multiplying it by the ratio of the radii to the third power (D = 8 mm was used as reference), the data corresponding to the different diameters should collapse into a mastercurve. Furthermore, if one multiplies the dynamic data at the different temperatures by the corresponding horizontal shift factors, then a single torque mastercurve is obtained. It is worth mentioning that the lowest value of the measured torque for all the dynamic measurements (7.2 μNm) is well above the minimum torque measurable with the rheometer (0.001 μNm). The data in Figure 4a show that indeed the normalized torque collapses into a mastercurve as expected, and the largest discrepancies between the different signals are observed at 130 °C. Moreover, Figure 4b reports the phase angle as a function of frequency. The phase angle δ is evaluated as $\tan^{-1}(\tan(\delta))$. The signal becomes noisy around the minimum due to the large difference between the elastic and the viscous contribution to the viscoelastic response [36]. The noise is more relevant with plates having small diameters because the magnitude of torque and angular displacement are smaller compared to those obtained with plates having larger diameters.

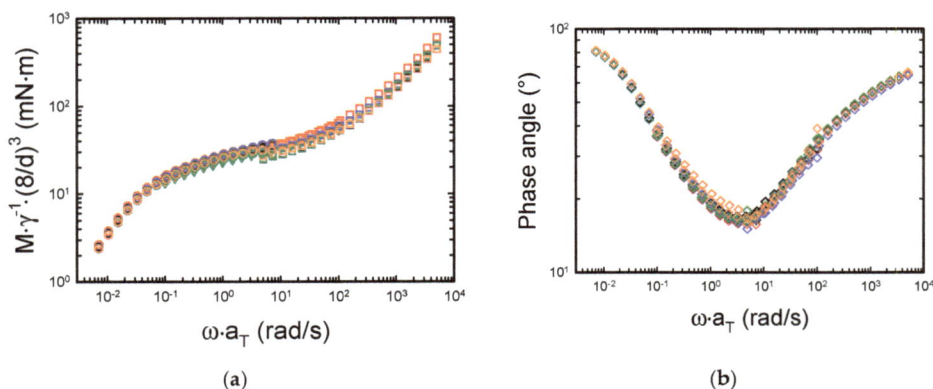

Figure 4. (a) Torque and (b) phase angle mastercurves. The color legend is the same as in Figure 2.

From the mastercurves in Figure 3, the low frequency crossover, corresponding to the terminal relaxation time of the polymer, can be extracted. However, inspection of Figure 3 shows that the terminal region, that is, the low frequency range where the storage modulus has a slope of 2 and the loss modulus has a slope of 1, is not fully attained. Measuring at temperatures higher than 170 °C with polystyrene is not recommended, as thermal degradation may occur [37]. In addition, with the smallest diameter plate and at higher temperatures, the torque can reach values close to the sensitivity limits of the instrument. An alternative is to conduct a creep measurement and convert the creep compliance into dynamic data. Figure 5a shows the creep experiments performed on the samples at 170 °C.

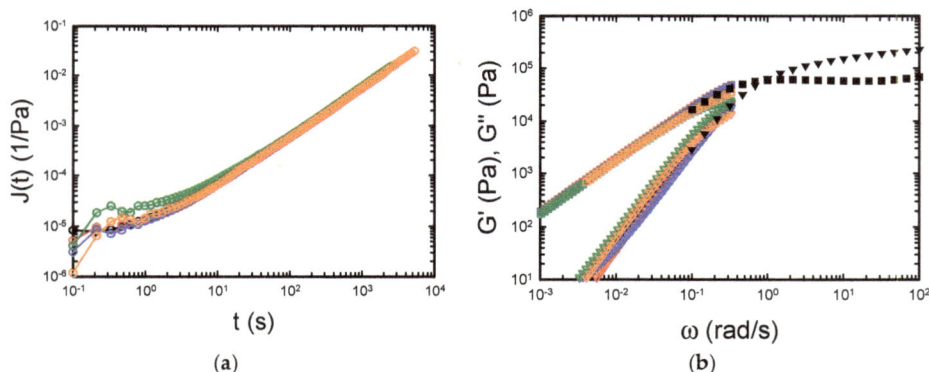

Figure 5. (a) Creep data at T = 170 °C; applied stress = 50 Pa. (b) Results of the conversion of the creep compliance into dynamic moduli. The color legend for both panels is the same as in Figure 2; squares = loss modulus; triangles = elastic modulus. The dynamic frequency sweep at 170 °C with PP8 is also reported for reference (black symbols).

Figure 5a shows the excellent agreement between the data measured with different geometries. Discrepancies are observed only at short times, which are not relevant for the determination of the terminal behavior. The typical creep ringing at early times due to inertia is also observed with the smallest geometries (3 mm and 4 mm) [14].

Figure 5b shows the conversion of the different compliance curves in Figure 5a into dynamic moduli obtained with the NLreg software. The very good agreement for the loss modulus is evident. Larger, but still acceptable deviations are observed in the elastic modulus for measurements with different diameters. In addition, the agreement of the converted dynamic data with the measured data (black symbols) confirms the validity of the conversion of creep compliance into dynamic moduli and, indirectly, the reliability of the creep measurements.

Hitherto, our study focused on linear measurements, demonstrating that the relaxation spectrum of polymer melts can be safely measured even with very small sample amounts. However, to the end of polymer processing, information on nonlinear shear flow behavior is also important. The point here is again about the minimum amount of sample that could be used in order to obtain reliable data in the nonlinear regime. To address this question, we performed nonlinear start-up shear tests on PS200k at 160 °C and different shear rates. To this end, we used two different partitioned cone geometries with an inner diameter of 6 mm (CPP6) and 4 mm (CPP4), respectively. For the CPP6 setup we used 22 mg of sample, whereas with CPP4 we used only 10 mg of sample. The results of the test are reported in Figure 6a.

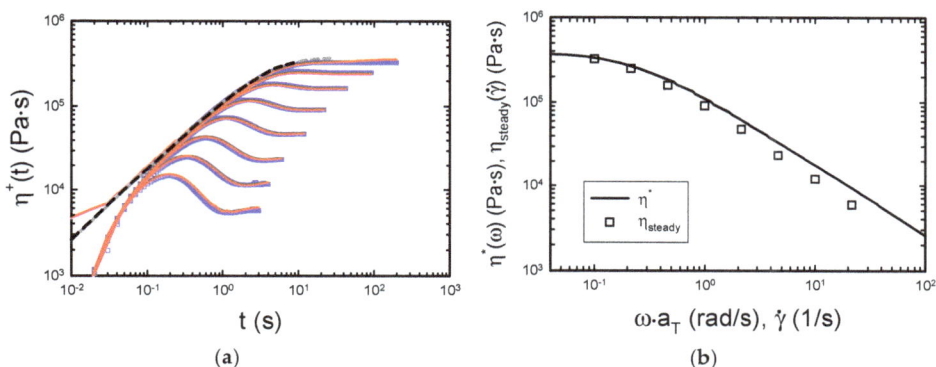

Figure 6. (**a**) Transient viscosity measurements performed with CPP6 (empty blue squares) and CPP4 (red lines) at T = 160°C; shear rates from top to bottom: 0.1, 0.215, 0.464, 1, 2.15, 4.64, 10, 21.5 s^{-1}. The linear viscoelastic envelope is also reported (dashed grey line and dashed black line). (**b**) Cox-Merz rule applied to the transient data.

Figure 6a demonstrates the excellent agreement between the data obtained with CPP6 and CPP4. The quality of the data is confirmed by the fact that the short-time nonlinear response overlaps with the linear viscoelastic (LVE) envelopes evaluated from a combination of the Cox-Merz rule [38] and the Gleissle rule [39]. The dashed black line is the LVE envelope calculated from the LVE response of the sample measured at 160 °C before starting the transient measurements. The grey dashed line, instead, is the mastercurve at 150 °C obtained with the 8-mm sample on the MCR702 and shifted to 160 °C. The experimental shift factor is $a_T = 0.189$, whereas the shift factor predicted from the WLF fit is $a_T = 0.234$. According to the WLF fit constants reported above, the latter value would correspond to a temperature of 161 °C. The mismatch could be due to the different thermal calibration of the two rheometers. To further confirm the quality of the data, we also checked the validity of the Cox-Merz rule (Figure 6b) [38]. We obtained a good agreement between the dynamic and transient data, consistently with previous literature reports [24,26]. The slope of the last four points is −0.91, slightly larger compared with the previously reported value of −0.82 [26]. It is worth noting that the steady-state viscosity data obtained from the transient tests are systematically lower than the dynamic ones and that this mismatch is higher with increasing shear rate. The reasons for this discrepancy are multiple and could be, for example, related to wall-slip. Wall-slip of polystyrene melts was demonstrated in capillary and parallel plate geometries [40,41]. Concerning our data, we find a satisfactory agreement between transient and dynamic viscosity values in the transition from the Newtonian regime to the shear thinning region. However, we cannot completely exclude slip at high shear rates. On the other hand, in such a range, edge fracture can also partly affect the data. In fact, as it was pointed out above, the CPP can delay the effect of edge fracture into the measurement volume but cannot completely avoid it. The link between edge fracture and wall-slip in CPP geometry is an open question that deserves further investigation. As far as this paper is concerned, we aimed to demonstrate that few milligrams of sample are enough to obtain reproducible nonlinear data with state-of-the-art instrumentation on rotational rheometers.

4. Conclusions

The laboratory synthesis of new materials is carried out on a small scale. This often poses the challenge to obtain reliable rheological measurements both in linear and nonlinear regimes working with very small quantities of samples. Therefore, we investigated the rheological behavior of a well-entangled polystyrene melt, both in linear and nonlinear regimes, working with relatively small geometries with different diameters and gap thicknesses. We showed that reliable data are obtained in the linear regime with good accuracy, measuring samples as small as less than 2 mg with parallel plates. We also demonstrated that artifact-free measurements are possible in transient shear working with cone-partitioned plate geometries that house samples as small as 10 mg.

Author Contributions: Data curation, S.C.; Investigation, S.C. and R.P.; Methodology, S.C., R.P., J.L. and N.G.; Supervision, N.G.; Validation, S.C. and J.L.; Writing—original draft, S.C.; Writing—review & editing, R.P. and N.G.

Funding: This research received no external funding.

Acknowledgments: We are grateful to Dimitris Vlassopoulos and Antonis Mavromanolakis for having kindly donated the PS sample and the homemade setups used in this work.

Conflicts of Interest: The authors declare no conflict of interest.

References

1. Malmström, E.E.; Hawker, C.J. Macromolecular engineering via 'living' free radical polymerizations. *Macromol. Chem. Phys.* **1998**, *199*, 923–935.
2. Sumerlin, B.S.; Vogt, A.P. Macromolecular Engineering through Click Chemistry and Other Efficient Transformations. *Macromolecules* **2010**, *43*, 1–13. [CrossRef]
3. Hutchings, L.R.; Kimani, S.M.; Hoyle, D.M.; Read, D.J.; Das, C.; McLeish, T.C.B.; Chang, T.; Lee, H.; Auhl, D. Correction to In Silico Molecular Design, Synthesis, Characterization, and Rheology of Dendritically Branched Polymers: Closing the Design Loop. *ACS Macro Lett.* **2012**, *1*, 742. [CrossRef]
4. Park, S.; Cho, D.; Ryu, J.; Kwon, K.; Lee, W.; Chang, T. Fractionation of Block Copolymers Prepared by Anionic Polymerization into Fractions Exhibiting Three Different Morphologies. *Macromolecules* **2002**, *35*, 5974–5979. [CrossRef]
5. Lee, H.C.; Lee, H.; Lee, W.; Chang, T.; Roovers, J. Fractionation of Cyclic Polystyrene from Linear Precursor by HPLC at the Chromatographic Critical Condition. *Macromolecules* **2000**, *33*, 8119–8121. [CrossRef]
6. Schlüter, A.D.; Rabe, J.P. Dendronized Polymers: Synthesis, Characterization, Assembly at Interfaces, and Manipulation. *Angew. Chem. Int. Ed.* **2000**, *39*, 864–883. [CrossRef]
7. Hirao, A.; Matsuo, A. Synthesis of Chain-End-Functionalized Poly(methyl methacrylate)s with a Definite Number of Benzyl Bromide Moieties and Their Application to Star-Branched Polymers. *Macromolecules* **2003**, *36*, 9742–9751. [CrossRef]
8. Scherz, L.F.; Costanzo, S.; Huang, Q.; Schlüter, A.D.; Vlassopoulos, D. Dendronized Polymers with Ureidopyrimidinone Groups: An Efficient Strategy to Tailor Intermolecular Interactions, Rheology, and Fracture. *Macromolecules* **2017**, *50*, 5176–5187. [CrossRef]
9. Costanzo, S.; Scherz, L.F.; Schweizer, T.; Kröger, M.; Floudas, G.; Schlüter, A.D.; Vlassopoulos, D. Rheology and Packing of Dendronized Polymers. *Macromolecules* **2016**, *49*, 7054–7068. [CrossRef]
10. Turton, R.; Bailie, R.; Whiting, W.; Shaeiwtz, J. *Analysis, Synthesis, and Design of Chemical Processes*; Prentice Hall: Englewood Cliffs, NJ, USA, 2008.
11. Busico, V.; Cipullo, R.; Mingione, A.; Rongo, L. Accelerating the Research Approach to Ziegler-Natta Catalysts. *Ind. Eng. Chem. Res.* **2016**, *55*, 2686–2695. [CrossRef]

12. Vittoria, A.; Meppelder, A.; Friederichs, N.; Busico, V.; Cipullo, R. Demystifying Ziegler-Natta Catalysts: The Origin of Stereoselectivity. *ACS Catal.* **2017**, *7*, 4509–4518. [CrossRef]

13. Macosko, C. *Rheology: Principles, Methods and Application*; Wiley-VCH: New York, NY, USA, 1994.

14. Ewoldt, R.; Johnston, M.; Caretta, L. Experimental challenges in shear rheology: How to avoid bad data. In *Complex Fluids in Biological Systems*; Springer: New York, NY, USA, 2015; pp. 207–241.

15. Ewoldt, R.H.; Winegard, T.M.; Fudge, D.S. Non-linear viscoelasticity of hagfish slime. *Int. J. Non-Linear Mech.* **2011**, *46*, 627–636. [CrossRef]

16. Griffiths, D.F.; Walters, K. On edge effects in rheometry. *J. Fluid Mech.* **1970**, *42*, 379–399. [CrossRef]

17. Laun, H.; Meissner, J. A Sandwich-Type Creep Rheometer for the measurement of rheological properties of polymer melts at low shear stresses. *Rheol. Acta* **1980**, *19*, 60–67. [CrossRef]

18. Sharma, V.; Jaishankar, A.; Wang, Y.-C.; McKinley, G.H. Rheology of globular proteins: Apparent yield stress, high shear rate viscosity and interfacial viscoelasticity of bovine serum albumin solutions. *Soft Matter* **2011**, *7*, 5150–5160. [CrossRef]

19. Davies, G.A.; Stokes, J.R. On the gap error in parallel plate rheometry that arises from the presence of air when zeroing the gap. *J. Rheol.* **2005**, *49*, 919–922. [CrossRef]

20. Kravchuk, O.; Stokes, J.R. Review of algorithms for estimating the gap error correction in narrow gap parallel plate rheology. *J. Rheol.* **2013**, *57*, 365–375. [CrossRef]

21. Costanzo, S.; Ianniruberto, G.; Marrucci, G.; Vlassopoulos, D. Measuring and assessing first and second normal stress differences of polymeric fluids with a modular cone-partitioned plate geometry. *Rheol. Acta* **2018**, *57*, 363–376. [CrossRef]

22. Meissner, J.; Garbella, R.W.; Hostettler, J. Measuring Normal Stress Differences in Polymer Melt Shear Flow. *J. Rheol.* **1989**, *33*, 843–864. [CrossRef]

23. Schweizer, T.; Schmidheiny, W. A cone-partitioned plate rheometer cell with three partitions (CPP3) to determine shear stress and both normal stress differences for small quantities of polymeric fluids. *J. Rheol.* **2013**, *57*, 841–856. [CrossRef]

24. Snijkers, F.; Vlassopoulos, D. Cone-partitioned-plate geometry for the ARES rheometer with temperature control. *J. Rheol.* **2011**, *55*, 1167–1186. [CrossRef]

25. Honerkamp, J.; Weese, J. A nonlinear regularization method for the calculation of relaxation spectra. *Rheol. Acta* **1993**, *32*, 65–73. [CrossRef]

26. Costanzo, S.; Huang, Q.; Ianniruberto, G.; Marrucci, G.; Hassager, O.; Vlassopoulos, D. Shear and Extensional Rheology of Polystyrene Melts and Solutions with the Same Number of Entanglements. *Macromolecules* **2016**, *49*, 3925–3935. [CrossRef]

27. Yan, Z.-C.; Costanzo, S.; Jeong, Y.; Chang, T.; Vlassopoulos, D. Linear and Nonlinear Shear Rheology of a Marginally Entangled Ring Polymer. *Macromolecules* **2016**, *49*, 1444–1453. [CrossRef]

28. Franck, A. *Understanding Instrument Compliance Correction in Oscillation*; TA Instruments: New Castle, DE, USA, 2006.

29. Laukkanen, O.-V. Small-diameter parallel plate rheometry: A simple technique for measuring rheological properties of glass-forming liquids in shear. *Rheol. Acta* **2017**, *56*, 661–671. [CrossRef]

30. Gottlieb, M.; Macosko, C.W. The effect of instrument compliance on dynamic rheological measurements. *Rheol. Acta* **1982**, *21*, 90–94. [CrossRef]

31. Ferry, J.D. *Viscoelastic Properties of Polymers*, 3rd ed.; Wiley: New York, NY, USA, 1980; ISBN 978-0-471-04894-7.

32. Dealy, J.; Plazek, D. Time—Temperature Superposition—A Users Guide. *Rheol. Bull.* **2009**, *78*, 16.

33. Van Ruymbeke, E.; Kapnistos, M.; Vlassopoulos, D.; Huang, T.; Knauss, D.M. Linear Melt Rheology of Pom-Pom Polystyrenes with Unentangled Branches. *Macromolecules* **2007**, *40*, 1713–1719. [CrossRef]

34. Farrington, P.J.; Hawker, C.J.; Fréchet, J.M.J.; Mackay, M.E. The Melt Viscosity of Dendritic Poly(benzyl ether) Macromolecules. *Macromolecules* **1998**, *31*, 5043–5050. [CrossRef]

35. Zoller, P.; Walsh, D. *Standard Pressure-Volume-Temperature Data for Polymers*; Technomic Publishing Company: Lancaster, PA, USA, 1995.

36. Velankar, S.S.; Giles, D. How do I know my phase angles are correct? *Rheol. Bull.* **2007**, *76*, 8.

37. Peterson, J.D.; Vyazovkin, S.; Wight, C.A. Kinetics of the Thermal and Thermo-Oxidative Degradation of Polystyrene, Polyethylene and Poly(propylene). *Macromol. Chem. Phys.* **2001**, *202*, 775–784. [CrossRef]

38. Cox, W.P.; Merz, E.H. Correlation of dynamic and steady flow viscosities. *J. Polym. Sci.* **1958**, *28*, 619–622. [CrossRef]

39. Gleissle, W. Two simple time-shear rate relations combining viscosity and first normal stress coefficient in the linear and non-linear flow range. In *Rheology*; Plenum: New York, NY, USA, 1980; pp. 457–462.

40. Ebrahimi, M.; Ansari, M.; Hatzikiriakos, S.G. Wall slip of polydisperse linear polymers using double reptation. *J. Rheol.* **2015**, *59*, 885–901. [CrossRef]

41. Henson, D.J.; Mackay, M.E. Effect of gap on the viscosity of monodisperse polystyrene melts: Slip effects. *J. Rheol.* **1995**, *39*, 359–373. [CrossRef]

![fluids logo] *fluids*

MDPI

Article

Saffman–Taylor Instability in Yield Stress Fluids: Theory–Experiment Comparison

Oumar Abdoulaye Fadoul and Philippe Coussot *

University of Paris-Est, Laboratoire Navier (ENPC-IFSTTAR-CNRS), 77420 Champs sur Marne, France;
fadoul.oumar-abdoulaye@ifsttar.fr
* Correspondence: philippe.coussot@ifsttar.fr

Received: 12 February 2019; Accepted: 12 March 2019; Published: 16 March 2019

Abstract: The Saffman–Taylor instability for yield stress fluids appears in various situations where two solid surfaces initially separated by such a material (paint, puree, concrete, yoghurt, glue, etc.) are moved away from each other. The theoretical treatment of this instability predicts fingering with a finite wavelength at vanishing velocity, and deposited materials behind the front advance, but the validity of this theory has been only partially tested so far. Here, after reviewing the basic results in that field, we propose a new series of experiments in traction to test the ability of this basic theory to predict data. We carried out tests with different initial volumes, distances and yield stresses of materials. It appears that the validity of the proposed instability criterion cannot really be tested under such experimental conditions, but at least we show that it effectively predicts the instability when it is observed. Furthermore, in agreement with the theoretical prediction for the finger size, a master curve is obtained when plotting the finger number as a function of the yield stress times the sample volume divided by the square initial thickness, in wide ranges of these parameters. This in particular shows that this traction test could be used for the estimation of the material yield stress.

Keywords: Saffman–Taylor instability; yield stress fluid; traction test

---

1. Introduction

The Saffman–Taylor instability (STI) is observed when a fluid pushes a more viscous fluid in a confined geometry. The term confined here means that the distance between the solid walls is much smaller than the characteristic length in the flow direction. Such boundary conditions are typically encountered in porous media or between two parallel plates (i.e., Hele–Shaw cell). Under so-called "stable conditions", the length of the interface between the two fluids remains minimal, so that it is straight for a flow in a single direction, or circular for a radial flow. When the STI develops, the interface evolves in the form of fingers. For viscous fluids, the origin of the instability is as follows: if the pressure along the interface is uniform, any perturbation or unevenness (local curvature) of the interface tends to develop further; this is so because the viscous fluid tends to advance faster in front of a curvature in the flow direction as the fluid volume to be pushed is smaller. The development of this perturbation may only be damped if surface tension, which, on the contrary, works against the deformation of the initial interface, and is sufficient to counterbalance the above viscous effect. This instability has been widely studied for simple fluids [1–3].

Experiments with radial Hele–Shaw cells using non-Newtonian fluids have shown striking qualitative differences in the fingering pattern (see e.g., [4,5]). It was discovered that, when the high-viscosity fluid is viscoelastic, the interface grows along a narrow and very tortuous finger leading to branched, fractal patterns [6]. It was also shown that this viscous fingering pattern can be replaced by a viscoelastic fracture pattern for appropriate Deborah numbers [7,8]. On the theoretical side, the treatment of the Saffman–Taylor instability problem was revisited for viscoelastic or shear-thinning fluids. Wilson [9] considered an Oldroyd-B fluid that exhibits elasticity and the case of power-law fluids was treated by Wilson [9] for unidirectional flows and by Sader et al. [10] and Kondic et al. [11] for radial flows. However, except in the case of fluids with a negative viscosity for which slip layers may form [11] or for strongly viscoelastic fluids [8], the corresponding theoretical results did not show strong changes in the basic process of instability as it appears for Newtonian fluids. For viscoelastic fluids, Wilson [9] found a kind of resonance that can produce sharply increasing (in fact unbounded) growth rates as the relaxation time of the fluid increases. Sader et al. [10] mainly showed that decreasing the power-law index dramatically increases the growth rates of perturbation at the interface and provides effective length compression for the formation of viscous-fingering patterns, thus enabling them to develop much more rapidly. For non-elastic weak shear-thinning fluids, Lindner et al. [12] showed that, during the evolution of the Saffman–Taylor instability in a rectangular Hele–Shaw cell, the width of the fingers as a function of the capillary number collapse onto the universal curve for Newtonian fluids, provided the shear-thinning viscosity is used to calculate the capillary number. For stronger shear-thinning, narrower fingers are found. Further observations on shear-thinning elastic materials were provided by Lindner et al. [13].

As far as we know, the theoretical description of STI with yield stress fluids (YSF) can flow only beyond a critical stress; otherwise, they behave as solids [14], starting with the work of Coussot [15], for both longitudinal and radial flows in Hele–Shaw cells. This approach is based on the use of an approximate Darcy's law for yield-stress fluids, which leads to a dispersion equation for both flow types similar to equations obtained for ordinary viscous fluids, except that now the viscous terms in the dimensionless numbers conditioning the instability contain the yield stress. As a consequence, the wavelength of maximum growth can be extremely small even at vanishing velocities, so that the STI can still exist and we have an original situation: a "hydrodynamic" instability at vanishing velocity. Another original aspect of this instability for YSF is that, at a sufficiently low flow rate, the fingering process leaves arrested fluid volumes behind the advancing front [15]. Miranda [16] presented a theoretical analysis that goes beyond the above theory by using a mode-coupling approach to examine the morphological features of the fluid–fluid interface at the onset of nonlinearity, and finally proposed mechanisms for explaining the rising of tip-splitting and side-branching events. However, this approach relies on a Darcy-law-like equation valid in the regime of high viscosity compared to yield stress effects, which is precisely not the scope of the present paper. On the contrary, as we are interested in the specific effect of yielding, we focus on situations for which there is a major impact of the yield stress. On another side, a numerical approach was also developed to study the standard problem of penetration of a finger in a Hele–Shaw cell (for Newtonian fluids a stationary finger forms), first for a simple YSF [17], and then for a thixotropic fluid [18].

Experimentally, the SFI instability of YSF has been studied in a rectangular Hele–Shaw cell with Carbopol gels [19,20]. This relies on the injection of air at a given point in the middle of the cell, which then propagates through the fluid. For a Newtonian viscous fluid, when the instability criterion is fulfilled, some fingers develop in the cell, but, after some distance, one finger becomes dominant while the others stop and this single finger advances steadily along the main cell direction, with a size equal to half the cell width. The result with a YSF is strongly different: at some time, there can be one finger, but with a size possibly much smaller than the cell width. This finger, however, will soon destabilize in secondary fingers, which are finally stopped, leaving again one finger and so on. A comparison with theory is hardly possible in this context, but the details of the evolution and the different regimes have been described [20]. Similar approaches were also developed for thixotropic YSF [21], which obviously gives rise to effects more complex to predict due to the time-dependency of the fluid behavior.

There is a situation in which the STI of YSF is currently observed: the separation of two plates initially in contact with a thin layer of YSF; as the plates are moved away, the layer thickness increases, which induces a radial flow towards some central position; if the distance between the plates is sufficiently small, the radial velocity is much larger than the axial one, so that the flow approximately corresponds to a radial flow driven by the air entering the gap, which corresponds to the conditions under which the STI can be considered. This is the most frequent situation under which the STI for YSF can be observed in our everyday life: as soon as some thin layer of paint, glue, puree, or yoghurt is squeezed between two solid surfaces (a tool, a spoon, etc.) are then separated, one observes a characteristic fingering shape. Note that it is possible to observe such pictures because the fluid leaves arrested regions behind the flow front, which finally give this definitive shape. This contrasts with simple liquids for which the fingers soon relax under the action of wetting effects and a uniform layer rapidly reforms.

Finally, most of the theory–experiment comparisons concern the observations from traction tests. In that case, a reasonable agreement between the fingering wavelength and the theoretical predictions was found [19,22], but this was done in relatively narrow range of parameters, as essentially the gap was varied. In addition, somewhat problematic are the observations of Barral et al. [23], which showed that there is a strong discrepancy between the theoretical conditions and the experimental data concerning the onset of the instability. The problem is that this appears to be the only experimental approach of the onset of this instability with YSF, and it is in complete disagreement with existing theory, which might suggest that something is missing in the theory.

Our present objective is to attempt to clarify the situation through new experiments and further discussion of the experimental criterion of instability and the fingering wavelength. We rely on new systematic traction tests under different conditions (fluid volume, initial aspect ratio, interaction with the solid surface) and an analysis of these data with a critical eye, allowing for reaching some clearer conclusions about the validity of the theory.

2. Materials and Methods

2.1. Materials

We used oil-in-water (direct) emulsions made of silicone oil (viscosity 0.35 Pa·s) as dispersed phase, and a continuous phase (viscosity 5 mPa·s) made of distilled water and 3 wt% myristyltrimethylammonium bromide (TTAB, Sigma-Aldrich, St. Louis, MO, USA). A Silverson mixer (model L4RT), equipped with a rotating steel blade inside a punched steel cylinder, was also used as an emulsifier. During the preparation, the fluids are sheared and the oil phase is broken into small droplets while the water or water/glycerol phase fill the surrounding environment, and the interface is stabilized by surfactants (TTAB). The rotation velocity of the mixer is progressively increased to reach the maximum rate of 6000 rpm. A part of the bubbles incorporated in the mixture during this process can be removed by tapping the container, and the rest of the bubbles are removed by centrifugation. The droplet size is approximately uniform, around 5 microns. Different emulsions with different oil volume fractions were prepared. The resulting yield stress of the emulsions prepared at 76%, 78%, 82% and 84% (volume concentration of oil) was, respectively, 20 Pa, 30 Pa, 40 Pa and 50 Pa, within 1 Pa.

We also used a Carbopol (U980) gel. It has been observed that this material is essentially a glass made of a high concentration of individual, elastic sponges (with a typical element size of 2 μm to 20 μm) [24], which gives rise to its yielding behavior. The preparation of Carbopol gel begins with the introduction of some water in a mortar mixer. The rotation velocity is set at 90 rpm and the appropriate amount of raw Carbopol powder (1 wt%) is slowly added to the stirring water. After about one hour, the incorporation of the powder is done and the appropriate amount of Sodium Hydroxide (1 mol/L) is quickly added to the solution, which increases its pH. The mixing is then maintained for approximately one day to allow a full homogenization of the mixture.

2.2. Rheological Characterization

Rheological tests were performed with a Kinexus Malvern-stress-controlled rheometer equipped with two circular, rough plates (diameter: 40 mm). The sample was carefully set up and the gap was fixed at 2 mm taking care not to entrap air bubbles. A logarithmically increasing and then decreasing stress ramp test was then applied over a total time of four minutes. Except for the first part of the increasing curve associated with deformations in the solid regime, the increasing and decreasing shear stress vs. shear rate curves superimpose. We retain, here, the decreasing part as the flow curve of the material. For similar emulsions, it has been shown that this apparent flow curve obtained from macroscopic observations correspond to the effective, local constitutive equation observed at a local scale with imaging technique [25]. The emulsions and the Carbopol gel exhibit a simple yield stress fluid behaviour and their flow curve can be well fitted by a Herschel–Bulkley (HB) model (see typical results in Figure 1):

$$\tau > \tau_c \Rightarrow \tau = \tau_c + k\dot{\gamma}^n, \tag{1}$$

in which τ is the shear stress, $\dot{\gamma} > 0$ the shear rate, τ_c the yield stress, k the consistency factor and n the power-law exponent.

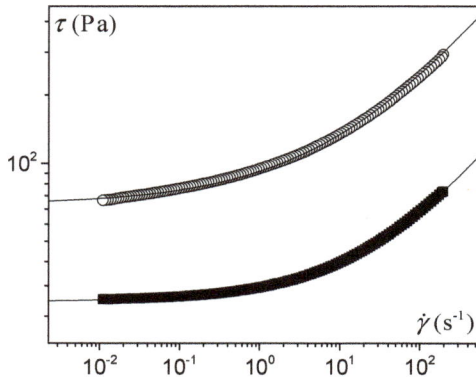

Figure 1. Typical flow curve of a one of our emulsion and of the Carbopol gel. The continuous lines are the Herschel–Bulkley model fitted to data with the parameters: (emulsion emulsion (78%)) $\tau_c = 30$ Pa, $k = 4.5$ Pa·s^n and $n = 0.45$; (Carbopol) $\tau_c = 70$ Pa, $k = 23.5$ Pa·s^n and $n = 0.4$.

2.3. Set up for Traction Tests

For the adhesion tests, a dual-column testing system (*Instron* model 3365, Instron, Norwood, MA, USA) with a position resolution of 0.118 µm was used. The column was equipped with either a 10 or 500 N static load cell, which were able to measure the force to within a relative value of $\pm 10^{-6}$ of the maximum value. Waterproof sandpaper (average particle diameter 82 µm, a dimension much larger than the typical droplet size) was attached to the top and bottom plates. Since the volume loss in the roughness could be significant in some cases, a generous amount of extra sample was applied to the surface of the sandpaper before each test and the excess removed by scraping the surface with a palette knife. This also ensured reproducible wetting conditions of the fluid onto the solid surface. However, qualitatively similar results were obtained with initially dry or wet surfaces. Between two successive tests, both plates were removed and cleaned. A given volume (Ω_0) of material was then collected with a syringe, put at the center of the bottom plate, and the upper plate was decreased at a fixed (initial) height (h_0), thus squeezing the material. The adhesion test then consisted of lifting the upper plate at a constant velocity (0.01 mm/s) while monitoring the force (F) applied to the upper plate. The initial distance was varied between 0.2 mm and 5 mm. The initial volume was varied between 0.3 mL and 3 mL.

3. Theoretical

3.1. Instability in a Straight Hele–Shaw Flow

The instability of radial flows of Newtonian fluids in Hele–Shaw cells has been studied [3,26,27] by using the vectorial form of Darcy's law. The treatment below summarizes the assumptions and results of Coussot [15], whose approach has some similarity with the one adopted by Wilson [9] or Sader et al. [10] who considered power-law fluids and could not directly use the standard (Newtonian) Darcy's law.

We consider a yield stress fluid pushed by an inviscid fluid (say, air) so that it tends to flow in a given direction x between two parallel plates separated by a distance $h = 2b$, with a mean fluid velocity U. The initial interface is assumed to be uniform and straight (along the z direction). A stable flow corresponds to a fluid motion along the x direction, uniform along the z direction. For an unstable flow, this interface does not remain straight (see Figure 2).

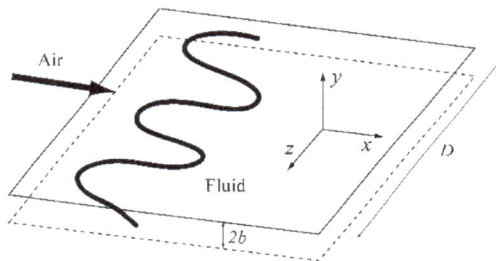

Figure 2. Scheme of main flow characteristics in a Hele–Shaw cell for a destabilized interface (thick line).

The linear stability analysis of this flow [15] relies on several assumptions: (i) the constitutive equation of the fluid can be well represented by a HB behaviour; (ii) the lubrication assumption is valid, i.e., the velocity component perpendicular to the cell plan can be neglected; and (iii) the shear stress at the wall, even around the front of the flow, can be approximated by a value close to the exact one for a stable uniform flow through this cell (see [15]):

$$\tau_w = \tau_c \left[1 + c \left(\frac{kU^n}{\tau_c b^n} \right)^d \right],$$ (2)

where c and d two parameters which depend on n. For example, for $n = 1/3$, $c = 1.93$, $d = 0.9$ [15].

Under these conditions, the linear stability analysis of the flow, for negligible gravity effects, predicts that the unidirectional flow above described is fundamentally unstable as soon as the inviscid fluid pushes the yield stress fluid. Moreover, the wavelength of maximum growth is

$$\lambda_m = 2\pi \sqrt{\frac{3\sigma b}{\tau_w}}$$ (3)

in which σ is the surface tension. Note that the Newtonian case is recovered from this approach: by using in Equation (3) the wall stress expression for a stable uniform flow of a Newtonian fluid in such a cell, i.e., $\tau_w = 3\mu U / b$, we find $\lambda_m = 2\pi b \sqrt{\sigma / \mu U}$, which is the standard expression found from a complete theoretical analysis in the Newtonian case [2].

From Equation (3), we also deduce that the instability will be apparent only if $\lambda_m < D$, where D is the width of the flow. This implies that the flow will be apparently unstable if

$$\tau_w > \frac{12\pi^2 \sigma b}{D^2}.$$ (4)

Finally, note that, for a yield stress fluid, τ_w tends to τ_c when $U \to 0$ or, more precisely, $\tau_w \approx \tau_c$ when $kU^n/\tau_c b^n \ll 1$. Thus, at vanishing velocity, the wavelength tends to a finite value, i.e., $\lambda_m = 2\pi\sqrt{3\sigma b/\tau_c}$. This strongly contrasts with the result of the Saffman–Taylor instability for simple fluids (i.e., without yield stress) for which the wavelength tends to infinity when the velocity tends to zero. Thus, for yield stress fluid, if the front width is sufficiently large, we will see the development of a hydrodynamic instability at vanishing velocity. Note that, more precisely, due to the square root of the stress in the wavelength expression, the approximation above leading to neglect the flow rate dependent term in the stress expression, leads to an approximation to within 10% of the exact value of the wavelength if $kU^n/\tau_c b^n$ is smaller than 0.2.

Moreover, in the case of small front velocity, the stress should slightly overcome the yield stress in the regions with highest velocities and, as a consequence, intuitively, the stress might be smaller than the yield stress in regions with lowest velocities (see further demonstration in [15]). As a consequence, the regions left behind should remain static just after the beginning of the unstable process. As long as the fingers grow, the pressure drop applied to these regions therefore decreases so that they should remain static even after a long time.

3.2. Instability in a Radial Hele–Shaw Flow

We consider now the case of a radial flow, with an inviscid fluid pushing the yield stress fluid towards the center. This assumes that, if the plates remain at the same distance, the YSF for example escapes through a central hole. Using again expression (2) for the wall shear stress (which neglects orthoradial components), a linear stability analysis [15] leads to

$$\lambda_m = 2\pi R \left(\frac{3\sigma b}{\sigma b + \tau_w R^2} \right)^{1/2}, \tag{5}$$

in which R is the radius of the circular interface. Once again, this expression allows for recovering the Newtonian case, $\lambda_m = 2\pi R/\sqrt{\mu U R^2/\sigma b^2 + 1/3}$ [3], by introducing in Label (5) the expression for the wall shear stress of the stable, and the uniform flow of a Newtonian fluid (see above).

Finally, for a YSF, the criterion for the apparent onset of instability ($\lambda_m < 2\pi R$) is:

$$\tau_w > \frac{2\sigma b}{R^2}. \tag{6}$$

The above remarks concerning the finite wavelength at vanishing velocity and the tracks left behind still apply in this case.

3.3. Flow Induced by a Traction Test

We now consider the flow induced by a traction test, in which the material initially forming a cylindrical layer situated between two plates, is then deformed as a result of the relative motion of the two plates away from each other along their common axis. As the distance between the plates increases, since the material remains in contact with the plate, the thickness of the sample increases. As a result, the material tends to gather towards its central axis. Let us consider the ideal case where the sample shape remains cylindrical during this process, i.e., the flow is stable and we neglect the deposits of material along its motion along the plates. In that case, as a result of mass conservation, the mean radial velocity (U) is related to the velocity of separation of the plates (V) through

$$U = \frac{R}{4b} V. \tag{7}$$

From Label (7), we see that, as soon as the aspect ratio (i.e., $R/2b$) of the sample is sufficiently large, the radial velocity is much larger than the separation velocity. In that case, the lubrication assumption, i.e., the velocity components parallel to the plates are much larger than the perpendicular ones, is relevant, and we can consider that the flow is similar to that resulting from a pure radial flow between plates at a fixed distance from each other. Obviously, this assumption will start to fail at some point during the process, as the aspect ratio progressively decreases toward smaller values when the plates are moved away from each other. In the following, we will a priori assume that the lubrication assumption is valid, and discuss its possible non-validity as an artefact of the tests.

On the other side, for such a traction test, we can easily estimate the normal force needed to separate the plates under the lubrication assumption for stable and sufficiently slow flows (i.e., $kU^n/\tau_c b^n << 1$) [28]. In that case, the radial flow along the plate induces a shear stress equal to the material yield stress. The momentum balance applied to the sample volume between R and r assuming no surface tension effect and negligible atmospheric pressure leads to:

$$p(r) = \frac{\tau_c}{b}(r - R).$$

(8)

The net normal force exerted onto the plate in that case is then found by integrating the pressure (8) over the surface of contact:

$$F = \frac{2\pi \tau_c R^3}{3b}.$$

(9)

Equation (9) thus provides an expression for the force applied in the case of slow flows. Since the assumed constitutive equation is continuous, i.e., it predicts a continuous transition from rest to slow flows around the yield stress, Equation (9) also provides an expression for the minimum force to induce some motion for a given separation distance and a given radius.

Note that, for a given volume of material ($\Omega_0 = 2\pi R^2 b$), this force may be rewritten as

$$F = \frac{4\tau_c \Omega_0^{3/2}}{3\sqrt{\pi}h^{5/2}},$$

(10)

which gives the force variation as a function of the distance ($h = 2b$) between the plates.

4. Results and Discussion

4.1. General Trends

The typical result of a traction test is the formation of an approximately symmetrical deposit over each solid surface, associated with a tendency to a gathering of the material towards the central axis, as proved by the larger thickness of material towards the central part. Depending on the experimental conditions, the final deposit has different aspects, from a simple conical shape to a fingered structure (see [23]). Since the initial shape is cylindrical, a stable flow would maintain a cylindrical interface. As a consequence, a final fingered structure (see Figure 3) is the hallmark of flow instability.

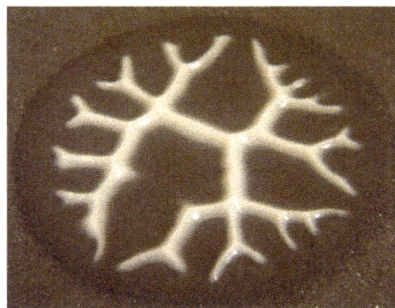

Figure 3. Fingering aspect of the remaining deposit of yield stress fluid after separation of two plates initially squeezing a thin fluid layer. (Photo Q. Barral).

4.2. Force vs. Distance

The force during such a test strongly decreases with the distance and approximately follows a slope of −2.5 in logarithmic scale (see Figure 4), which tends to confirm the validity of Equation (10). However, we can remark that the force curve is shifted towards smaller values when the initial distance is smaller, in contradiction with the above theory since expression (10) only depends on the sample volume and the current distance. This is explained by the development of fingering, which implies that some significant parts of the material do not flow anymore in the radial direction.

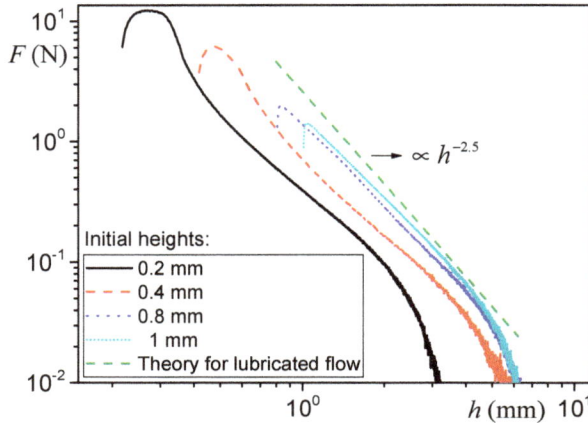

Figure 4. Force vs. distance during a traction experiment for an emulsion (82%) (yield stress of 40 Pa) for different initial aspect ratios (corresponding to first point of curves on the left) (Ω = 3 mL). The dashed line is the lubrication model (see text).

Let us try to take this phenomenon into account. We assume that, during the withdrawal of dR, the fingers well develop so that half the material is left behind as deposited material, while the central flowing region is "plain", with a current volume Ω. Thus, we have a variation of the current volume of material still in the flowing region as $d\Omega = \pi R h dR$, since, by definition of this volume: $\Omega = \pi R^2 h$, we deduce $d\Omega/\Omega = dR/R$ and by integration $\Omega = \Omega_0 R/R_0 = \Omega_0^2/\pi h R_0^2$. We finally find $F_c = \frac{2\tau_c \Omega_0^{3/2}}{3\pi^{1/2}} h_0^{3/2} h^{-4}$. Although this expression now effectively predicts a decrease of the force with the initial distance, it also predicts a decrease of this force with the current distance as a power −4, in contradiction with the data. Thus, we can conclude that, although we are able to reproduce some qualitative trends through different approaches, we still lack a full theory for describing the force evolution with distance when fingering develops significantly.

4.3. Characteristics of the Instability

4.3.1. Instability Criterion

We now discuss the characteristics of this instability. In order to better discuss the origin of the evolution of the final shape of the deposit with the material and process parameters, we consider the theoretical prediction of the instability criterion (i.e., Equation (6)) under negligible "additional viscous effects". Note that we checked that for all the tests with the emulsions $kU^n/\tau_c b^n$ was smaller than 0.2, which means that the above simplified expression for the finger width is relevant. This was not the case for the Carbopol gel, for which $kU^n/\tau_c b^n$ was as large as 0.5 at the beginning of the test in some cases, but, in the following, we neglect this aspect and it appears that this does not affect the consistency of our results and analysis. In the instability criterion (Equation (6)), we can then use the approximation $\tau_w \approx \tau_c$, so that this criterion may be rewritten as $\tau_c \Omega_0/h^2 > 2\pi\sigma$. From this expression, we see that, if it is to occur, the instability will occur at the beginning of the withdrawal, when the height is the smallest. As a consequence, the instability criterion writes: $X = \tau_c \Omega_0 / h_0^2 > 2\pi\sigma$. Under these conditions, we can expect that the instability will be "more developed" for increasing values of X.

In Figure 5, we show the different final shapes observed for different values of X as a function of the initial distance between plates. We see that h_0 does not determine solely the intensity of the instability: various deposit aspects are found for a given h_0 value. On the contrary, as expected from the theory, the aspect of the deposits seems to be close for a given value of X: we get approximately similar branched structures along each horizontal line in this representation (see Figure 5). Then, we can determine the limit between the unstable and stable regimes, by considering that the absence of apparent fingering is the hallmark of stable flows. Note that, for small values of h_0, this is an extrapolation, since we were unable to get experimental data in this region of the graph as it required too small sample volumes. We thus find that this limit corresponds to $X \approx 10$ Pa·m. (Note that a similar approach from the data of Barral et al. [15] would lead to $X \approx 80$ Pa·m.) On the other side, using for the surface tension the value of the interstitial liquid (water) [29], i.e., $\sigma = 0.07$ Pa·m, the left hand-side of the instability criterion (6), i.e., $2\pi\sigma$ is equal to 0.4 Pa·m. Thus, we find that experiments give stable flows in a wide range where unstable flows are expected from theory, namely between say $X \approx 10$ Pa·m and $X \approx 0.4$ Pa·m.

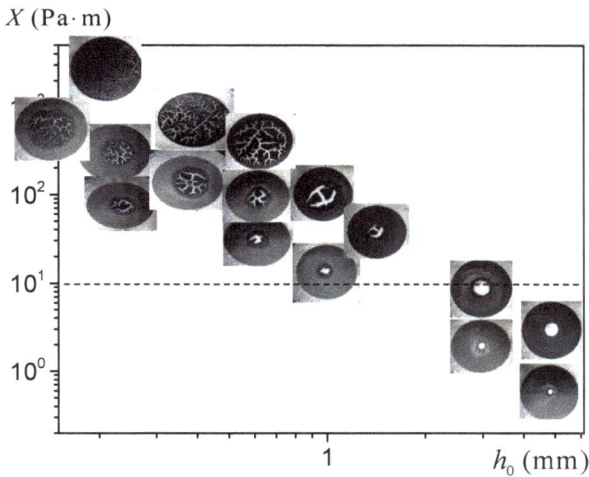

Figure 5. Final aspects of the deposit after plate separation as a function of the parameter X and the initial thickness, for a 40 Pa yield stress emulsion and different sample volumes (the volume corresponding to each picture may be estimated from the value of X and h_0 through $\Omega_0 = X h_0^2 / \tau_c$). The diameter of the dark disk in all the photos is 10 cm.

At first glance, this result may be seen as a strong discrepancy between experiments and theory. Actually, this is not so obvious. Indeed, looking at the flows considered as "stable", we see that they correspond to a rather limited radial flow. Under such conditions, the STI, or more precisely fingering, would simply not have enough time or distance to develop. More precisely, this is the validity of the lubrication assumption, which should be discussed, as we expect that, if this assumption is valid, during the traction test the distance will now increase to large values, i.e., at least of the same order as the initial radius, which implies a significant radial flow. If we compute the ratio h_0/R_0, we find that the stable flows correspond exactly to those for which h_0/R_0 is larger than 0.1. This suggests that here we in fact find stable flows because the lubrication assumption is not valid. In that case, we have indeed a more complex flow than assumed in the theory, in particular there is now likely a significant component of elongation along the vertical axis, which might dampen the instability. In fact, looking further at the initial and final sample shape, we see that the material essentially transforms from a circular disk layer to a cone of same basis, and does not significantly flow radially. Actually, in similar experiments carried out with smooth plates, with roughness of less than one nanometer, no instability at all is observed in a wide range of initial distances [30]. In that case, we have a pure elongation flow along most of the flow. This further confirms the above suggestion that, when the elongational component is significant, no instability can be expected.

This suggests that we cannot really test the validity of the instability criterion because well before the range for which stability is expected, the lubrication assumption is not valid. We can just say that, under the proper assumptions, i.e., lubrication assumption, the flow is unstable, in agreement with the theoretical criterion.

4.3.2. Fingering Wavelength

Let us now try another approach to test the theory, by looking at the wavelength of the fingering. With that aim, we need to look at the fingering characteristics at the onset of the instability, since this is the only aspect relevant within the frame of a linear stability analysis, which considers the flow evolution for a slight perturbation of the initial stable flow. We assume that the corresponding wavelength corresponds to the fingers apparent at the periphery of the initial sample layer and we simply count the number of deposited fingers around the sample, which corresponds to $N/2$. The theoretical prediction for the finger number N at the onset of the instability, i.e., $2\pi R/\lambda_m$, as deduced from Label (5) is:

$$N = \frac{1}{\sqrt{3}} \left(1 + \frac{2Y}{\pi} \right)^{1/2} \tag{11}$$

in which $Y = X/\sigma$.

All the data are shown in Figure 6, where we can see that globally the theoretical prediction is in agreement with the experiments: all the data for the different fluids, different initial distances, and different sample volumes, fall along a master curve which corresponds to an increase of the finger number following on average the theoretical curve. Nevertheless, we can note some discrepancy: at low Y values, the finger number is systematically below (by a factor about 1.5) the theoretical prediction, which suggests that, in that case, the elongational component of the flow plays some significant role and slightly dampens the instability; at large Y, the finger number is slightly above the theoretical prediction, but we have no explanation for that.

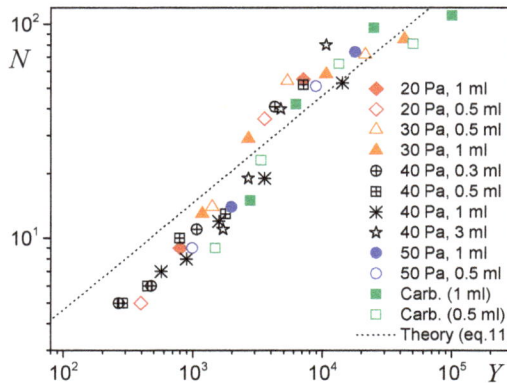

Figure 6. Saffman instability during traction tests with emulsions of various yield stresses, for various sample volumes and various initial thicknesses: Number of fingers as a function of the parameter Y. Caption: emulsions, except when mentioned (Carbopol).

5. Conclusions

We have shown that it is in fact not possible to test the theoretical criterion for the instability onset of yield stress fluids in traction tests: as soon as the radial flow induced is sufficiently developed, an unstable flow is expected and effectively observed. On the contrary, in agreement with the theoretical prediction for the finger size appears, i.e., a master curve is obtained when plotting N as a function of X for the different volumes, yield stresses and initial thicknesses in rather large ranges. However, there remains a discrepancy between theory and experiments: the finger number is smaller by a factor about 1.5 for $X < 25$ Pa·m, and larger by a factor about 1.5 for $X > 25$ Pa·m. As a consequence, the basic theory can be used to estimate and predict the fingering aspect for any application, as soon as one knows the material yield stress, sample volume and initial thickness. On the other side, this traction test could be used for the estimation of the material yield stress, from an analysis of the fingering characteristics. Further work in that field could focus on the exact flow characteristics during such a traction test, in particular when the instability has begun.

Author Contributions: Conceptualization, P.C.; data curation, investigation and visualization, O.A.F.; formal analysis, O.A.F. and P.C.; writing P.C.

Funding: This research was partly funded by BID (Islamic Development Bank).

Conflicts of Interest: The authors declare no conflict of interest.

References

1. Saffman, P.G.; Taylor, G. The penetration of a fluid into a porous medium or Hele-Shaw cell containing a more viscous liquid. *Proc. R. Soc. Lond.* **1958**, *A245*, 312–329.
2. Homsy, G.M. Viscous fingering in porous media. *Ann. Rev. Fluid Mech.* **1987**, *19*, 271–311. [CrossRef]
3. Paterson, L. Radial fingering in a Hele-Shaw cell. *J. Fluid Mech.* **1981**, *113*, 513–529. [CrossRef]
4. Van Damme, H.; Lemaire, E.; Abdelhaye, O.M.; Mourchid, A.; Levitz, P. Pattern formation in particulate complex fluids: A guided tour. In *Non-Linearity and Breakdown in Soft Condensed Matter*; Bardan, K.K., Chakrabarti, B.K., Hansen, A., Eds.; Lecture Notes in Physics; Springer: Berlin/Heidelberg, Germany, 1994; Volume 437.
5. McCloud, K.V.; Maher, J.V. Experimental perturbations to Saffman-Taylor flow. *Phys. Rep.* **1995**, *260*, 139–185. [CrossRef]
6. Nittmann, J.; Daccord, G.; Stanley, E. Fractal growth of viscous fingers: Quantitative characterization of a fluid instability phenomenon. *Nature* **1985**, *314*, 141–144. [CrossRef]
7. Lemaire, E.; Levitz, P.; Daccord, G.; Van Damme, H. From viscous fingering to viscoelatic fracturing in colloidal fluids. *Phys. Rev. Lett.* **1991**, *67*, 2009–2012. [CrossRef]
8. Foyart, G.; Ramos, L.; Mora, S.; Ligoure, C. The fingering to fracturing transition in a transient gel. *Soft Matter* **2013**, *32*, 7775–7779. [CrossRef]
9. Wilson, S.D.R. The Saffman-Taylor problem for a non-Newtonian liquid. *J. Fluid Mech.* **1990**, *220*, 413–425. [CrossRef]
10. Sader, J.E.; Chan, D.Y.C.; Hughes, B.D. Non-Newtonian effects on immiscible viscous fingering in a radial Hele-Shaw cell. *Phys. Rev. E* **1994**, *49*, 420–432. [CrossRef]
11. Kondic, L.; Palffy-Muhoray, P.; Shelley, M.J. Models for non-Newtonian Hele-Shaw flow. *Phys. Rev. E* **1996**, *54*, R4536–R4539. [CrossRef]
12. Lindner, A.; Bonn, D.; Meunier, J. Viscous fingering in a shear-thinning fluid. *Phys. Fluids* **2000**, *12*, 256–261. [CrossRef]
13. Lindner, A.; Bonn, D.; Poire, E.C.; Ben Amar, M.; Meunier, J. Viscous fingering in non-Newtonian fluids. *J. Fluid Mech.* **2002**, *469*, 237–256. [CrossRef]
14. Coussot, P. Yield stress fluid flows: A review of experimental data. *J. Non-Newton. Fluid Mech.* **2014**, *211*, 31–49. [CrossRef]
15. Coussot, P. Saffman-Taylor instability for yield stress fluids. *J. Fluid Mech.* **1999**, *380*, 363–376. [CrossRef]
16. Fontana, J.V.; Lira, S.A.; Miranda, J.A. Radial viscous fingering in yield stress fluids: Onset of pattern formation. *Phys. Rev. E* **2013**, *87*, 013016. [CrossRef] [PubMed]

17. Ebrahimi, B.; Mostaghimi, P.; Gholamian, H.; Sadeghy, K. Viscous fingering in yield stress fluids: A numerical study. *J. Eng. Math.* **2016**, *97*, 161–176. [CrossRef]
18. Ebrahimi, B.; Seyed-Mohammad, T.; Sadeghy, K. Two-phase viscous fingering of immiscible thixotropic fluids: A numerical study. *J. Non-Newton. Fluid Mech.* **2015**, *218*, 40–52. [CrossRef]
19. Maleki-Jirsaraei, N.; Lindner, A.; Rouhani, S.; Bonn, D. Saffman-Taylor instability in yield stress fluids. *J. Phys. Cond. Matter* **2005**, *17*, S1219–S1228. [CrossRef]
20. Eslamin, A.; Taghavi, S.M. Viscous fingering regimes in elasto-visco-plastic fluids. *J. Non-Newton. Fluid Mech.* **2017**, *243*, 79–94. [CrossRef]
21. Maleki-Jirsaraei, N.; Erfani, M.; Ghane-Golmohamadi, F.; Ghane-Motlagh, R. Viscous fingering in laponite and mud. *J. Test. Eval.* **2015**, *43*, 11–17. [CrossRef]
22. Lindner, A.; Bonn, D.; Coussot, P. Viscous fingering in a yield stress fluid. *Phys. Rev. Lett.* **2000**, *85*, 314–317. [CrossRef]
23. Barral, Q.; Ovarlez, G.; Chateau, X.; Boujlel, J.; Rabideau, B.D.; Coussot, P. Adhesion of yield stress fluids. *Soft Matter* **2010**, *6*, 1343–1351. [CrossRef]
24. Piau, J.M. Carbopol gels: Elastoviscoplastic and slippery glasses made of individual swollen sponges Meso- and macroscopic properties, constitutive equations and scaling laws. *J. Non-Newton. Fluid Mech.* **2007**, *144*, 1–29. [CrossRef]
25. Coussot, P.; Tocquer, L.; Lanos, C.; Ovarlez, G. Macroscopic vs. local rheology of yield stress fluids. *J. Non-Newton. Fluid Mech.* **2009**, *158*, 85–90. [CrossRef]
26. Bataille, J. Stabilité d'un déplacement radial non miscible. *Rev. Inst. Français Pétr.* **1968**, *XXIII*, 1349–1364.
27. Wilson, S.D.R. A note on the measurement of dynamic contact angles. *J. Colloid Interface Sci.* **1975**, *51*, 532–534. [CrossRef]
28. Coussot, P. *Rheometry of Pastes, Suspensions and Granular Materials*; Wiley: New York, NY, USA, 2005.
29. Boujlel, J.; Coussot, P. Measuring the surface tension of yield stress fluids. *Soft Matter* **2013**, *9*, 5898–5908. [CrossRef]
30. Zhang, X.; Fadoul, O.; Lorenceau, E.; Coussot, P. Yielding and flow of soft-jammed systems in elongation. *Phys. Rev. Lett.* **2018**, *120*, 048001. [CrossRef] [PubMed]

Article

Inhomogeneous Flow of Wormlike Micelles: Predictions of the Generalized BMP Model with Normal Stresses

J. Paulo García-Sandoval [1], Fernando Bautista [1], Jorge E. Puig [1] and Octavio Manero [2,*]

[1]   Departamentos de Física e Ingeniería Química, Universidad de Guadalajara, Blvd. M. García Barragán 1451, Guadalajara 44430, Jal., Mexico; paulo.garcia@cucei.udg.mx (J.P.G.-S.); ferbautistay@yahoo.com (F.B.); puig_jorge@hotmail.com (J.E.P.)

[2]   Instituto de Investigaciones en Materiales, Universidad Nacional Autónoma de México, Ciudad Universitaria, Ciudad de México 04510, Mexico

*   Correspondence: manero@unam.mx

Received: 26 December 2018; Accepted: 2 March 2019; Published: 8 March 2019

Abstract: In this work, we examine the shear-banding flow in polymer-like micellar solutions with the generalized Bautista-Manero-Puig (BMP) model. The couplings between flow, structural parameters, and diffusion naturally arise in this model, derived from the extended irreversible thermodynamics (EIT) formalism. Full tensorial expressions derived from the constitutive equations of the model, in addition to the conservation equations, apply for the case of simple shear flow, in which gradients of the parameter representing the structure of the system and concentration vary in the velocity gradient direction. The model predicts shear-banding, concentration gradients, and jumps in the normal stresses across the interface in shear-banding flows.

Keywords: shear-banding flow; BMP model; normal stresses

1. Introduction

Beyond the local equilibrium hypothesis, the extended irreversible thermodynamics (EIT) provides a consistent methodology to derive constitutive equations for systems far from equilibrium. These equations, together with the conservation laws, predict flow-induced concentration changes produced by inhomogeneous stresses in complex fluids [1–4].

Flow produces changes in the internal structure of complex fluids and induces fluctuations in concentration and in the rheological properties. In some analyses of the rheology of these complex fluids, the stress constitutive equation couples with an evolution equation of a scalar representing the flow-induced modifications on the internal structure of the fluid (a variable such as the fluidity or micellar length, in the particular case of giant micellar solutions). Simultaneously, the stress coupling with the diffusion equation of the dispersed phase explains the phenomena arising from concentration changes, as suggested in reports on flow-induced concentration fluctuations and diffusive interfaces [5,6]. An additional coupling of diffusion and structural changes closes this scheme.

Rheological measurements in complex fluids, in particular those performed in wormlike micellar systems, demonstrate that a unique selected shear stress exists independently of flow history [7]. The steady-state flow curve has a well-defined, reproducible plateau. Coexistence of low and high viscosity bands has been observed by nuclear magnetic resonance (NMR) spectroscopy [8], small-angle neutron scattering (SANS) [9], and from flow birefringence [10], which reveals a highly oriented band coexisting with an isotropic one.

The Johnson-Segalman model (JS) predicts a history dependence of banded solutions after imposing several flow histories, i.e., the apparent flow curves and the stress plateau depend on flow history. To find a unique stress selection, non-local gradient terms have been heuristically added to the JS constitutive equation [11,12], although diffusion in the stress arises naturally in kinetic theory, in particular in dumbbell models [13].

Although the non-local JS model has been useful to understand some features of the kinetics and stability of band formation, nevertheless there are two important setbacks of this model. The first one is that this model cannot describe the breaking and reformation processes of the micellar systems under flow to enable an understanding of the relation between shear-band formation and microstructural evolution. The second one refers to the inability of the model to describe the evolution of the stress and normal stress differences under step-strain experiments in shear flow and it gives wrong responses in extensional flow [14]. Furthermore, the non-local JS model may predict reversal in the band ordering in Couette flows [9] in contrast to experimental data.

Relevant alternative approaches, particularly that by Yuan and Jupp [15] using a 2-fluid J-S model, apparently give unique stress selection, even though interfacial terms were only present in the equation of motion for the concentration dynamics, and not in the constitutive viscoelastic equation. Vasquez et al. developed a two-species reptation-reaction network model [16] that captures the continuous breakage and reformation of long entangled chains that forms an entangled viscoelastic network, and the enhanced breaking that takes place during imposed shear deformation. The same group compared the homogeneous flow predictions of their model with steady and transient shear flow experiments performed in concentrated cetylpyridium chloride/sodium salicylate solutions [17]. In general, the predictions for nonlinear shear flow agree quite well with the rheological behavior for shear rates below the stress plateau, including the first normal stress difference, but it cannot predict a non-zero value of the second normal stress difference.

Concentration-coupling in the shear-banding transition of wormlike micellar systems has been a generic explanation of the slightly upward slope of the plateau stress observed with increasing shear rates [18–20]. This effect implies that a micellar concentration difference is established between the bands as the high-shear band grows to fill the gap. A model of concentration-coupled shear banding [21] was introduced by combining the diffusive (spatially non-local) J-S model [22] with a two-fluid approach to concentration fluctuations [23]. This model does not address the microscopic features of any particular viscoelastic system, but it should be regarded as a minimal model that combines an unstable constitutive curve, such as that of semi-dilute wormlike micelles, with spatially non-local terms in the viscoelastic constitutive equation and a simple approach to concentration coupling. In the two-fluid approach, the viscoelastic stress causes the micelles to diffuse up in stress gradients, and so it couples stress with concentration. If the viscoelastic stress then increases with concentration, positive feedback occurs, causing net diffusion of micelles up their own concentration gradient. This mechanism causes shear-enhanced concentration fluctuations and shear-induced de-mixing (concentration coupling) in systems that shear-band.

Normal stresses may be the reason for vorticity structuring that can emerge in complex fluids. A banding state may undergo a secondary linear stability due to the action of normal stresses across the interface between bands. Recent experiments of gradient-banded solutions of wormlike micelles show that they are unstable with respect to interfacial undulations with the wave vector in the vorticity direction [11,12]. The underlying mechanism includes normal stresses and shear-rate jumps across the interface.

In a previous work [24], we examined the non-homogeneous shear-banded flow of giant micelles with the BMP (Bautista-Manero-Puig) model. Results included the phase portraits around the flow curve and for the confined fluid, predictions were given for the velocity, stress, and fluidity fields as functions of both space and time. It was found that the same stress plateau and critical shear rates are approached independently of the initial conditions and shear history for a given applied shear rate or shear stress. The flow histories included forward and backward sweeps under strain-controlled and stress-controlled conditions.

In this work, the shear-banded flow predictions of the generalized BMP (Bautista-Manero-Puig) model [5], which contains the above-mentioned couplings, including normal stresses derived from EIT for complex fluids, are analyzed here. In reference [4], we present the formulation of the governing equations for flows that include normal stresses.

The derivation of the model's main concepts is given in the Appendix A and in [5]. The predictions of the phase portraits of the dynamics before approaching steady state and the variation of normal stresses in space and time of the resulting banded state are exposed. As in the previous paper, it is found that the same plateau is reached for various flow conditions; in this case, a stepping-up variation of shear rate is imposed. The diffusion equation allows for concentration gradients and gradients in the structural variable, and predicts stress gradient terms in the equation for the stress [13], which naturally arise in the constitutive equation, without the need to include them in an ad-hoc manner. Depleted regions where the concentration decreases are found near the moving boundary. It is shown that jumps in the normal stresses across the interface are also predicted.

2. Theoretical Description

The set of equations of the generalized BMP model are [5]:

$$\frac{d\varphi}{dt} = \frac{1}{\lambda}(\varphi_0 - \varphi) + k_0(1 + \vartheta(II_D))(\varphi_\infty - \varphi)\underline{\underline{\sigma}} : \underline{\underline{D}} + \varphi_0\beta_0'\nabla \cdot \bar{J} \tag{1}$$

$$\bar{J} + \tau_1\frac{\varphi_0}{\varphi}\overset{\nabla}{\bar{J}} = -\frac{\mathcal{D}\varphi_0}{\varphi}\nabla c - \frac{\beta_0}{\varphi}\nabla\phi + \frac{\beta_2\varphi_0}{\varphi}\nabla \cdot \underline{\underline{\sigma}} \tag{2}$$

$$\underline{\underline{\sigma}} + \frac{1}{G_0\phi}\overset{\nabla}{\underline{\underline{\sigma}}} = \frac{2}{\varphi}\underline{\underline{D}} + \frac{\psi_2}{\varphi}\underline{\underline{D}}\cdot\underline{\underline{D}} + \frac{\beta_2'\varphi_0}{\varphi}(\nabla\bar{J})^s \tag{3}$$

where $(\nabla\bar{J})^s$ stands for the symmetric part of $\nabla\bar{J}$ and the upper-convected derivatives of the diffusive concentration flux vector $\bar{J}$ and of the stress tensor $\underline{\underline{\sigma}}$ are defined, respectively, as:

$$\overset{\nabla}{\bar{J}} = \frac{d\bar{J}}{dt} - \underline{\underline{L}}\cdot\bar{J}, \tag{4}$$

$$\overset{\nabla}{\underline{\underline{\sigma}}} = \frac{d\underline{\underline{\sigma}}}{dt} - \left(\underline{\underline{L}}\cdot\underline{\underline{\sigma}} + \underline{\underline{\sigma}}\cdot\underline{\underline{L}}^T\right) \tag{5}$$

here d/dt is the material-time derivative, $\underline{\underline{D}}$ is the symmetric part of the velocity gradient tensor $\underline{\underline{L}}$, and II_D is its second invariant. φ is the inverse of the shear viscosity (η) is the fluidity, $\varphi_0(\equiv \eta_0^{-1})$ is the fluidity at zero shear rate, G_0 is the plateau shear modulus, λ is a structure relaxation time, and k_0 can be interpreted as a kinetic parameter for structure breaking. τ_1 is a relaxation time for the mass flux, $\mathcal{D}$ is the Fickean diffusion coefficient, and ψ_2 is the second normal stress coefficient; c is the local equilibrium concentration and ϑ, β_0, β_0', β_2, and β_2' are phenomenological parameters. The structural variable σ has been identified with the normalized fluidity $\phi = \varphi/\varphi_0$ [5].

Equations (1)–(3), together with the conservation equations, represent a closed set of time evolution equations for all the independent variables chosen to describe the behavior of complex fluids. Note the mutual coupling of these equations.

For simple-shear (where x is the direction of the macroscopic flow velocity, y is the direction of the velocity gradient, and z is the vorticity direction), we assume small inertia and that the mass flux relaxation time is negligible compared to the stress relaxation time, i.e., $(G_0\varphi)^{-1} \gg \tau_1$ (which a plausible assumption for wormlike micelles). Equations (1)–(3) become:

$$\frac{d\varphi}{dt} = \frac{1}{\lambda}(\varphi_0 - \varphi) + k_0(1 + \vartheta(II_D))(\varphi_\infty - \varphi)\underline{\underline{\sigma}} : \underline{\underline{D}} + \varphi_0\beta_0'\left[\frac{\partial J_x}{\partial x} + \frac{\partial J_y}{\partial y}\right] \tag{6}$$

$$J_x = -\frac{\mathcal{D}\varphi}{\varphi_0}\frac{\partial c}{\partial x} - \frac{\beta_0}{\varphi}\frac{\partial \varphi}{\partial x} + \frac{\beta_2\varphi_0}{\varphi}\left[\frac{\partial \sigma_{xy}}{\partial y} + \frac{\partial \sigma_{xx}}{\partial x}\right] \tag{7}$$

$$J_y = -\frac{\mathcal{D}\varphi}{\varphi_0}\frac{\partial c}{\partial y} - \frac{\beta_0}{\varphi}\frac{\partial \varphi}{\partial y} + \frac{\beta_2\varphi_0}{\varphi}\frac{\partial \sigma_{xy}}{\partial y} \tag{8}$$

$$\sigma_{xy} + \frac{1}{G_0\varphi}\left[\frac{\partial \sigma_{xy}}{\partial t} - \dot{\gamma}\sigma_{yy}\right] = \frac{\dot{\gamma}}{\varphi} + \frac{\beta_2'\varphi_0}{\varphi}\left[\frac{\partial J_x}{\partial y} + \frac{\partial J_y}{\partial x}\right] \tag{9}$$

$$\sigma_{xx} + \frac{1}{G_0\varphi}\left[\frac{\partial \sigma_{xx}}{\partial t} - 2\dot{\gamma}\sigma_{yy}\right] = \frac{\beta_2'\varphi_0}{\varphi}\frac{\partial J_x}{\partial x} \tag{10}$$

$$\sigma_{yy} + \frac{1}{G_0\varphi}\left[\frac{\partial \sigma_{yy}}{\partial t}\right] = \psi_2\dot{\gamma}^2\frac{\varphi_0}{\varphi} + \frac{\beta_2'\varphi_0}{\varphi}\frac{\partial J_y}{\partial y} \tag{11}$$

$$\sigma_{zz} + \frac{1}{G_0\varphi}\left[\frac{\partial \sigma_{zz}}{\partial t}\right] = 0 \tag{12}$$

where $\dot{\gamma}$ is the shear rate. Equations (6)–(12) are the ones given particular attention in this work. To close the system of equations, the conservation of mass, concentration, and momentum are:

$$\rho\frac{\partial v_x}{\partial x} = 0 \tag{13}$$

$$\frac{\partial c}{\partial t} = -\left[\frac{\partial J_x}{\partial x} + \frac{\partial J_y}{\partial y}\right] \tag{14}$$

$$\rho\frac{\partial v_x}{\partial t} = \frac{\partial \sigma_{xy}}{\partial y} + \eta_s\frac{\partial^2 v_x}{\partial y^2} + \frac{\partial \sigma_{xx}}{\partial x}, \tag{15}$$

where η_s is the solvent viscosity. In Equations (7), (8), and (15), the derivatives of the normal stresses involve terms of third order in the derivatives of the fluidity and concentration, which we neglect. In addition, a solution for Equations (13) and (15) can be obtained by taking the derivative of each term of Equation (15) with respect to x. This leads to:

$$\frac{\partial^2 \sigma_{xy}}{\partial x \partial y} + \frac{\partial^2 \sigma_{xx}}{\partial x^2} = 0 \tag{16}$$

Next, the normal stress differences are defined in the usual form:

$$N_1 = \sigma_{xx} - \sigma_{yy}, \quad N_1 = \sigma_{yy} - \sigma_{zz} \tag{17}$$

From Equations (10)–(12) we obtain:

$$N_1 + \frac{1}{G_0\varphi}\left[\frac{\partial N_1}{\partial t} - 2\dot{\gamma}\sigma_{xy}\right] = \frac{\beta_2'\varphi_0}{\varphi}\left[\frac{\partial J_x}{\partial x} - \frac{\partial J_y}{\partial y}\right] \tag{18}$$

$$N_2 + \frac{1}{G_0\varphi}\left[\frac{\partial N_2}{\partial t}\right] = \psi_2\frac{\varphi_0}{\varphi}\dot{\gamma}^2 + \beta_2'\frac{\varphi_0}{\varphi}\frac{\partial J_y}{\partial y} \tag{19}$$

Until now, Equations (6), (9), (18), and (19) have been the more general expressions in two dimensions (the shear plane), including normal stresses. They preserve the tensorial character of the equations, upon the assumptions of small inertia and negligible mass flux relaxation time. Following the translational symmetry of the flow, we address the particular case where the derivatives in the direction of flow are negligible. In such case, we have:

$$\frac{\partial J_x}{\partial x} = \frac{\partial J_y}{\partial x} = 0 \tag{20}$$

$$\frac{\partial J_x}{\partial y} = \beta_2 \varphi_0 \frac{\partial}{\partial y} \left[\frac{1}{\phi} \frac{\partial \sigma_{xy}}{\partial y} \right] \tag{21}$$

$$\frac{\partial J_y}{\partial y} = \frac{\partial}{\partial y} \left[\frac{1}{\phi} \left(-\beta_0 \frac{\partial \phi}{\partial y} + \beta_2 \varphi_0 \frac{\partial \sigma_{yy}}{\partial y} \right) \right] + \frac{\partial}{\partial y} \left(\frac{-\mathcal{D} \phi_0}{\phi} \frac{\partial c}{\partial y} \right) \tag{22}$$

$$N_1 + \frac{1}{G_0 \phi} \left[\frac{\partial N_1}{\partial t} - 2\dot{\gamma} \sigma_{xy} \right] = -\psi_2 \frac{\phi_0}{\phi} \dot{\gamma}^2 - \beta_2' \frac{\phi_0}{\phi} \left[\frac{\partial J_y}{\partial y} \right] \tag{23}$$

while (19) remains equal. Written in terms of the non-dimensional variables:

$$\phi = \varphi / \varphi_0, \ \phi_\infty = \varphi_\infty / \varphi_0,$$

Equation (9) becomes:

$$\phi \sigma_{xy} + \tau_\sigma \left[\frac{\partial \sigma_{xy}}{\partial t} \right] = \eta_0 \dot{\gamma} + \dot{\gamma} \tau_\sigma N_2 + \frac{\beta_2 \beta_2'}{2} \frac{\partial}{\partial y} \left[\frac{1}{\phi} \frac{\partial \sigma_{xy}}{\partial y} \right] \tag{24}$$

where τ_σ is the stress relaxation time ($\tau_\sigma = (G_0 \phi_0)^{-1}$) and η_0 is the zero shear-rate viscosity. In the limit of creeping flow $\phi \to 1$, Equation (24) reduces to:

$$\sigma_{xy} + \tau_\sigma \left[\frac{\partial \sigma_{xy}}{\partial t} \right] = \eta_0 \dot{\gamma} + \frac{1}{2} \beta_2 \beta_2' \frac{\partial^2 \sigma_{xy}}{\partial y^2} \tag{25}$$

Equation (25) is similar to the diffusion equation for the stress analyzed in the current literature to predict diffusion of interfaces [12,13,25]. In fact, following similar assumptions (simple shear, small inertia), Equation (25) is equal to that derived from the constitutive equation:

$$\underline{\underline{\sigma}} + \tau_\sigma \underline{\underline{\overset{\circ}{\sigma}}} = 2\eta_0 \underline{\underline{D}} + \tau_\sigma \mathcal{D} \nabla^2 \underline{\underline{\sigma}} \tag{26}$$

where $\underline{\underline{\overset{\circ}{\sigma}}}$ is the (Gordon-Schowalter) convected time derivative defined in the Johnson-Segalman model [26]. In creeping shear flows, Equation (26) reduces to Equation (25), providing the phenomenological coefficients $\beta_2 \beta_2'$ to be identified with $\tau_\sigma \mathcal{D}$. This identification of the coefficients allows a physical interpretation and measurement of their magnitudes. Similarly, the coefficients $\beta_0 \beta_0'$ may be identified with the structure diffusion coefficient $\mathcal{D}'$, and $\beta_0 \beta_2'$ can be identified with $\rho \mathcal{D} \mathcal{D}'$. With these identifications, and considering that Equation (23) is decoupled, Equations (6), (9), (19), and (23) become:

$$\frac{\partial J_y}{\partial y} = \frac{\partial}{\partial y} \left[\frac{1}{\phi} \left(-\rho \mathcal{D}' \frac{\partial \phi}{\partial y} + \tau_\sigma \frac{\partial N_2}{\partial y} \right) \right] + \frac{\partial}{\partial y} \left(\frac{-\mathcal{D}}{\phi} \frac{\partial c}{\partial y} \right) \tag{27}$$

$$\frac{\partial \phi}{\partial t} = \frac{1}{\lambda}(1 - \phi) + k_0(1 + \vartheta \dot{\gamma})(\phi_\infty - \phi)\sigma_{xy} \dot{\gamma} + \frac{1}{\rho} \frac{\partial J_y}{\partial y} \tag{28}$$

$$\tau_\sigma \left[\frac{\partial \sigma_{xy}}{\partial t} \right] = \eta_0 \dot{\gamma} + \dot{\gamma} \tau_\sigma N_2 + \frac{\mathcal{D} \tau_\sigma}{2} \frac{\partial}{\partial y} \left[\frac{1}{\phi} \frac{\partial \sigma_{xy}}{\partial y} \right] \tag{29}$$

$$\phi N_1 + \tau_\sigma \left[\frac{\partial N_1}{\partial t} - 2\dot{\gamma} \sigma_{xy} \right] = -\psi_2 \dot{\gamma}^2 + \mathcal{D} \left[\frac{\partial J_y}{\partial y} \right] \tag{30}$$

$$\phi N_2 + \tau_\sigma \left[\frac{\partial N_2}{\partial t} \right] = \psi_2 \dot{\gamma}^2 - \left[\frac{\partial J_y}{\partial y} \right] \tag{31}$$

Equations (28)–(31) are the main results of this section. It is noticeable that this formulation leads to structure-dependent variables, i.e., viscosity (η_0 / ϕ), stress relaxation time (τ_σ / ϕ), and diffusion coefficients ($\mathcal{D} / \phi$ and $\rho \mathcal{D}' / \phi$). The structure itself follows an evolution Equation (28). In the limit $\phi \to 1$ (constant structure) with no normal stresses, Equation (29) reduces to the simple-shear version of Equation (26). These equations contain seven constants. Five of them (λ, k_0, η_0, ϕ_∞, τ_σ) can be evaluated from independent rheological experiments [4]. Under heterogeneous (shear banding) flow, ϑ (the shear-banding intensity parameter) is related to the position of the stress plateau (set by the equal-areas criterion or equal minima in the dissipated energy) and $\mathcal{D}$ may be evaluated from data of interface diffusion. Dhont [27] has suggested an expression for the relaxed stress under simple shear (without normal stresses) similar to that of Equation (25), in which the term $(\mathcal{D}/\phi)$ is identified with the "curvature viscosity", which actually follows the same form of the shear-thinning viscosity observed in worm-like micellar solutions. According to the magnitudes shown in [27], usual values for the zero-shear rate viscosity are around 20 Pa·s. In the absence of normal stresses, the reaction-diffusion character of equation is preserved

$$\sigma_{xy} + \tau_\sigma \left[\frac{\partial \sigma_{xy}}{\partial t} \right] = \eta_0 \dot{\gamma} + \frac{\mathcal{D}\tau_\sigma}{2} \frac{\partial}{\partial y} \left[\frac{1}{\phi} \frac{\partial \sigma_{xy}}{\partial y} \right]$$

which reduces to the simple shear version of Equation (25) as $\phi \to 1$. It is worth mentioning that the inclusion of diffusion in the constitutive equations leads to a finite thickness of the interface between the bands, as shown in the results presented in the next section.

2.1. Steady-State Solution

Under steady state, the conservation Equations (14) and (15) lead to

$$\left[\frac{\partial J_y}{\partial y} \right] = 0, \frac{\partial \sigma_{xy}}{\partial y} = 0$$

which means that both J_y and σ_{xy} are independent of the coordinates, and since there is no flux at the boundaries, then $J_y = 0$ for all y. Furthermore, Equation (8) leads to:

$$\rho \mathcal{D}' \left[\frac{\partial \phi}{\partial y} \right] = -\mathcal{D} \left[\frac{\partial c}{\partial y} \right] + \tau_\sigma \left[\frac{\partial N_2}{\partial y} \right] \tag{32}$$

In the particular case when the mass and viscosity diffusion coefficients and the stress relaxation times are constant, Equation (32) can be integrated to give:

$$\phi = \phi_c - (\mathcal{D}/\rho\mathcal{D}')c + (\tau_\sigma/\rho\mathcal{D}')N_2$$

For a given constant ϕ_c, the ratio of the diffusion coefficients is positive, and hence the fluidity increases with decreasing concentration. In addition, under steady state, Equations (28)–(31) become

$$0 = \frac{1}{\lambda}(1 - \phi) + k_0(1 + \vartheta\dot{\gamma})(\phi_\infty - \phi)\sigma_{xy}\dot{\gamma} \tag{33}$$

$$\phi\sigma_{xy} = \eta_0\dot{\gamma} + \dot{\gamma}\tau_\sigma N_2 \tag{34}$$

$$\phi N_1 = (2\tau_\sigma\sigma_{xy} - \psi_2\dot{\gamma})\dot{\gamma} \tag{35}$$

$$\phi N_2 = \psi_2\dot{\gamma}^2 \tag{36}$$

Substitution of Equations (36) and (34) in (33) leads to a fifth and third order equation for the shear rate and fluidity, respectively:

$$0 = (1 - \phi)\phi^2 + k_0\lambda(1 + \vartheta\dot{\gamma})(\phi_\infty - \phi)\left(\eta_0\phi + \tau_\sigma\psi_2\dot{\gamma}^2\right)\dot{\gamma}^2$$

In most cases, the order of magnitude of ψ_2 is negligible in comparison with $\eta_0\phi$, and the previous equation reduces to a third and second order equation for the shear rate and fluidity, respectively, which has been previously analyzed [4].

2.2. Numerical Method

To analyze the transient behavior at the inception of flow predicted by the model, Equations (28)–(31) were numerically solved together with equations

$$\frac{\partial c}{\partial t} = -\frac{\partial J_y}{\partial y} \tag{37}$$

$$\rho \frac{\partial v_x}{\partial t} = \frac{\partial \sigma_{xy}}{\partial y} \tag{38}$$

which are, respectively, the simplified version, where the derivatives in the direction of flow (x direction) as well as the solvent viscosity, η_s, are negligible. The moving plate is located at position $y = L$, while the fixed plate is at $y = 0$, with boundary conditions:

$$v_x(t, 0) = 0, \; v_x(t, L) = v_L(t) \tag{39}$$

$v_L(t) \geq 0$ is the upper plate velocity, which follows the following dynamics:

$$v_L(t) = v_{L,0} + (v_{L,ss} - v_{L,0})(1 - e^{-t/\tau_c}),$$

where $v_{L,0}$ is the initial velocity of the moving plate, $v_{L,ss}$ is the steady state velocity, and τ_c is a characteristic time. This time is linked to a controller, which regulates the velocity of the moving plate, which in general is of the same order of magnitude of the characteristic time of the system τ_σ. In addition to the conservation law (Equation (38)) and mass conservation, we have to satisfy the following conditions:

$$\frac{\partial \sigma(t, 0)}{\partial y} = 0, \; \frac{\partial \sigma(t, L)}{\partial y} = \frac{1}{\rho} \frac{dv_L(t)}{dt} \tag{40}$$

$$J_y(t, 0) = 0, \; J_y(t, L) = 0 \tag{41}$$

As pointed out before [24], due to the conservation of momentum and the flow history $v_L(t)$, the spatial derivative of the stress on the upper plate cannot be zero. The boundary conditions in Equation (40) imply that the spatial derivative of the shear stress at the upper plate is zero only when steady state is reached. Thus, the boundary conditions at the upper plate are also history-dependent. In some cases, these boundary conditions give rise to a stable three-band state, similar to that found for Dirichlet boundary conditions, wherein the stress distribution in the gap depends on both the boundary condition and stress gradient [28,29].

Equations (28)–(31), (37), and (38), together with boundary conditions (39)–(41), are solved using the numerical method described in reference [24], where the ordinary differential equations resulting in discretization in space are integrated to obtain numerical solutions as a function of time.

3. Results

Figure 1 illustrates the variation of the stress, first normal stress difference (N_1), and the absolute value of the second normal stress difference (N_2) with shear rate, under shear-rate controlled flow starting from rest. The shear rate was increased in a stepwise mode allowing attainment of steady-state at each step. At the highest shear rate, N_2 approaches within a tenth of the value of N_1; this is the upper bound of N_2 as suggested by the scarce experimental data. N_1 shows a behavior similar to that of the stress-shear rate curve depicting, in fact, a shear-banding unstable region. In contrast, N_2 grows monotonically. For low shear rates, the slope of N_1 tends to a value of 2, although this range is not shown in Figure 1.

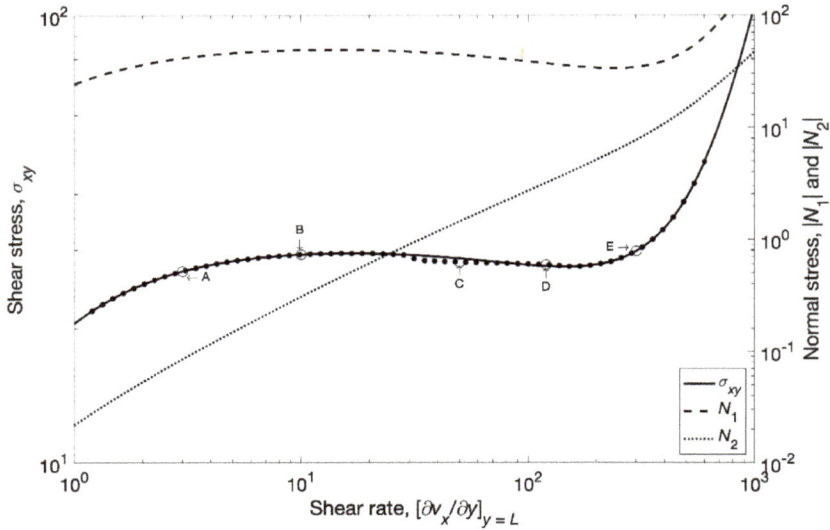

Figure 1. Steady state constitutive curves obtained under controlled shear rate starting from rest. Parameters used in the simulations are: $c_0 = 5$ wt %, $\varphi_\infty = 847$ m·s·kg^{-1}, $\tau_\sigma = 1.06$ s, $\eta_0 = 36.56$ kg m^{-1}s^{-1}, $k_0 = 3.28 \times 10^{-4}$ m s^{-1} kg^{-1}, $\lambda = 0.136$ s, $\vartheta = 0.0061$ s, $\rho = 1000$ kg m^{-3}, $\mathcal{D} = 1 \times 10^{-5}$ m^2 s^{-1}, $\mathcal{D}' = 1 \times 10^{-11}$ m^2 s^{-1}, $\psi_2 = 0.0388$ s.

Figure 2a–g depict the dynamics and attainment of steady state under controlled shear rate when the reference shear rate is 3 s^{-1} (see point A in Figure 1). At this shear rate, the flow is homogeneous, and the steady state is reached at the time scale of the Maxwell relaxation time. The position $y = L$ corresponds to the moving plate and $y = 0$ refers to the fixed boundary. Figure 2g depicts uniform concentration throughout the geometry. The shear stress increases monotonically (Figure 2e) in the same form observed in the fluidity, while first and second normal stress differences are depicted in Figure 2b,c,f, respectively, and a velocity profile corresponding to a single shear rate is observed in Figure 2d.

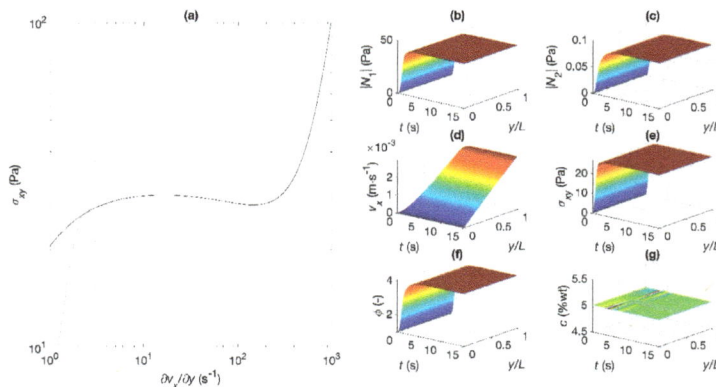

Figure 2. Dynamics and steady state under controlled shear rate. Reference shear rate is 3 s^{-1}. Temporal trajectory of the stress and shear rate for various spatial positions (**a**), evolution of the normal stress differences N_1 (**b**) and N_1 (**c**), velocity (**d**), shear stress (**e**), fluidity (**f**), and concentration (**g**), in space and time.

In Figure 3a–g, the reference shear rate has been increased to 10 s^{-1} (point B in Figure 1) corresponding to the top-jumping stress. The dynamics shown in Figure 3a are different to that in the homogeneous region, although the attainment to steady state is fast. The trajectory before steady-state includes an overshoot before landing at the top jumping stress. Once again, the concentration is uniform (Figure 2g) and the fluidity and second normal stress difference grow monotonically (Figure 2c,f, respectively). In contrast, the shear stress (Figure 2e) and the first normal stress difference (Figure 2b) present an overshoot at the inception of flow before they reach steady state. The velocity profile corresponds to a single shear rate (Figure 2d).

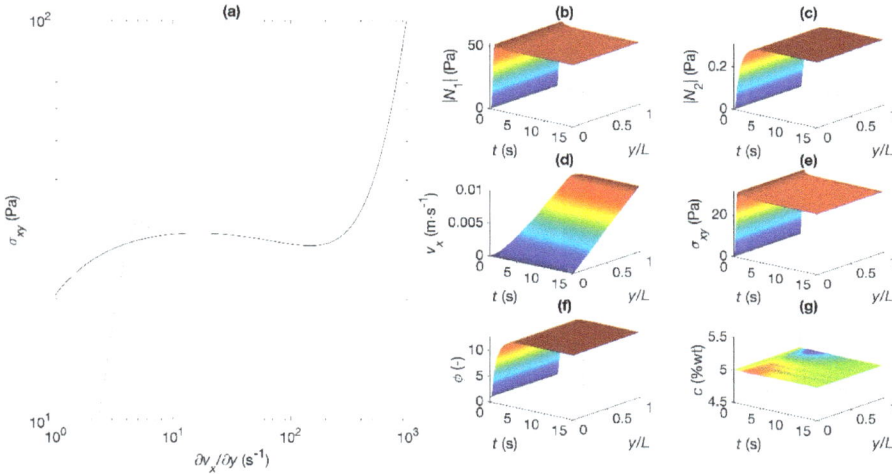

Figure 3. Dynamics and steady state under controlled shear rate. Reference shear rate is 10 s^{-1}. Temporal trajectory of the stress and shear rate for various spatial positions (**a**), evolution of the normal stress differences N_1 (**b**) and N_2 (**c**), velocity (**d**), shear stress (**e**), fluidity (**f**), and concentration (**g**), in space and time.

A different situation is shown when the reference shear rate increases to 50 s^{-1}, within the shear banding region (point C). The dynamics now oscillate between the two attractors located at the critical shear rates (at the binodals or extremes of the plateau stress) after describing an overshoot in the stress (Figure 4a). The concentration is not uniform (Figure 4g) and transient and steady state banding is predicted as the velocity profile changes as a function of time from a single into a two-banded profile corresponding to two shear rates, with the steepest one located next to the moving plate (Figure 4d), within the region where the concentration decreases. A depletion zone then appears near the moving plate in the high shear rate region resulting in a concentration gradient; the fluidity in turn increases next to the moving plate (Figure 4f). As before, the shear stress (Figure 4e) and N_1 (Figure 4b) describe an overshoot at the inception of flow, but rapidly they attain steady state. A remarkable result in both stress differences is the decrease in N_1 simultaneous to an increase in N_2 in the region next to the moving plate (Figure 4b,c, respectively). In fact, a jump in both stress differences across the interfaces is revealed in these predictions.

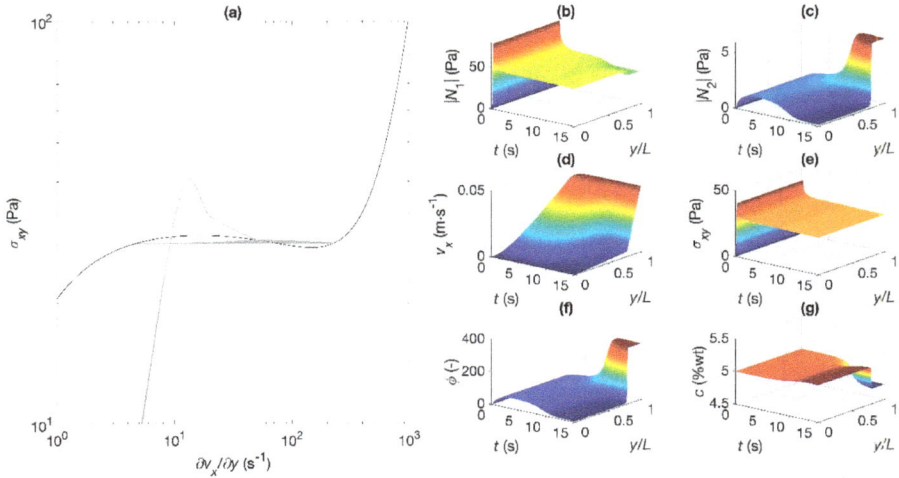

Figure 4. Dynamics and steady state under controlled shear rate. Reference shear rate is 50 s^{-1}. Temporal trajectory of the stress and shear rate for various spatial positions (**a**), evolution of the normal stress differences N_1 (**b**) and N_2 (**c**), velocity (**d**), shear stress (**e**), fluidity (**f**), and concentration (**g**), in space and time.

In Figure 5, the reference shear rate is now 120 s^{-1}, which is in the region near the minimum of the flow curve (high shear rate attractor, point D in Figure 1). The dynamics rapidly converge to the extreme of the plateau stress (Figure 5a) after an overshoot and oscillations along the plateau. Once again, concentration gradients are predicted, including a sudden decrease of concentration at the interface (Figure 5g) inducing a sudden rise in the fluidity (Figure 5f) next to the moving wall. The high shear rate band covers most of the flow region (Figure 5d) and past a pronounced maximum the total stress is uniform along the flow cell (Figure 5e). As found in the shear banding region, there is a jump in the normal stresses across the interface (Figure 5b,c), but N_1 develops this sudden change after a short overshoot, in contrast to N_2, which monotonically increases in the high shear rate band and decreases to almost zero in the low shear rate band.

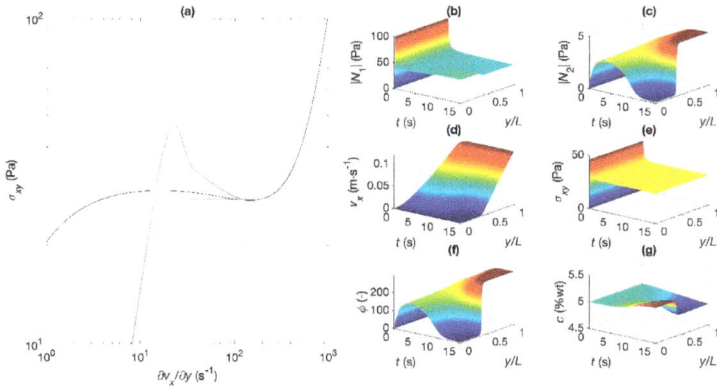

Figure 5. Dynamics and steady state under controlled shear rate. Reference shear rate is 120 s^{-1}. Temporal trajectory of the stress and shear rate for various spatial positions (**a**), evolution of the normal stress differences N_1 (**b**) and N_2 (**c**), velocity (**d**), shear stress (**e**), fluidity (**f**), and concentration (**g**), in space and time.

Finally, in Figure 6a–g, the reference shear rate is 300 s^{-1} located in the high shear-rate branch (point E in Figure 1). Past an overshoot, the transient dynamics ends at the high shear-rate branch of the flow curve. The flow is again homogeneous, since a single velocity gradient is predicted (Figure 6d). Concentration gradients are absent at steady state and fluidity is uniform again (Figure 6f,g). The shear stress and N_1 rapidly attain steady state after maxima following the inception of flow (Figure 6b); N_2 attains steady-state monotonically (Figure 6c).

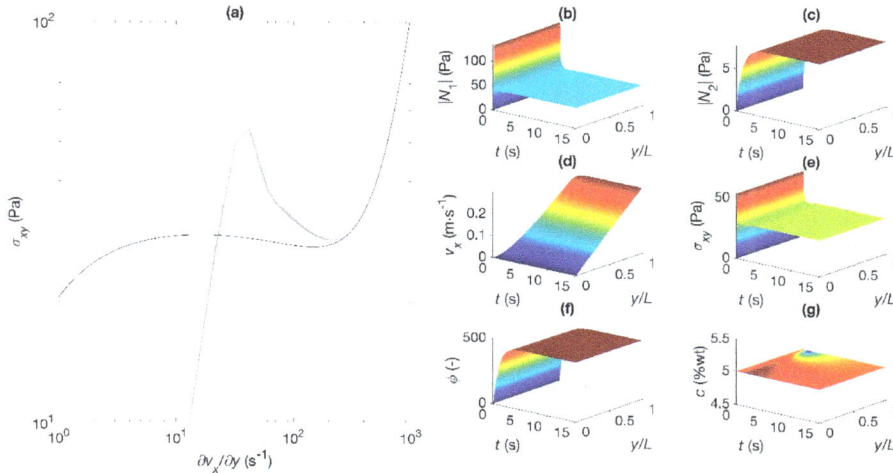

Figure 6. Dynamics and steady state under controlled shear rate. Reference shear rate is 300 s^{-1}. Temporal trajectory of the stress and shear rate for various spatial positions (**a**), evolution of the normal stress differences N_1 (**b**) and N_2 (**c**), velocity (**d**), shear stress (**e**), fluidity (**f**), and concentration (**g**), in space and time.

4. Discussion and Conclusions

The model presented here contains three constitutive equations: the equation for the stress, for the structural parameter, and for the diffusion of mass. They are mutually-coupled by phenomenological coefficients, and their physical significance arises as they identify with the mechanisms acting on the system. The mechanisms included here are the kinetics involved in the attainment of steady state, after which a banding state is induced by the non-monotonic nature of the constitutive equation. The existence of a banded state induces jumps in the normal stresses and concentration differences amidst the bands, which constitute important predictions of the model. The diffusion of structure and mass are governed by the diffusion coefficients, as it should be in situations where concentration gradients or "structure gradients" arise. The latter is a consequence of the existence of ordered phases or bands coexisting with more disordered or isotropic bands in the flow cell.

Instabilities leading to undulations of the interface [9,10] appear from the action of normal stresses across the interface between the bands. Accordingly, the mechanism of instability is not fully understood, but it is likely to stem from steep gradients in the normal stress and shear rate across the interface. Various authors have found instabilities due to the jump of normal stresses across the interface in polymeric systems [30,31]. Alternating vorticity bands [32] and concentration gradients in polymer solutions that exhibit shear banding have been predicted.

One of the advantages of the present model is that it describes the breakage-reformation process of the micelles under flow so as to enable an understanding of the relation between shear band formation and microstructure evolution. In the BMP model, the relation between the fluidity, relaxation time τ, and micellar length n is [24]:

$$\tau = \frac{1}{G_0 \varphi} = \tau_0 \left(\frac{n}{n_0} \right)$$

where τ_0 is the relaxation time when $\varphi = \varphi_0$. Substitution of this equation into the evolution equation for the fluidity (see Equation (1)) yields an equation expressed in terms of the average micellar length, which is the microstructural variable. The physical interpretation is that the micellar length n follows an evolution equation related to the breakage and reformation process of the micelles.

Results of the present model and those contained in reference [24] (see Figure 5 therein), where no normal stresses are included, reveal predictions of multiple bands in the absence of normal stresses. In fact, a comparison at similar imposed shear rates (those near the high shear-rate extreme of the plateau) of the velocity and stress fields are similar, but the fluidity presents two regions with high fluidity next to the walls (without normal stresses). With normal stresses, we observe a single region of high fluidity next to the moving wall (see Figure 5 of the present paper).

We have shown that the EIT formalism described here is consistent with previous works on shear-banding inhomogeneous flows. In fact, in the two-fluid approach, the viscoelastic stress causes the micelles to diffuse up in gradients of the stress, and so it couples stress with concentration. In the present work, predictions of a depleted layer next to the moving surface reveal agreement with this underlying mechanism, i.e., the existence of a positive feedback, which causes diffusion of micelles up their own concentration gradient. As an explanation to micellar migration [21], the strain component W_{yy} is more negative in the high shear rate phase than in the lower shear phase, then micelles migrate to the low-shear band. This corresponds to the strongly sheared micelles stretched strongly along the flow direction.

A model for wormlike micellar solutions involving scission and reforming of chains based on non-affine network theory and a discrete version of the Cates theory was forwarded [16]. Although the model does not predict N_2, one of the variants of the model (PEC + M, partial extended convected derivative with two interacting species) predicts a behavior of N_1 as a function of shear rate quantitatively similar to that predicted by the present model. In reference [16], the first normal stress difference also shows a banded structure and a sudden drop at the interface. Predictions of the overshoot at the onset for flow follow the same manner as predictions by the BMP model.

Further concordance with predictions of other models arises. The stress overshoot predicted in the banded state in Figure 4 agrees with the three stages predicted by the non-local JS model after a step growth in shear rate [33], i.e., band destabilization, interface reconstruction, and interface traveling. As indicated in this reference, the instability and reconstruction of the interface in the first two stages end when the interface between stable bands sharpens. Front propagation is controlled by the diffusion constant D, as in the BMP model.

Predictions of the concentration profiles in entangled polymeric systems [34] depict a sudden decrease (quasi-step like) of concentration at a given position in the flow cell, and this change occurs nearer to the moving wall when the shear rate is smaller. These predictions agree qualitatively with those depicted in Figures 4 and 5 of the present paper. Band migration and band shapes are similar in both systems, illustrating that this phenomenon is common to wormlike micelles and entangled polymeric systems

In summary, the relevant predictions of the present model are the depleted concentration region near the moving boundary and the jumps in normal stresses across the interface. Stress diffusion arises naturally in the constitutive equations. Two diffusion mechanisms are involved, the mass and the structural diffusion, which arise in the equations for the stress and stress differences, but in addition, they are present in the equation for the reformation-breakage of the structure.

Author Contributions: All four authors (J.P.G.-S., F.B., J.E.P., O.M.) participated in the theoretical descriptions. J.E.P. and O.M. undertook the editing process.

Funding: This research was funded by CONACYT (National Council for Science and Technology, Mexico).

Conflicts of Interest: The authors declare no conflict of interest. Sponsors had no role in this study, nor in the writing of the manuscript, or decision to publish the results.

Appendix A

Extended Thermodynamic description of complex fluids (see [4]). The non-equilibrium thermodynamic state of a complex fluid is described by a formulation contained in the usual procedure of Extended Irreversible Thermodynamics (EIT). The thermodynamic state shall be described by taking as conserved variables {C} the internal energy density (e), the mass density (ϱ), and the relative concentration of the dispersed phase (c) that is embedded in a Newtonian liquid of concentration c', i.e., $c + c' = 1$. As for the set {R} of non-conserved state variables, a scalar representing the internal structure of the fluid ς, the diffusive concentration flux **J**, and the traceless symmetric part of the stress tensor σ are included. Hence, the space of state variables for this system is given by the set G = C U R = {e, ς, c; ρ, $\bar{J}$, $\underline{\sigma}$}. As the first basic assumption of the theory, EIT assumes the existence of a sufficiently continuous and differentiable function η_E, defined over a complete space G: $\eta_E = \eta_E$ {e, ς, c; ρ, $\bar{J}$, $\underline{\sigma}$}. This assumption aims to generate a differential form, which, in a strictly formal sense, will generalize the Gibbs relation of local equilibrium thermodynamics. For an incompressible fluid at constant temperature, applying the usual procedure of EIT to the given generalized-entropy function and restricting the scheme to the lowest order in the non-conserved variables, we obtain the following generalized Gibbs relation:

$$T\frac{d\eta_E}{dt} = -\mu\frac{dc}{dt} + \frac{1}{\rho}\alpha_0\frac{d\varsigma}{dt} + \frac{1}{]\rho}\bar{\alpha}_1 \cdot \frac{d\bar{J}}{dt} + \frac{1}{\rho}\underline{\alpha}_2 : \frac{d\underline{\sigma}}{dt} \tag{A1}$$

here T and μ are the local equilibrium values of the temperature and the chemical potential of the dispersed phase; α_0, $\bar{\alpha}_1$, and $\underline{\alpha}_2$ are phenomenological coefficients that are defined as the partial derivatives of η_E with respect to the state variables, and hence, depend on the equilibrium value of c. The scalar α_0, the vector $\bar{\alpha}_1$, and the tensor $\underline{\alpha}_2$ should be constructed as the most general scalar, vector, and tensor expressions that may be obtained in terms of all independent variables in G. Thus, according to the theory of invariants in space G and to the first order in the non-conserved variables, they are given by:

$$\underline{\alpha}_0 = \alpha_{00}\zeta, \quad \underline{\alpha}_1 = \alpha_{10}\underline{J}, \quad \underline{\alpha}_2 = \alpha_{20}\underline{\sigma} \tag{A2}$$

where α_{i0} (i = 0, 1, 2) are scalar coefficients. It should be stressed that these phenomenological coefficients can only be determined from experiment or from a microscopic theory.

The second postulate of EIT assumes that the function η_E satisfies a balance equation, namely,

$$\rho\frac{d\eta_E}{dt} + \nabla \cdot \bar{J}_\eta = S_\eta \tag{A3}$$

$\bar{J}_\eta$ and S_η denote the flux and source term associated to η_E, respectively. They should be expressed as the most general vector and scalar in G. Consistency with the order considered in arriving at Equation (A1) requires that

$$\bar{J}_\eta = \beta_0\varsigma\bar{J} + \beta_1\bar{J} + \beta_2\bar{J} \cdot \underline{\sigma} \tag{A4}$$

$$S_\eta = X_0\varsigma + \bar{X}_1 \cdot \bar{J} + \underline{X}_2 : \underline{\sigma} \tag{A5}$$

where the phenomenological coefficients β_i depend on the local equilibrium value of c. Furthermore, we consider X_i, up to first order, as the most general quantities in G. It is important to point out that in order to recover the usual results of the linear irreversible thermodynamics (LIT) near equilibrium, η_E, $\bar{J}_\eta$, and S_η should reduce to the entropy production, the entropy flux, and the entropy production, respectively. Therefore, $\beta_1 = -\mu/T$.

By computing the divergence of Equation (A4) and using Equation (A1) and the mass conservation equation, we get:

$$\rho \frac{dc}{dt} = -\nabla \cdot \bar{J} \tag{A6}$$

The following explicit expression is obtained from Equation (A3):

$$\rho \frac{d\eta_E}{dt} + \nabla \cdot \bar{J}_\eta = \frac{\varsigma}{T} \left[\alpha_{00} \frac{dc}{dt} + \beta_0 T \nabla \cdot \bar{J} \right] + \frac{1}{T} \bar{J} \cdot \left[\alpha_{10} \frac{d\bar{J}}{dt} - T\nabla(\frac{\mu}{T}) + \beta_0 T \nabla \varsigma + \beta_2 T \nabla \cdot \underset{=}{\sigma} \right]$$
$$+ \frac{1}{T} \underset{=}{\sigma} : \left[\alpha_{20} \frac{d\sigma}{dt} + \beta_2 T \nabla \bar{J} \right] \tag{A7}$$

By substituting Equation (A5) into Equation (A3) and using Equation (A7), the following coupled relaxation equations for the non-conserved variables are obtained:

$$\tau_0 \frac{d\varsigma}{dt} = -X_0 + \beta_0 \nabla \cdot \bar{J} \tag{A8}$$

$$\tau_1 \frac{d\bar{J}}{dt} = -\overline{X}_1 - \nabla \mu + \beta_0 \nabla \varsigma + \beta_2 \nabla \cdot \underset{=}{\sigma} \tag{A9}$$

$$\tau_2 \frac{d\underset{=}{\sigma}}{dt} = -\underline{X}_2 + \beta_2 \nabla \bar{J} \tag{A10}$$

It is important to stress that in Equation (A5), S_η may also depend on the parameters that lie outside G, but which are essential to specify the non-equilibrium state of the system. For the model under consideration, the velocity gradient tensor $\nabla \bar{v}$ is required to formulate the constitutive equations for the stress tensor. Therefore, the expressions for X_0, $\overline{X}_1$, and $\underline{X}_2$ are written in terms of the non-conserved variables and $\nabla \bar{v}$. With these considerations, the explicit form of Equations (A8)–(A10) is given in Equations (1)–(3).

References

1. Jou, D.; Criado-Sancho, M.; Del Castillo, L.F.; Casas-Vázquez, J. A thermodynamic model for shear-induced concentration banding and macromolecular separation. *Polymer* **2001**, *42*, 6239–6245. [CrossRef]
2. Jou, D.; Casas-Vázquez, J.; Criado-Sancho, M. *Thermodynamics of Fluids under Flow*; Springer: Berlin, Germany, 2000.
3. Manero, O.; Rodríguez, R.F. A thermodynamic description of coupled flow and diffusion in a viscoelastic binary mixture. *Non-Equilib. Thermodyn.* **1999**, *24*, 177–195. [CrossRef]
4. Bautista, F.; Soltero, J.F.A.; Manero, O.; Puig, J.E. Irreversible thermodynamics approach and modeling of shear-banding flow of wormlike micelles. *J. Phys. Chem. B* **2002**, *106*, 13018–13026. [CrossRef]
5. Manero, O.; Pérez-López, J.H.; Escalante, J.I.; Puig, J.E.; Bautista, F. A thermodynamic approach to rheology of complex fluids: The generalized BMP model. *J. Non-Newton. Fluid Mech.* **2007**, *146*, 22–29. [CrossRef]
6. García-Rojas, B.; Bautista, F.; Puig, J.E.; Manero, O. A thermodynamic approach to the rheology of complex fluids: Flow-concentration coupling. *Phys. Rev. E* **2009**, *80*, 036313. [CrossRef] [PubMed]
7. Puig, J.E.; Bautista, F.; Soltero, J.F.A.; Manero, O. *Giant Micelles. Properties and Applications*; Surfactant Science Series; CRC Press: Boca Raton, FL, USA, 2007.
8. Callaghan, P.T.; Cates, M.E.; Rofe, C.F.; Smeulders, J.B.A.F. A study on the "spurt" effect in wormlike micelles using nuclear magnetic resonance microscopy. *J. Phys. II* **1996**, *6*, 375–393. [CrossRef]
9. Berret, J.F.; Roux, D.C.; Porte, G.; Lindler, P. Shear-induced isotropic–to-nematic phase transition in equilibrium polymers. *Europhys. Lett.* **1994**, *25*, 521–526. [CrossRef]
10. Decrupe, J.P.; Cressely, R.; Makhloufi, R.; Cappelaere, E. Flow birefringence experiments showing a shear banding structure in a CTAB solution. *Colloid Polym. Sci.* **1995**, *273*, 346–351. [CrossRef]
11. Fielding, S.M. Linear instability of planar shear banded flow. *Phys. Rev. Lett.* **2005**, *95*, 134501. [CrossRef]
12. Fielding, S.M. Vorticity structuring and velocity rolls triggered by gradient shear bands. *Phys. Rev. E* **2007**, *76*, 016311. [CrossRef]
13. El-Kareh, A.W.; Leal, L.G. Existence of solutions for all Deborah numbers for a non-Newtonian model modified to include diffusion. *J. Non-Newton. Fluid Mech.* **1989**, *33*, 257–287. [CrossRef]

14. Zhou, L.; Vasquez, P.A.; Cook, L.P. Modeling the inhomogeneous response and formation of shear bands in steady and transient flows of entangled liquids. *J. Rheol.* **2008**, *52*, 591–623. [CrossRef]

15. Yuan, X.S.F.; Jupp, L. Interplay of flow-induced phase separations and rheological behavior of complex fluids in shear banding flow. *Europhys. Lett.* **2002**, *60*, 691–697. [CrossRef]

16. Vasquez, P.A.; McKinley, G.H.; Cook, L.P. A network scission model for wormlike micellar solutions. *J. Non-Newton. Fluid Mech.* **2007**, *144*, 122–139. [CrossRef]

17. Pipe, C.J.; Kim, N.J.; Vasquez, P.A.; Cook, L.P.; McKinley, G.H. Wormlike micellar solutions: II. Comparison between experimental data and scission model predictions. *J. Rheol.* **2010**, *54*, 881. [CrossRef]

18. Olmsted, P.D.; Lu, C.-Y. Coexistence and phase separation in sheared complex fluids. *Phys. Rev. E* **1997**, *56*, R55–R58. [CrossRef]

19. Lerouge, S.; Decruppe, J.P.; Berret, J.F. Correlations between rheological and optical properties of a micellar solution under shear banding flow. *Langmuir* **2000**, *16*, 6464–6474. [CrossRef]

20. Schmitt, V.; Marques, C.M.; Lequeux, F. Shear induced phase separation of complex fluids: The role of flow-concentration coupling. *Phys. Rev. E* **1995**, *52*, 4009–4015. [CrossRef]

21. Fielding, S.M.; Olmsted, P.D. Flow phase diagrams for concentration-coupled shear banding. *Eur. Phys. J. E* **2003**, *11*, 65–83. [CrossRef]

22. Olmsted, P.D. Perspectives on shear banding in complex fluids. *Rheol. Acta* **2008**, *47*, 283–300. [CrossRef]

23. Milner, S.T. Hydrodynamics of semidilute polymer solutions. *Phys. Rev. Lett.* **1991**, *66*, 1477–1480. [CrossRef] [PubMed]

24. García Sandoval, J.P.; Bautista, F.; Puig, J.E.; Manero, O. Inhomogeneous flows and shear banding formation in micellar solutions: Predictions of the BMP model. *J. Non-Newton. Fluid Mech.* **2012**, *179*, 43–54.

25. Lu, C.Y.D.; Olmsted, P.D.; Ball, R.C. Effects of nonlocal stress on the determination of shear banding flow. *Phys. Rev. Lett.* **2000**, *84*, 642–645. [CrossRef] [PubMed]

26. Johnson, M.; Segalman, D. A model for viscoelastic fluid behavior which allows non-affine deformation. *J. Non-Newton. Fluid Mech.* **1977**, *2*, 255–270. [CrossRef]

27. Dhont, J.K.G. A constitutive relation describing the shear banding transition. *Phys. Rev. E* **1999**, *60*, 4534–4544. [CrossRef]

28. Adams, J.M.; Fielding, S.M.; Olmsted, P.D. The interplay between boundary conditions and flow geometries in shear-banding: Hysteresis, band configurations and surface transitions. *J. Non-Newton. Fluid Mech.* **2008**, *155*, 101–118. [CrossRef]

29. Skeel, R.D.; Berzins, M. A Method for the Spatial Discretization of Parabolic Equations in One Space Variable. *SIAM J. Sci. Stat. Comput.* **1990**, *11*, 1–32. [CrossRef]

30. McLeish, T.C.B. Stability of the interface between two dynamic phases in capillary flow of linear polymer melts. *J. Polym. Sci. B* **1987**, *25*, 2253–2264. [CrossRef]

31. Hinch, E.J.; Harris, O.J.; Rallison, J.M. The instability mechanism for two elastic liquids being co-extruded. *J. Non-Newton. Fluid Mech.* **1992**, *43*, 311–324. [CrossRef]

32. Herle, V.; Kohlbrecher, J.; Pfister, B.; Fischer, P.; Windhab, E.J. Alternating Vorticity Bands in a Solution of Wormlike Micelles. *Phys. Rev. Lett.* **2007**, *99*, 158302. [CrossRef]

33. Radulescu, O.; Olmsted, P.D.; Decruppe, J.P.; Lerouge, S.; Berret, J.P.; Porte, G. Time scales in shear banding of wormlike micelles. *Europhys. Lett.* **2003**, *62*, 230–236. [CrossRef]

34. Peterson, J.D.; Cromer, M.; Fredrickson, G.H.; Leal, L.G. Shear Banding Predictions for the Two-Fluid Rolie-Poly Model. *J. Rheol.* **2016**, *60*, 927–951. [CrossRef]

fluids

MDPI

Article

A 3D Numerical Study of Interface Effects Influencing Viscous Gravity Currents in a Parabolic Fissure, with Implications for Modeling with 1D Nonlinear Diffusion Equations

Eden Furtak-Cole [1,*,†,‡] and Aleksey S. Telyakovskiy [2,‡]

1   Department of Mathematics & Statistics, Utah State University, Logan, UT 84321, USA
2   Department of Mathematics & Statistics, University of Nevada-Reno, Reno, NV 89557, USA;
    alekseyt@unr.edu
*   Correspondence: efurtakc@cityu.edu.hk
†   Current address: School of Energy and Environment, City University of Hong Kong, Kowloon Tong,
    Kowloon, Hong Kong.
‡   These authors contributed equally to this work.

Received: 15 December 2018; Accepted: 22 May 2019; Published: 28 May 2019

Abstract: Although one-dimensional non-linear diffusion equations are commonly used to model flow dynamics in aquifers and fissures, they disregard multiple effects of real-life flows. Similarity analysis may allow further analytical reduction of these equations, but it is often difficult to provide applicable initial and boundary conditions in practice, or know the magnitude of effects neglected by the 1D model. Furthermore, when multiple simplifying assumptions are made, the sources of discrepancy between modeled and observed data are difficult to identify. We derive one such model of viscous flow in a parabolic fissure from first principals. The parabolic fissure is formed by extruding an upward opening parabola in a horizontal direction. In this setting, permeability is a power law function of height, resulting in a generalized Boussinesq equation. To gauge the effects neglected by this model, 3D Navier-Stokes multiphase flow simulations are conducted for the same geometry. Parameter variations are performed to assess the nature of errors induced by applying the 1D model to a realistic scenario, where the initial and boundary conditions can not be matched exactly. Numerical simulations reveal an undercutting effect observed in laboratory experiments, but not modeled when the Dupuit-Forchheimer assumption is applied. By selectively controlling the effects placed on the free surface in 3D simulations, we are able to demonstrate that free surface slope is the primary driver of the undercutting effect. A consistent lag and overshoot flow regime is observed in the 3D simulations as compared to the 1D model, based on the choice of initial condition. This implies that the undercutting effect is partially induced by the initial condition. Additionally, the presented numerical evidence shows that some of the flow behavior unaccounted for in the 1D model scales with the 1D model parameters.

Keywords: viscous gravity spreading; nonlinear diffusion equation; Navier-Stokes equations; volume of fluid method; OpenFOAM; generalized Boussinesq equation; porous medium equation; Dupuit-Forchheimer assumption

1. Introduction

Multiple mechanisms can be responsible for the movement of fluids. Gravity currents occur when a difference in the density between two or more fluids induces movement. Applications of this type of movement include both natural [1–4] and industrial [5,6] settings. Examples of such flows are the propagation of fog banks [7], the phreatic surface in unconfined aquifers [8], C02 injection in brine aquifers [9], and the pouring of molten metal or glass [10]. Gravity currents can be used to model miscible and immiscible flows, which may be laminar or turbulent.

Traditionally, modeling approaches for viscous gravity currents rely on analytical 1D models. In many instances, similarity reductions are possible [8,11]. It is harder to extend such techniques to more realistic 2D and 3D settings. Few solutions [12,13] are available for models using the 2D potential flow equations, which can be reduced to a 1D non-linear diffusion equation by using the Dupuit-Forchheimer (DF) assumption [14]. This assumption allows velocity to be calculated from the slope of the free surface and restricts the velocity to the horizontal direction. This results in a non-divergence free velocity field for incompressible fluids, with the continuity equation imposed on the height of the free surface [15]. For uniform permeability, these assumptions result in the Boussinesq equation with constant coefficients. This equation is commonly used in problems involving the spreading of unconfined groundwater [8,16,17], or flow between parallel plates [1]. Similarity solutions exist for a variety of initial and boundary conditions. For the free spreading of a finite pulse of fluid, Barenblatt [11] constructed a solution using a point mass at initial time. Using a symmetry argument, this solution can be applied to both boundary and initial value problems. Barenblatt's solution was extended to the fractal dimensions in work of Hayek [18]. Domains with spatially variable permeability can be modeled using the generalized Boussinesq equation, also called the porous medium equation (PME). Similarity solutions exist for cases where the permeability takes the form of a power law in the horizontal or vertical directions [19–22]. Hele-Shaw cells can be used to study these flows experimentally, with viscous drag serving as a proxy for permeability [23]. Cells constructed in a wedge configuration or with curved walls can be used to validate the PME experimentally. Ciriello et al. [22] constructed similarity solutions to injection problems in horizontal and vertical wedge configurations. For drainage from an initially full aquifer, Zheng et al. [24] developed similarity solutions corresponding to similarly configured experiments. Curved plates resulting in power-law permeability are difficult to manufacture, but have been used for horizontally oriented experiments by Zheng et al. [5], as well as vertical experiments filled with porous media by Longo et al. [25].

Numerical simulations of gravity currents are motivated by the limited scope of similarity solutions, and limitations of the 1D model. Few researchers have conducted numerical simulations of the experiments described above in detail or high dimension. Frolkovič [26] developed a level set solution to the potential flow equations that is applicable to various groundwater problems, including seepage and aquifer drawdown. Numerical modeling efforts specific to the Hele-Shaw cell include Tsay and Hoopes [6] conducting simulations of the free surface of a groundwater mounding problem, Chesnokov and Liapidevskii [27] conducting 2D simulations of viscous fingering, and Bernal and Kindelan [28] simulating 2D non-Newtonian injection. Many authors have pointed out localized discrepancies between experiments and similarity solutions [1,5,22,29], and have offered a variety of case-dependent explanations for the mechanism of error. We have found no examples of attempts to use highly detailed numerical simulations to further investigate these discrepancies. In this work, we pursue this objective.

Numerical investigations of viscous spreading between two non-parallel plates necessitates 3D modeling. For 1D and 2D models, permeability is approximated as a function of space. Our approach is to conduct laminar flow simulations using the 3D Navier-Stokes equations (NSE) with multiphase flow modeled using a volume of fluid (VOF) method. This provides a more direct modeling of the viscous layer that acts in place of permeability, resolves the interface between the fluid phases, and allows for the inclusion of various effects on the interface. To our knowledge, this approach has not yet been applied in this setting. Parabolic prisms provide interesting geometries that are compatible with both 1D analytical solutions and 3D numerical simulations. We derive a flow model for viscous gravity spreading in a parabolic fissure using the DF assumption, resulting in a generalized Boussinesq type equation in Section 2.1. We then numerically investigate different flow scenarios using the 3D model outlined in Section 2.2. This approach allows us to isolate the effects of flow assumptions and parameters in a way that would be intractable using physical experiments. Results of the numerical simulations are outlined in Section 3. From these, mechanisms for the discrepancies between the two models is discussed in Section 4.

2. Governing Equations and Boundary Conditions

2.1. 1D Flow Model

We derive a flow model for a viscous fluid in a parabolic fissure. Consider a fluid element formed by vertically cutting through a parabolic prism, as seen in Figure 1. Conservation of mass dictates that the rate of change of fluid in the elementary volume is equal to the difference in volumetric fluxes through the faces at y and $y + dy$:

$$\frac{\partial}{\partial t}\left(\int_0^{h_v(y,t)} b(z)dzdy\right) = \left(\int_0^{h_v(y,t)} b(z)u_v(y,z,t)dz\right)$$
$$- \left(\int_0^{h_v(y+dy,t)} b(z)u_v(y+dy,z,t)dz\right), \quad (1)$$

where the width of the wedge for height z is given by $b(z)$, u_v is the fluid speed in the y direction, and $h_v(y,t)$ is the height of the free surface. We consider only the flux of an incompressible high viscosity fluid (e.g., glycerol), disregarding the flux of gaseous atmosphere being displaced. Neglecting the higher order terms of the Taylor expansion in the second term on the right hand side of Equation (1) yields,

$$\frac{\partial}{\partial t}\left(\int_0^{h_v(y,t)} b(z)dzdy\right) = -\frac{\partial}{\partial y}\int_0^{h_v(y,t)} b(z)u_v(y,z,t)dzdy. \quad (2)$$

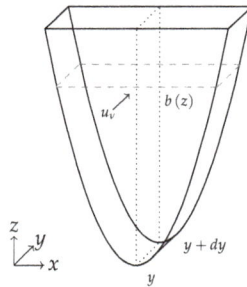

Figure 1. A fluid element within the parabolic Hele-Shaw cell. The cross-sectional width is denoted $b(z)$, and the flux through the element is u_v.

As seen in Figure 1, the parabolic cross section is oriented against gravity in the x-z plane. The equation of the fissure walls in the x-z plane is,

$$z = \omega x^2, \quad (3)$$

for some constant $\omega > 0$. Then the distance between the walls as a function of height is,

$$b(z) = 2\sqrt{\frac{z}{\omega}}. \quad (4)$$

Under the assumption of Poiseuille flow between parallel plates of width $b(z)$ for an arbitrary height z, permeability as a power law function of height is obtained:

$$k(z) = \frac{(b(z))^2}{12} = \frac{z}{3\omega}. \tag{5}$$

This permeability results in the Darcy velocity,

$$u_v(y, z, t) = -\frac{z}{3\omega} \frac{\rho g}{\mu} \frac{\partial h_v(y, t)}{\partial y}. \tag{6}$$

Integrating Equation (2), with substitutions for width (4) and Darcy velocity (6), yields

$$\frac{\partial}{\partial t} \left(\frac{4}{3\sqrt{\omega}} h_v^{\frac{3}{2}} \right) = \frac{\partial}{\partial y} \left(\frac{4}{15\sqrt{\omega^3}} \frac{\rho g}{\mu} h_v^{\frac{5}{2}} \frac{\partial h_v}{\partial y} \right), \tag{7}$$

which further simplifies to,

$$\frac{\partial}{\partial t} \left(h_v^{\frac{3}{2}} \right) = \frac{\rho g}{5\omega\mu} \frac{\partial}{\partial y} \left(h_v^{\frac{5}{2}} \frac{\partial h_v}{\partial y} \right). \tag{8}$$

The change of variables, $Y = h_v^{\frac{3}{2}}$, is implemented to obtain a special form of the PME [15]:

$$\frac{\partial}{\partial t} Y = \frac{2\rho g}{35\omega\mu} \frac{\partial^2}{\partial y^2} Y^{\frac{7}{3}}. \tag{9}$$

We are interested in case of the free spreading of a finite pulse of liquid under the force of gravity, which results in the boundary condition,

$$Y(0, t) = \sigma t^{-\frac{1}{7/3+1}}, \tag{10}$$

where $\sigma > 0$ for an initially empty fissure [11].

The problem defined by Equations (9) and (10) can be solved by introducing dimensionless similarity variables,

$$\xi = \frac{y(1 + \beta(m-1))^{\frac{1}{2}}}{(a\sigma^{(m-1)} t^{1+\beta(m-1)})^{\frac{1}{2}}}, \tag{11}$$

and

$$Y = \sigma t^{\beta} H(\xi), \tag{12}$$

where $\xi(y, t)$ is the similarity variable, $H(\xi)$ is a scaling function, $m = 7/3$ is the nonlinear diffusion exponent in the PME, and

$$a = \frac{2\rho g}{35\omega\mu}. \tag{13}$$

The parameter β is the time exponent of the boundary condition, in this case $-\frac{1}{7/3+1}$. Substitution of the similarity variables results in the ordinary differential equation,

$$\frac{d^2 H^m(\xi)}{d\xi^2} + \frac{\xi}{2} \frac{dH(\xi)}{d\xi} - \lambda H(\xi) = 0, \tag{14}$$

where $\lambda = \frac{\beta}{1+\beta(m-1)}$, and $H(0) = 1$, $H(\xi_0) = 0$. The front position in terms of similarity variables is ξ_0, and is found in the process of solution. The "mathematician's pressure" change of variables [15],

$$H = u^{\frac{1}{m-1}}, \tag{15}$$

performed on Equation (14) produces an ODE with integer powers:

$$\frac{m}{m-1}u\frac{d^2u}{d\xi^2} + \frac{m}{m-1}\left(\frac{m}{m-1}-1\right)\left(\frac{du}{d\xi}\right)^2 + \frac{\xi}{2}\frac{1}{m-1}\frac{du}{d\xi} - \lambda u = 0. \tag{16}$$

Boundary conditions are given by,

$$u(0) = 1, \tag{17}$$

and

$$u(\xi_0) = 0. \tag{18}$$

Equations (16) and (18) can be solved using a Runge-Kutta solver [30], or a power series of the form found in [29]:

$$u(\xi) = \sum_{n=0}^{\infty} a_n \left(1 - \frac{\xi}{\xi_0}\right)^n. \tag{19}$$

2.2. 3D Flow Model

Flow through the parabolic fissure can be modeled with the incompressible 3D Navier-Stokes equations,

$$\rho\frac{D\mathbf{u}}{Dt} = -\nabla p + \rho\mathbf{g} + \mu\Delta\mathbf{u}, \tag{20}$$

and

$$\nabla \cdot \mathbf{u} = 0, \tag{21}$$

where ρ is density, $\mathbf{u}$ is velocity, p is pressure, μ is viscosity, and $\mathbf{g}$ is the gravity vector. Phases are tracked using a VOF method [31,32]. The indicator function α takes on values in the range of $[0,1]$ and can be interpreted as the fraction of two phase types within a computational cell. For our purposes, 0 denotes air, and 1 denotes a viscous liquid. For any computational cell in the domain, the average fluid properties for density and viscosity are calculated by,

$$\rho = \sum_i \rho_i \alpha_i, \tag{22}$$

and

$$\mu = \sum_i \mu_i \alpha_i, \tag{23}$$

where α_i denotes the volume fraction of phase i. The indicator function evolves under the equation,

$$\frac{\partial\alpha}{\partial t} + \nabla \cdot \mathbf{u}\alpha + \nabla \cdot \mathbf{u}_c\alpha\left(1 - \alpha\right) = 0, \tag{24}$$

where the counter gradient transport term $\nabla \cdot \mathbf{u}_c\left(1 - \alpha\right)$ acts as a limiter based on the maximum velocity $\mathbf{u}_c$ on interface. Surface tension effects are calculated by,

$$S = \gamma K \nabla\alpha, \tag{25}$$

where γ is the surface tension, and K is the curvature $-\nabla \cdot \mathbf{n}$. S is added to the momentum Equation (20) to obtain,

$$\rho\frac{D\mathbf{u}}{Dt} = -\nabla p + \rho\mathbf{g} + \mu\Delta\mathbf{u} + \gamma K \nabla\alpha. \tag{26}$$

This method has been successfully used in applications at a variety of scales [33–36]. Our use of this method is very conservative, in that the flow is laminar and surface effects are not the main driver of flow. We use the *interFoam* multiphase flow solver included in the finite volume method toolbox *openFOAM* [37] for all 3D flow simulations. A hexahedral base mesh was constructed with non-uniform grading decreasing cell size towards the y-z plane and base of the parabola. This mesh was snapped to a 3D parabolic geometry using the *snappyHexMesh* utility. The parabolic geometry was constructed to the specifications given in Section 2.3. The same mesh was utilized for all simulations, with time-step limiting based on a maximum Courant number of 0.25. With the available computing resources, each simulation required under two days to complete 3 s of simulated time.

Boundary conditions are chosen to mimic a lab experiment, with the top of the parabola open to the atmosphere. For velocity, a no-slip condition,

$$\mathbf{u} = 0, \tag{27}$$

is applied to all solid walls. To allow fluxes of air at the top of the experiment, a zero gradient condition is applied with respect to the patch normal vector:

$$\frac{\partial \mathbf{u}}{\partial \mathbf{n}} = 0. \tag{28}$$

A no-flow pressure condition is applied at the solid walls,

$$\frac{\partial p}{\partial \mathbf{n}} = 0, \tag{29}$$

while at the top of the experiment total pressure P_{tot} is calculated using the gauge pressure p_{atm},

$$p_{tot} = p_{atm} - .5\rho \|\mathbf{u}\|^2. \tag{30}$$

Unless otherwise specified, the indicator function α is fixed at the top of the experiment and zero gradient at solid walls,

$$\frac{\partial \alpha}{\partial \mathbf{n}} = 0. \tag{31}$$

2.3. Flow Assumptions, Initial Conditions and Parameter Values

The 1D flow model uses the DF assumption, and does not include the effects of air flow, surface tension or contact angle. The 3D model includes a complete velocity field and optional effects at the liquid-gas-wall interface, but requires large amounts of computational power to solve. This is largely due to the aspect ratio of the domain, which requires very small finite volumes between the parabola walls. The parabolic geometry defined by Equation (4) is extruded along the y axis for one meter with a value $\omega = 2000$. This corresponds to a gap width of $x = 2$ cm at a maximum height of $z = 0.2$ m.

An initial condition for the 3D simulation was chosen from the 1D analytical solution using the parameter values in Table 1. The 1D viscosity is held constant at $\mu = 1.4$ Pa · s for generating the initial condition, though this is effectively varied as part of a (Equation (13)) in the generation of results. Choosing a solution curve from the 1D model at a nonzero starting time is required, as the solution collapses to a singularity at $t = 0$ s and the height of the solution is above the $z \in [0, 0.2]$ m of computational domain at early time. The resulting curve at time $t = 0.3$ s was chosen as a suitable initial condition containing enough potential energy to produce interesting results. Additional 3D numerical experiments were performed under different values of the parameter a using the same initial condition by exploiting the scaling nature of the similarity solution. After the ODE system Equations (16) and (18) is solved and ζ_0 obtained, the propagation distance y of the 1D solution can be computed as a function of time:

$$y = \xi_0 \frac{\left(a\sigma^{(m-1)} t^{1+\beta(m-1)}\right)^{1/2}}{(1+\beta(m-1))^{1/2}}. \tag{32}$$

To compare between 1D and 3D results, the various moments in time from the 1D model to which the initial condition used in 3D simulations corresponds must be calculated. Let the propagation distance of the aforementioned numerical initial condition be y_0, and a' be an alternate parameter value calculated from a 3D simulation parameter set. Then the start time of the numerical initial condition under the parameter set that produced a' is,

$$t_0 = \left(\frac{y_0^2}{\xi_0^2} \frac{1+\beta(m-1)}{a'\sigma^{(m-1)}}\right)^{\frac{1}{1+\beta(m-1)}}. \tag{33}$$

Table 1. Parameter values for the 1D and 3D models.

1D Parameter	Value	3D Parameter	Value
σ	0.06 m	-	-
ρ	1261 kg $\cdot$ m^{-3}	ρ_{liquid}	1261 kg $\cdot$ m^{-3}
μ	0.8–2.4 Pa $\cdot$ s	μ_{liquid}	0.8–2.4Pa $\cdot$ s
g	9.8 m $\cdot$ s^{-2}	g	9.8 m $\cdot$ s^{-2}
ω	2000	-	-
-	-	ρ_{air}	1 kg $\cdot$ m^{-3}
-	-	μ_{air}	1.48×10^{-5} Pa $\cdot$ s
-	-	γ	0.0634 N $\cdot$ m^{-1}

3. Results

In Section 3.1, the models are validated against one another, and the mesh quality of the numerical simulations is shown to be adequate. The mechanisms of discrepancy between the 1D and 3D models is established. Section 3.2 explores the less prominent effects of contact angle and surface tension at the free surface. Additionally, we examine the transition to the similarity flow regime from an initial condition that is not in the similarity solution set. Finally, Section 3.3 examines the behavior of the 3D solution under variations in the parameter a, which is compared against the 1D solution for the analytical initial condition.

3.1. Profile Comparison and Validation

A comparison between a baseline 3D simulation with $\gamma = 0$ and the 1D similarity solution is shown in Figure 2. Excellent agreement is observed between the two models, both in propagation distance and shape. The poorest fit is seen at the earliest time, Figure 2a. The simulated front lags behind the analytical solution at the base of the parabola, and undercutting is observed at the bottom of the advancing front. At later times, the free surface of the 1D analytical solution lags behind the NSE simulation at the "nose" of the advancing front.

The flow profiles observed in this simulation indicate that a steep free surface is responsible for the undercutting observed at early time. The effects of surface tension and contact angle are not included in the simulation used to produce Figure 2, a situation that would be difficult to replicate with experimental techniques.

The 1D flow model presented in Section 2.1 stems from a larger family of nonlinear diffusion equations which have been validated for flow through various fissures. These equations can be derived by substituting an alternative expression for Equation (5). In the case of parallel walls, the resulting Boussinesq equation was experimentally validated by Huppert and Woods [1] using a Hele-Shaw cell and glycerol. Experiments with angled planar walls were validated by Furtak-Cole et al. [29] as well as Ciriello et al. [22] using similar methods. The permeability profile from Equation (5) is bounded on either side by these two classes of experiments, making experimental validation unnecessary.

The interFoam solver presented in Section 2.2 is a type of VOF method [31] which has been used and validated for a variety of applications [33,36]. The cases presented in this paper represent a conservative use of this method, as the maximum Reynolds number is less than one. The role of mesh quality was tested at two levels. Base meshes were constructed with 60 and 80 cells partitioning the width between the parabolic walls. This width is consistent with the x direction shown in the fluid element for the 1D model, see Figure 1. Cell counts in the remaining directions were calculated to maintain a 1-1-1 average aspect ratio. As seen in Figure 3, a consistent rate of propagation was reached at both levels. Propagation distance for the fine mesh lags slightly behind as expected, due to its ability to better resolve the free surface. The finer mesh was used for all simulations presented here. Validation of both 1D and 3D flow models in independent settings, as well as their agreement displayed in Figure 2, implies an acceptable level of accuracy for both solution types.

3.2. Effects of Surface Tension, Contact Angle, and Initial Conditions

The similarity solution to the 1D flow problem is a weak, continuous solution. By contrast, the interface resolved by the numerical simulations is double valued at the propagating edge. To quantify this early time behavior observed in experiments and simulations, we measure the height of the undercutting observed in Figure 2a. This is measured vertically from the base of the parabola to the "nose" of the advancing front, where the air-liquid interface is vertical. To further test the assertion that the vertical velocity is primarily responsible for the undercutting behavior, the height of undercutting for simulations with surface tension, dynamic contact angle, and an alternative initial condition are measured. Surface tension for the 3D air-liquid interface is listed in Table 1 and based on the value for glycerol, a viscous fluid commonly used in experiments. A dynamic contact angle θ was modeled using the function,

$$\theta = \theta_0 + (\theta_a - \theta_r) \tanh\left(\frac{u_{wall}}{u_\theta}\right), \tag{34}$$

where θ_0 is the static contact angle, u_{wall} the speed of the interface tangent the static wall, u_θ is a scaling factor, and θ_a and θ_r are the advancing and receding angles respectively. Given the hypothetical nature of this study and lack of physical data, we arbitrarily choose a neutral parameter set, $\theta_0 = 90°$, $\theta_r = 60°$, $\theta_a = 120°$, and scaling parameter $u_\theta = 1$.

To perform physical experiments, initial conditions that do not coincide with the similarity solution are often used out of necessity. Such initial conditions have been observed to converge quickly on the similarity solution [5,38]. Numerical experiments offer a previously unobtainable opportunity to compare the convergence of arbitrary initial conditions. A hypothetical mass-conserving initial condition of rectangular cross section was produced by solving,

$$\int_0^{y_1} \int_0^{z=0.18} 2\sqrt{\frac{z}{2000}}\, dz\, dy = 3.425\,48 \times 10^{-4}, \tag{35}$$

where the upper limit $z = 0.18$ was chosen arbitrarily. The value for the right hand side was produced by numerically computing the volume integral defined by the analytically-derived initial condition from the 1D flow model in the parabolic geometry. For practical purposes, such an initial condition could be used in an experiment by setting a lock gate to hold a volume of liquid at the initial time.

Undercut heights for simulations with no interface effects (baseline), contact angle, surface tension, and the alternative rectangular initial condition are shown in Figure 4. Propagation distance and speed for the baseline and rectangular initial condition 3D simulations are compared to the 1D model in Figure 5.

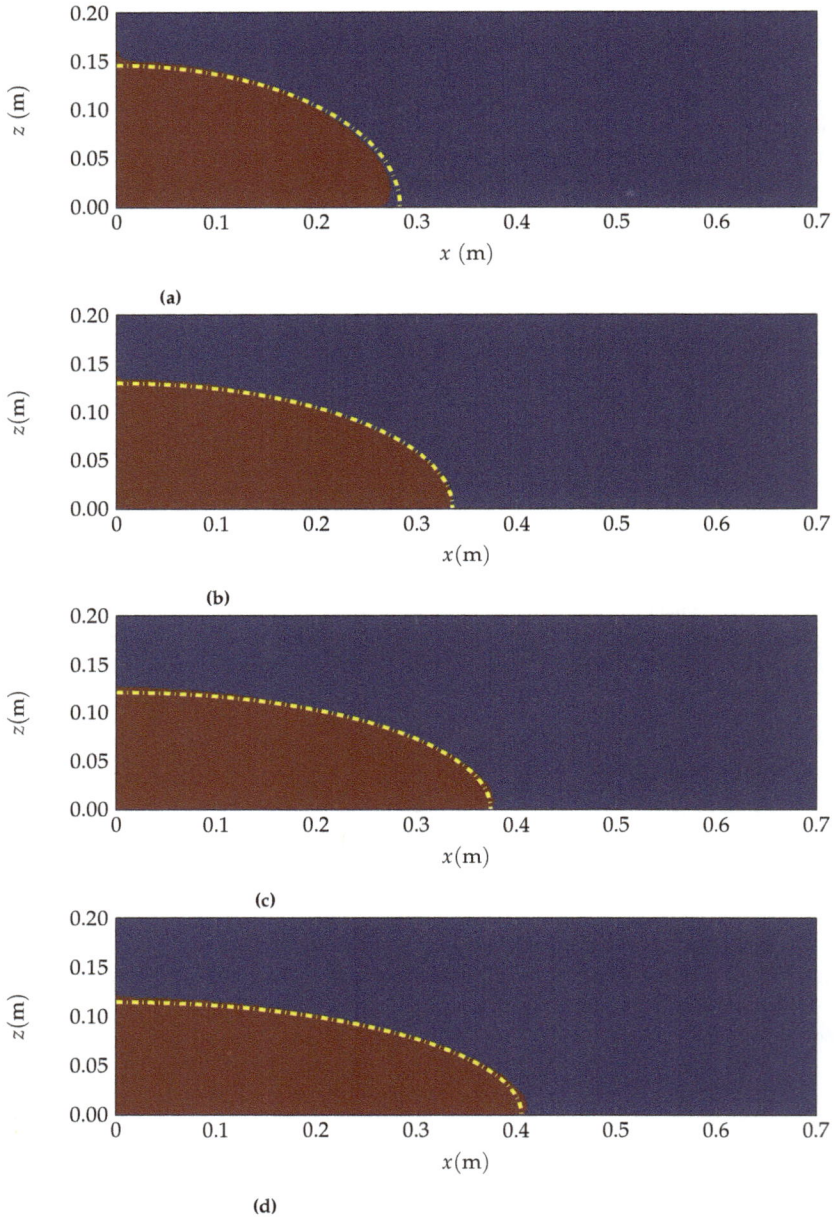

Figure 2. Comparisons in the $y - z$ plane cross section between the NSE simulation and 1D flow model at (**a**) 1.3 s, (**b**) 2.3 s, (**c**) 3.3 s, and (**d**) 4.3 s. Red represents glycerol in the simulation results, the yellow dotted line is the free surface from the 1D model.

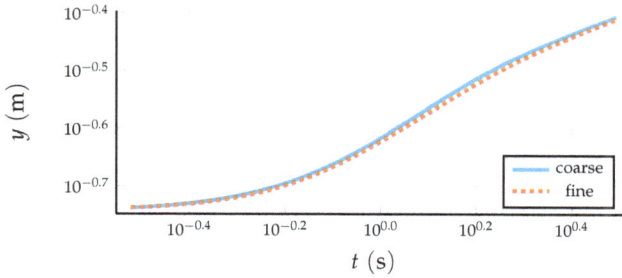

Figure 3. Front propagation for identical simulations of viscosity $\mu = 1.4$ with two mesh partitions. The coarse corresponds to 60 cells in the *x*-direction, while the fine has 80.

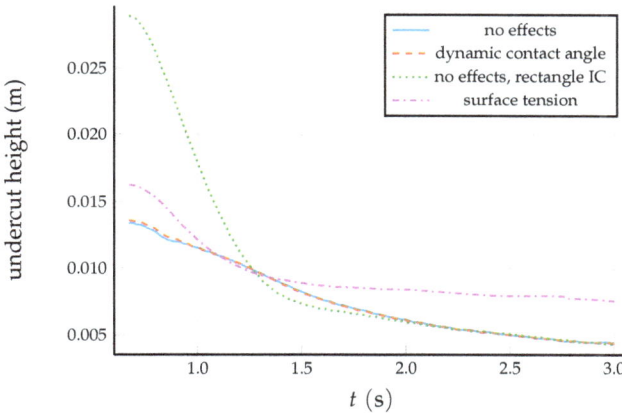

Figure 4. Height of undercutting plotted from its maximum as a function of time. The removal of surface tension results in a profile with less undercutting.

3.3. Scaling Effects

The previously presented results raise the question of whether or not early time inaccuracies of the 1D model can be predicted knowing only the parameters in Equation (9). Given that the coefficients of the right hand side of Equation (9) can be consolidated in a single constant a (Equation (13)), a single parameter can be varied without changing the ode problem given by Equation (14). A natural choice is to vary μ. This allows for all simulations to be performed with the same computational settings, preventing any false comparison of finite volume mesh-induced errors. For the same initial condition, simulations were conducted at the viscosities $\mu = 0.8 \, \text{Pa} \cdot \text{s}$, $1.2 \, \text{Pa} \cdot \text{s}$, $1.6 \, \text{Pa} \cdot \text{s}$, $2.0 \, \text{Pa} \cdot \text{s}$, and $2.4 \, \text{Pa} \cdot \text{s}$. This approach assumes that the parameter values result in a laminar flow, and that at the boundary $y = 0$ the fixed parameters of Equation (10) guarantee the liquid phase free surfaces change strictly due to gravity. For each simulation, Equation (33) was used to calculate the simulation start time for the single numerical initial condition chosen in Section 2.3, since the same initial profile for different values of μ corresponds to different initial moments in time. Propagation distances are shown in Figure 6, while undercut heights are presented in Figure 7.

(a)

(b)

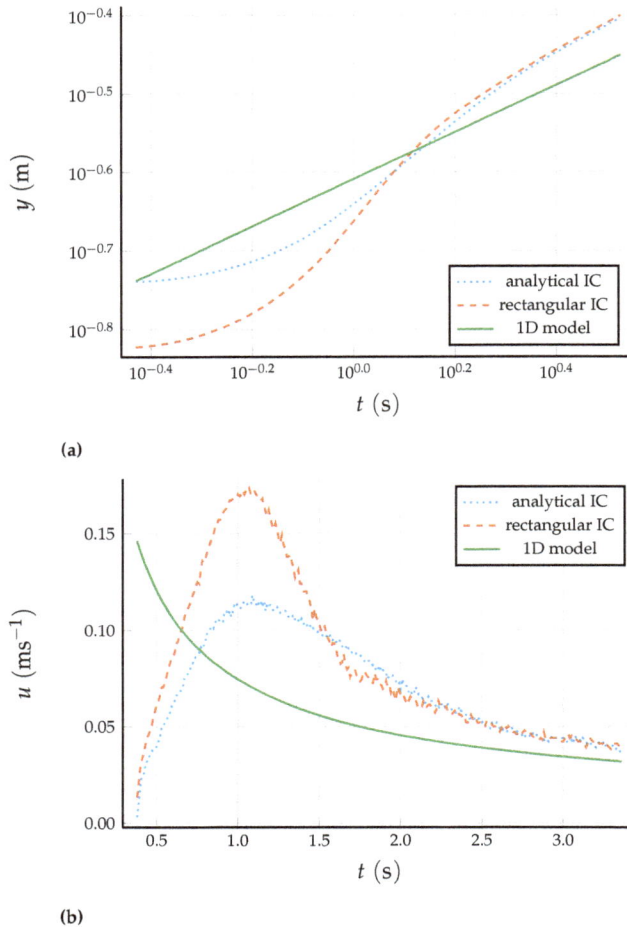

Figure 5. (**a**) Front propagation for two baseline 3D simulations and the 1D flow model. The analytical and rectangular initial conditions used have the same mass. (**b**) Propagation velocity *u* for the same 3D simulations and 1D model. A lag is observed as momentum builds, and results in overshoot at the nose of the advancing front where distance is measured in (**a**). Numerical calculation of front velocity reveals convergence in (**b**).

4. Discussion

The DF assumption is known to perform poorly for steep free surfaces and the data presented from the numerical simulations make it possible to quantify this deficiency. Use of the DF assumption is often preceded with stipulations such as the ratio of horizontal displacement over height being much larger than one [1]. A smaller ratio results in an increasingly steep free surface with more vertical velocity. Our numerical simulations indicate that vertical velocity components are not only important at the advancing front. Figure 8 shows velocity vectors and their magnitude at a cross section of a baseline simulation as given in Figure 2, after 0.4 s of simulation time. Within the fluid mound, velocities are large only near the interface, while fluid at the lower-left corner is nearly at rest. Primarily vertical velocities are seen at both steep and nearly horizontal segments of the free surface. We calculate continuity error numerically using two control volumes given by a modified version of Equation (1). Here, u_v is given by a surface integral of the spatially variable y component of the velocity in the liquid phase of the 3D simulation seen in Figure 8. Control volumes of 0.01 m width are placed 0.02 m from the left wall and 0.01 m to the left of the advancing front. The time derivative for the volume integral on the left hand side of Equation (1) is approximated with a forward derivative calculated over 0.01 s. The left control volume, positioned at a nearly horizontal free surface, results in a relative continuity error of 0.54 %. The right control volume, located on the steep free surface of the advancing front, gives a relative error of 5.06 %. In this sense, the 1D model holds well for vertical motions that do not appear near a vertical interface. This indicates that nose formation at the advancing front is at least partially due to a flow regime where fluid "skims" in a direction parallel to the free surface, perhaps better called "overtopping" than "undercutting". What we have measured as "undercut" is in fact a combination of mechanisms, including viscous drag unaccounted for in the 1D model. In some experiment configurations [1,5] viscous drag is not modeled at the base of the experiment and this effect can not be distinguished from vertical velocity due to the steep free surface. In these cases, it is likely that overtopping is actually compensating viscous drag at the base of experiments, resulting in good agreement between the 1D and experimental results at later times. Indeed, other experiments using vertical wedge configured plates [22,29] experience undercut profiles. No unaccounted drag exists at the base of experiments in these cases, though effects such as surface tension become immense where the plates meet and the cited experiments were performed at an empty initial state with flux boundaries. For the 1D model of a parabolic geometry there is no unaccounted drag at the base of the parabola, though an analogous problem exists: as height decreases to zero, Equation (6) is an increasingly poor approximation of the Darcy velocity because it is based on horizontal width only.

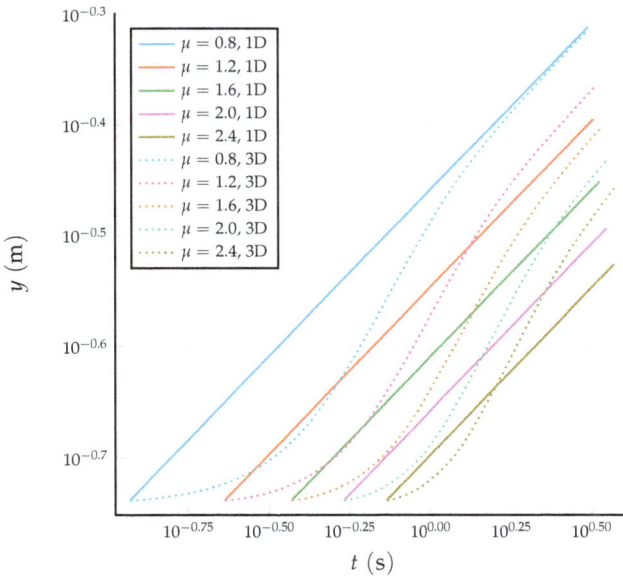

Figure 6. An overlay of propagation distances for the 1D model and 3D simulations at various viscosities.

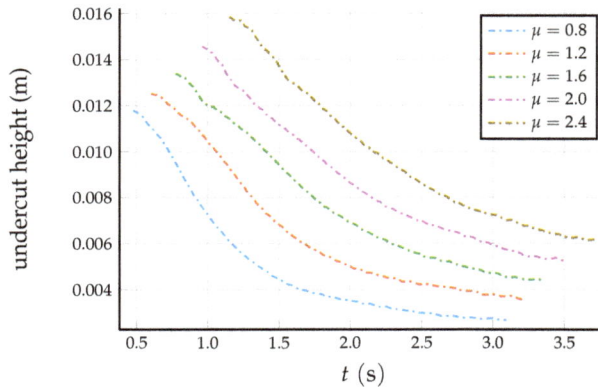

(a)

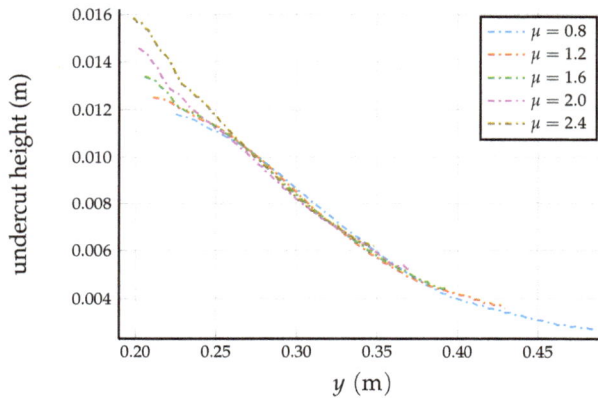

(b)

Figure 7. (**a**) Undercut heights plotted from peak value as a function of time. Shifting of the peaks is largely due to the start times calculated by Equation (33). (**b**) Undercut heights plotted from peak value as a function of space. In all cases, only μ is varied and no interfacial effects are present.

As shown in Figure 4, the magnitude of measured undercutting can be modified by small perturbations in the initial condition or effects acting at the free surface. The rectangular initial condition's infinite slope dramatically increased the magnitude of undercutting, but the solution rapidly converged on the same trajectory as the analytical initial condition within two seconds. Surface tension caused more extreme long term behavior. The extra force acting parallel to the free surface resists bending, allowing the fluid to override the lowest region of the parabolic wedge for an extended period of time. The Bond number is calculated $Bo = \Delta \rho g L^2 / \gamma$, and represents the interplay between gravitational and surface tension forces. Here, the characteristic length L^2 is determined by the width of the wedge at the height of undercutting in Figure 9a. In the surface tension case, undercut is consistently higher than the baseline case, with the exception of the curves meeting between 1 s and 1.5 s, as seen in the difference plot Figure 9b. This corresponds to large vertical velocities along the leading edge driving the liquid phase downward. Similar residuals are observed at 0.75 s and 2.5 s for an order of magnitude difference in Bo, emphasizing the complex relationship between dynamic velocity and surface tension at the interface. The additional simulation using fairly neutral dynamic contact angle parameters shows little change in undercutting as compared to the baseline case.

This could undoubtedly be altered by selecting more extreme parameters, however the complicated physics and numerical issues of this topic warrant a separate study. As with the undercutting results, the propagation plots of Figure 5 show the numerical simulations converging on one another rapidly from different initial conditions. While there appears to be discrepancy between the simulations and 1D solution at later times, it is important to note that propagation is measured as the furthest point at the nose of the advancing front. Thus the shape of the profile plays an out-sized role in the measurement of propagation distance, which is not an all-inclusive means of comparing a 3D interface to a 1D solution. More importantly, Figure 5b shows convergence between the propagation velocities for the 3D numerical simulations and 1D model.

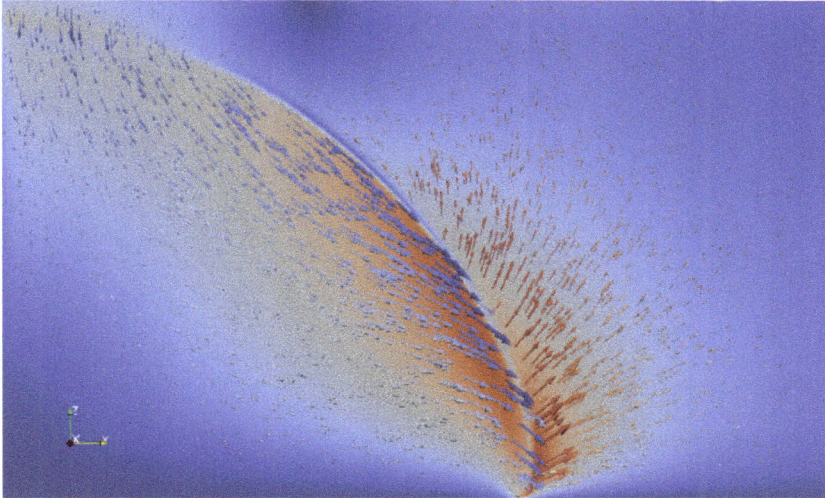

Figure 8. Velocity vectors from the cross section of a 3D simulation. Vectors are scaled by magnitude and colored by vertical component, with blue denoting downward velocity and red upward velocity. The background is colored by velocity magnitude, with blue denoting zero and red denoting high velocity. Strong downward motions are seen within the fluid mound (left), while strong updrafts are seen in the air being displaced by the advancing front (center right). The midpoint along the arc length of the interface is characterized by a nearly stagnant zone, where velocity changes from negative at left to positive at right with respect to the interface outward normal vector.

Simulations produced for different viscosities show that a period of velocity deficit and overcompensation occurs whenever a simulation is started, as compared to the 1D model solution. This presents a particular problem for experiments where it would be difficult to provide the initial condition with a velocity field from the 1D solution. The starting point of the initial condition relative to the similarity solution and lack of initial velocity are dominant factors in the lack of agreement between the two models. The simulations show increasingly poor convergence to the 1D model the later they are started in time, see Figure 6. In the case of our specific similarity solution, this corresponds to using a higher value of viscosity; a technique often employed to maintain laminar flow in experiments. Initial profiles close to the singular initial condition for the 1D model, and correct initial velocity, are far from the aspect ratio condition that makes the similarity solution applicable. Our results do not establish a general guideline for how severe this effect is; we have not generated a set of different initial profiles corresponding to the same initial moment in time for the different values of viscosity considered here. Still, Figure 6 does show that it would be easy to unknowingly choose a poor combination of initial condition and viscosity (or other parameters that make up *a*) prior to performing an experiment. Special attention should be paid to this issue.

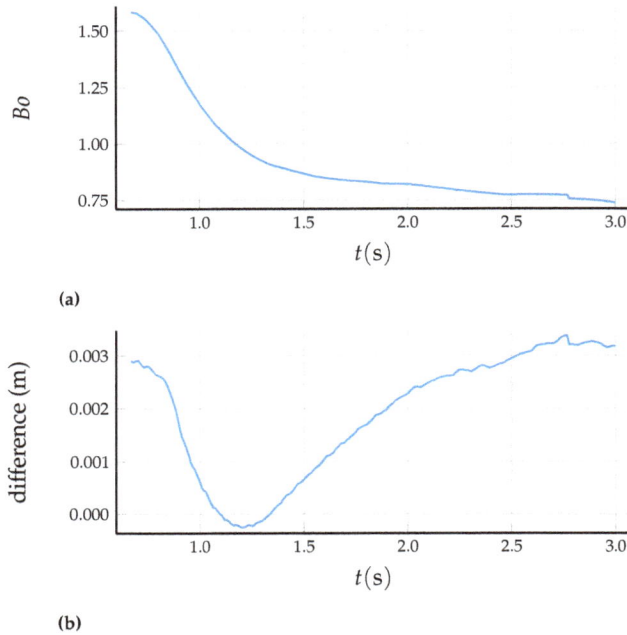

(a)

(b)

Figure 9. (a) Bond number calculated at the height of undercut for the surface tension case. (b) Difference between the surface tension and baseline cases. Bond number and differences from simulations presented in Figure 4. The lack of correlation between *Bo* and the difference occurs as velocity builds from the initial condition.

The dominant role of the initial condition in generating discrepancy is further underscored by plotting the undercut as a function of space in Figure 7b. For each viscosity the time corresponding to the initial condition is different, though each simulation quickly converges to the same trajectory through space. However, the peak value of undercut and its magnitude is shown to shift in space for variations in the value of *a*. Further analysis reveals a linear relationship between the peak *y* displacement and the parameter *a*, shown in Figure 10a. This relationship suggests realistic effects that are not accounted for in the 1D model are still intimately related to the similarity variables governing the 1D solution, and can be predicted as such. Similarly, the magnitude of undercut is plotted against *a* in Figure 10b, and shows equally linear behavior. The relationships established are purely empirical and further work is needed to understand their mechanism.

5. Conclusions

This paper investigated some of the effects that are neglected by 1D models of viscous gravity spreading under the DF assumption. To have the ability to gauge each effect separately, we conducted highly accurate 3D NSE VOF simulations. Effects neglected by the 1D model were included separately to assess the magnitude of their influence on fluid flow. These effects include vertical components of the flow, surface tension, and contact angle. Each simulation involved the spreading of a finite volume of liquid under the action of gravity.

(a)

(b)

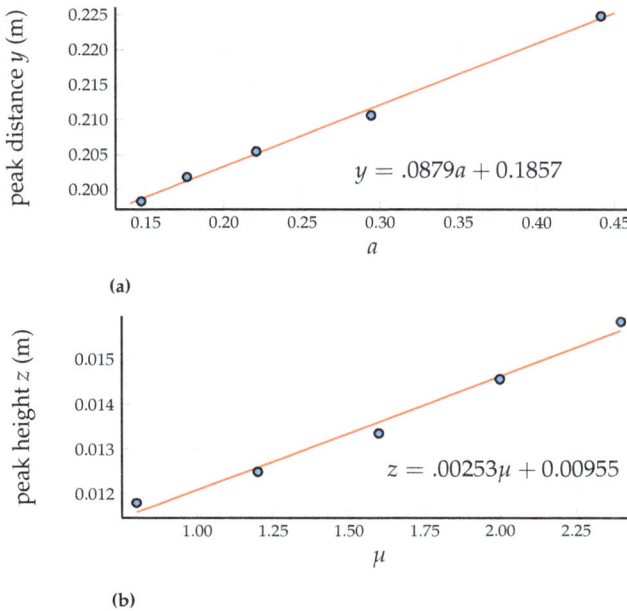

Figure 10. (**a**) A strong linear relationship is observed between the peak distance y and the parameter a, with a standard error of 0.0031 m. (**b**) Peak undercut height as a function of μ. Standard error for the linear fit is 0.0018 m. Both are measured from the baseline case when no modeled interface effects are present.

Undercutting was observed in both the 3D numerical simulations presented here, as well as in previous work by the authors and others. This effect can not be modeled if the DF assumption is used. Undercutting appears when the free surface of the fluid overruns the position of the wetting front at the bottom of the fissure. In this paper, numerical investigations have shown that the primary cause of this effect is vertical velocity induced by a steep free surface. Effects on the free surface may intensify this behavior. Undercutting is strongly influenced by the choice of initial condition. However, late time overall agreement between the two solutions was observed to be very good. This observation is corroborated by previous experimental work. At late times the similarity solution is known to perform well, and the observed height of undercutting takes a small value that changes very slowly for all initial conditions.

To summarize, the placement of the initial condition for numerical experiments relative to the similarity solution parameters produces a wide dilatation in both undercutting and in the position of the advancing front. Initial conditions for numerical simulations that are close quickly converge on each other. Generally, the 1D model predicted the correct speed of propagation, with shortcomings in modeling the shape of the free surface. The DF assumption and associated vertical velocity near steep segments of the free surface are largely responsible for discrepancies between the two models. Effects included at the interface have a secondary role to the above points for the cases considered here.

Author Contributions: Conceptualization, E.F.-C. and A.S.T.; methodology, E.F.-C. and A.S.T.; software, E.F.-C.; validation, E.F.-C.; formal analysis, E.F.-C.; resources, E.F.-C.; writing—original draft preparation, E.F.-C.; writing—review and editing, E.F.-C. and A.S.T.; visualization, E.F.-C.

Funding: This research received no external funding.

Acknowledgments: Eden Furtak-Cole thanks the Utah Center for High Performance Computing for the management of computing resources and access to their technical expertise.

Conflicts of Interest: The authors declare no conflict of interest.

Abbreviations

The following abbreviations are used in this manuscript:

PME   Porous medium equation
DF   Dupuit-Forcheimer
NSE   Navier-Stokes equation
VOF   Volume of Fluid
Bo   Bond number

References

1. Huppert, H.E.; Woods, A.W. Gravity-driven flows in porous layers. *J. Fluid Mech.* **1995**, *292*, 55–69. [CrossRef]
2. Huppert, H.E. Gravity currents: A personal perspective. *J. Fluid Mech.* **2006**, *554*, 299. [CrossRef]
3. Blanchette, F.; Strauss, M.; Meiburg, E.; Kneller, B.; Glinsky, M.E. High-resolution numerical simulations of resuspending gravity currents: Conditions for self-sustainment. *J. Geophys. Res.* **2005**, *110*. [CrossRef]
4. Simpson, J.E. Gravity Currents in the Laboratory, Atmosphere, and Ocean. *Annu. Rev. Fluid Mech.* **1982**, *14*, 213–234. [CrossRef]
5. Zheng, Z.; Christov, I.; Stone, H. Influence of heterogeneity on second-kind self-similar solutions for viscous gravity currents. *J. Fluid Mech.* **2014**, *747*, 218–246. [CrossRef]
6. Tsay, T.S.; Hoopes, J.A. Numerical simulation of ground water mounding and its verification by Hele–Shaw model. *Comput. Geosci.* **1998**, *24*, 979–990. [CrossRef]
7. Hrebtov, M.; Hanjalić, K. Numerical Study of Winter Diurnal Convection Over the City of Krasnoyarsk: Effects of Non-freezing River, Undulating Fog and Steam Devils. *Bound. Layer Meteorol.* **2017**, *163*, 469–495. [CrossRef]
8. Rupp, D.E.; Selker, J.S. Drainage of a horizontal Boussinesq aquifer with a power law hydraulic conductivity profile. *Water Resour. Res.* **2005**, *41*. [CrossRef]
9. Golding, M.J.; Neufeld, J.A.; Hesse, M.A.; Huppert, H.E. Two-phase gravity currents in porous media. *J. Fluid Mech.* **2011**, *678*, 248–270. [CrossRef]
10. Naaim, M.; Gurer, I. Two-phase Numerical Model of Powder Avalanche Theory and Application. *Nat. Hazards* **1998**, *17*, 129–145.:1008002203275. [CrossRef]
11. Barenblatt, G.I. *Scaling*; Cambridge University Press: Cambridge, UK, 2003.
12. Polubarinova-Kochina, P.Y. *Theory of Groundwater Movement*; Princeton University Press: Princeton, NJ, USA, 1962.
13. Longo, S.; Di Federico, V. Axisymmetric gravity currents within porous media: First order solution and experimental validation. *J. Hydrol.* **2014**, *519*, 238–247. [CrossRef]
14. McWhorter, D.B.; Sunada, D.K. *Ground-Water Hydrology and Hydraulics*; Water Resources Publications: Highlands Ranch, CO, USA, 1977.
15. Vazquez, J.L. *The Porous Medium Equation, Mathematical Theory*; Oxford University Press: Oxford, UK, 2007.
16. Lockington, D.A.; Parlange, J.Y.; Parlange, M.B.; Selker, J. Similarity solution of the Boussinesq equation. *Adv. Water Resour.* **2000**, *23*, 725–729. [CrossRef]
17. Telyakovskiy, A.S.; Braga, G.A.; Kurita, S.; Mortensen, J. On a power series solution to the Boussinesq equation. *Adv. Water Resour.* **2010**, *33*, 1128–1129. [CrossRef]
18. Hayek, M. An exact solution for a nonlinear diffusion equation in a radially symmetric inhomogeneous medium. *Comput. Math. Appl.* **2014**, *68*, 1751–1757. [CrossRef]
19. Barenblatt, G.I. On Some Unsteady-State Movements of Liquid and Gas in Porous Medium. *Prikl. Mat. Mekh.* **1952**, *16*, 67–78. (In Russian)
20. Barenblatt, G.I. On some problems of unsteady filtration. *Izv. AN SSSR* **1954**, *6*, 97–110. (In Russian)
21. Olsen, J.S.; Telyakovskiy, A.S. Polynomial approximate solutions of a generalized Boussinesq equation. *Water Resour. Res.* **2013**, *49*, 3049–3053. [CrossRef]
22. Ciriello, V.; Longo, S.; Chiapponi, L.; Di Federico, V. Porous gravity currents: A survey to determine the joint influence of fluid rheology and variations of medium properties. *Adv. Water Resour.* **2016**, *92*, 105–115. [CrossRef]

23. Helmig, R. *Multiphase Flow and Transport Processes in the Subsurface: A Contribution to the Modeling of Hydrosystems*; Springer: Berlin/Heidelberg, Germany, 1997.

24. Zheng, Z.; Soh, B.; Huppert, H.E.; Stone, H.A. Fluid drainage from the edge of a porous reservoir. *J. Fluid Mech.* **2013**, *718*, 558–568. [CrossRef]

25. Longo, S.; Ciriello, V.; Chiapponi, L.; Di Federico, V. Combined effect of rheology and confining boundaries on spreading of gravity currents in porous media. *Adv. Water Resour.* **2015**, *79*, 140–152. [CrossRef]

26. Frolkovič, P. Application of level set method for groundwater flow with moving boundary. *Adv. Water Resour.* **2012**, *47*, 56–66. [CrossRef]

27. Chesnokov, A.; Liapidevskii, V. Viscosity-stratified flow in a Hele–Shaw cell. *Int. J. Non Linear Mech.* **2017**, *89*, 168–176. [CrossRef]

28. Bernal, F.; Kindelan, M. RBF meshless modeling of non-Newtonian Hele–Shaw flow. *Eng. Anal. Bound. Elem.* **2007**, *31*, 863–874. [CrossRef]

29. Furtak-Cole, E.; Telyakovskiy, A.S.; Cooper, C.A. A series solution for horizontal infiltration in an initially dry aquifer. *Adv. Water Resour.* **2018**, *116*, 145–152. [CrossRef]

30. Shampine, L.F. Some Singular Concentration Dependent Diffusion Problems. *J. Appl. Math. Mech.* **1973**, *53*, 421–422. [CrossRef]

31. Hirt, C.W.; Nichols, B.D. Volume of fluid (VOF) method for the dynamics of free boundaries. *J. Comput. Phys.* **1981**, *39*, 201–225. [CrossRef]

32. Darwish, M.; Moukalled, F. Convective Schemes for Capturing Interfaces of Free-Surface Flows on Unstructured Grids. *Numer. Heat Transf. B Fund* **2006**, *49*, 19–42. [CrossRef]

33. Deshpande, S.S.; Anumolu, L.; Trujillo, M.F. Evaluating the performance of the two-phase flow solver interFoam. *Comput. Sci. Discov.* **2012**, *5*, 014016. [CrossRef]

34. Yin, X.; Zarikos, I.; Karadimitriou, N.; Raoof, A.; Hassanizadeh, S. Direct simulations of two-phase flow experiments of different geometry complexities using Volume-of-Fluid (VOF) method. *Chem. Eng. Sci.* **2019**, *195*, 820–827. [CrossRef]

35. Issakhov, A.; Imanberdiyeva, M. Numerical simulation of the movement of water surface of dam break flow by VOF methods for various obstacles. *Int. J. Heat Mass Transf.* **2019**, *136*, 1030–1051. [CrossRef]

36. Duguay, J.; Lacey, R.; Gaucher, J. A case study of a pool and weir fishway modeled with OpenFOAM and FLOW-3D. *Ecol. Eng.* **2017**, *103*, 31–42. [CrossRef]

37. Weller, H.G.; Tabor, G.; Jasak, H.; Fureby, C. A tensorial approach to computational continuum mechanics using object-oriented techniques. *Comput. Phys.* **1998**, *12*, 620. [CrossRef]

38. Hesse, M.; Tchelepi, H.; Cantwel, B.J.; Orr, F.M., Jr. Gravity currents in horizontal porous layers: Transition from early to late self-similarity. *J. Fluid Mech.* **2007**, *577*, 363–383. [CrossRef]

Article

A Continuum Model for Complex Flows of Shear Thickening Colloidal Solutions

Joseph A. Green [1], Daniel J. Ryckman [2] and Michael Cromer [3],*

[1] Department of Mechanical Engineering & Department of Electrical Engineering, Rochester Institute of Technology, Rochester, NY 14623, USA; jag2138@rit.edu
[2] Center for Imaging Science, Rochester Institute of Technology, Rochester, NY 14623, USA; djr9015@rit.edu
[3] School of Mathematical Sciences, Rochester Institute of Technology, Rochester, NY 14623, USA
* Correspondence: mec2sma@rit.edu

Received: 15 December 2018; Accepted: 21 January 2019; Published: 1 February 2019

Abstract: Colloidal shear thickening fluids (STFs) have applications ranging from commercial use to those of interest to the army and law enforcement, and the oil industry. The theoretical understanding of the flow of these particulate suspensions has predominantly been focused through detailed particle simulations. While these simulations are able to accurately capture and predict the behavior of suspensions in simple flows, they are not tractable for more complex flows such as those occurring in applications. The model presented in this work, a modification of an earlier constitutive model by Stickel et al. *J. Rheol.* **2006**, *50*, 379–413, describes the evolution of a structure tensor, which is related to the particle mean free-path length. The model contains few adjustable parameters, includes nonlinear terms in the structure, and is able to predict the full range of rheological behavior including shear and extensional thickening (continuous and discontinuous). In order to demonstrate its capability for complex flow simulations, we compare the results of simulations of the model in a simple one-dimensional channel flow versus a full two-dimensional simulation. Ultimately, the model presented is a continuum model shown to predict shear and extensional thickening, as observed in experiment, with a connection to the physical microstructure, and has the capability of helping understand the behavior of STFs in complex flows.

Keywords: shear thickening; colloids; continuum model; computational rheology

1. Introduction

Colloidal solutions consist of a liquid, Newtonian or viscoelastic, in which solid particles (with diameters ranging from 2 to 1000 nm) are suspended, so that the particles are large compared to the lengthscale of the fluid microstructure, but small enough to be affected by Brownian motion. At low concentrations, colloidal fluids are approximately Newtonian, however, with increasing concentrations they show complex rheological behavior including shear thinning or thickening and glass-like relaxation states. At these higher concentrations the complex rheological responses are due to hydrodynamic interactions between particles at high flow rates. The focus of this work is building towards to a continuum model that captures the shear thickening behavior.

Colloidal shear thickening fluids (STFs) have exciting applications ranging from commercial use (e.g., polymeric binders for paints) to those of interest to the army and law enforcement (e.g., soft body armor) and more [1]. For example, it has been shown experimentally through ballistics tests [2] and "stabbing" tests [3] that, in the application to body armor [4], the frequency response of certain shear thickening fluids (STFs) allows a fabric interwoven with an STF to stop a projectile or prevent the piercing of a knife while also being light and flexible enough to provide full body protection. These fluids also have a possible use in the oil recovery industry. A novel microfluidic device has recently been developed to simulate flow through a sandstone core similar to that which would occur during chemical injection [5]. In this work, a series of experiments were devised to investigate the efficiency of several different liquid materials on their ability to remove oil from capillaries and pores. Using water as the baseline, they investigated how different rheological properties affect oil recovery. In particular, it was shown that shear thickening fluids show great promise for enhanced oil recovery in the future [5].

A key feature of the flows in these applications is a mixture of shear and extensional kinematics. Although the shear rheology of thickening suspensions are well understood, the extensional rheology of these suspensions remains mostly unexplored. Only a very limited number of experimental studies have investigated the response of shear thickening suspensions to extensional flows [6–11]. Two of these studies [6,7] focused on measuring the steady state extensional response of shear thickening suspensions. In both cases, the steady state extensional viscosity as a function of extension rate shows sharp extensional thickening transition very similar to shear flows. In [6], they confirmed by small-angle light scattering measurements in a hyperbolic contraction that the increase in viscosity is likely due to the formation of strings and clusters ordered in the flow direction, which is one possible mechanism responsible for the shear thickening response. A more recent, and now widely accepted key ingredient for these responses, in particular for discontinuous shear thickening, is stress-induced frictional contacts between particles [12,13], which has been recently simulated under extensional flow using particle dynamics [14].

A key challenge of suspension mechanics is the generation of a constitutive model that can be used for complex flows [5,15–17], which are not practical using detailed particle simulations [4,13,18–23]. Such a constitutive model would need to take into account the underlying mechanisms of shear and extensional thickening, as well as the construction and destruction of microstructure by flow [24]. To tackle this challenge, recent work has focused on developing constitutive models that describe the evolution of the microstructure of particle suspensions coupled with general equations for fluid flow [25–33]

In early models [25–28], particle stress is made explicitly dependent on the microstructure through the consideration of a local conformation tensor that is inspired from the orientation distribution tensor defined for dilute fiber suspensions [34]. Hand [35] formulated a general representation theorem for the total Cauchy stress tensor in terms of the conformation tensor and the deformation rate tensor. Barthés-Biesel and Acrivos [36] show that Hand's approach is applicable for a number of different suspension types as long as a suitable microstructure tensor is available. For concentrated suspensions of spherical particles, Phan-Thien [25] proposed a differential constitutive equation for the conformation tensor based upon the unit vector joining two neighboring particles, thereby encoding a direct connection with the pair distribution function. Later, Phan-Thien et al. [26] went further with a micro-macro model inspired from statistical mechanics for the constitutive equation of the conformation tensor, but no quantitative comparisons were obtained. In 2006, Goddard [28] revisited this approach, and proposed a model involving 12 material parameters and two tensors for describing the anisotropy. Stickel et al. [27,37,38] defined the conformation tensor on the base of a directionally-dependent particle mean free path, and simplified the expression of the stress to be linear in the deformation rate and the conformation tensor, involving 13 free parameters. All of the tensorial models are, by construction, frame-invariant and potentially applicable to arbitrary flow geometries and conditions. One downside to these models is that they contain *ad hoc* terms and require the tuning of associated fit parameters with little physical meaning. As a result they provide limited insight into the dynamics of the microstructure and the link with the suspension stress, furthermore Chacko et al. [39] showed that these types of model fail to properly describe the dynamics of the fabric tensor under shear reversal. They attributed this to the fact that, while a second-rank tensor captures reasonably well the microstructure in steady flows, it gives a poor description during significant parts of the microstructural evolution following shear reversal. Other purely macroscopic continuum models have been developed that describe the evolution of the volume fraction, ϕ, of particles under flow with phenomenological descriptions of the stresses and viscosity as a function of ϕ [12,30,32,40–42].

While these models do not account, explicitly, for the microstructure, they could, coupled with a model describing the microstructure, help serve as the basis of a two-fluid model, e.g., [38].

More recent theories have focused on deriving constitutive models that reduce the number of free parameters and provide a more direct connection with the microstructure of non-colloidal suspensions [31,33]. Ozenda et al. [31] proposed a new, minimal tensorial model representing the role of microstructure on the viscosity of non-colloidal suspensions of rigid particles. Their model qualitatively reproduces several of the main rheological trends exhibited by concentrated suspensions: anisotropic and fore-aft asymmetric microstructure in simple shear and transient relaxation of the microstructure toward its stationary state. The model contains only a few model parameters, which have physical meaning, and can be identified from comparisons with experimental data. Gillissen and Wilson [33] developed a model for the microstructure and the stress of dense suspensions of non-Brownian, perfectly smooth spheres. These quantities are defined in terms of the second-order moment of the distribution function of the orientation unit vector between hydrodynamically interacting particles. The model is developed from first principles, and the evolution equation for the microstructure contains a source term that accounts for the association and the dissociation of interacting particle pairs.

All of these models have focused primarily on non-colloidal solutions, and have yet to capture the strong shear and extensional thickening behavior observed in many particulate suspensions. The goal of this paper is therefore to provide a continuum-level constitutive equation for the stress in colloidal suspensions that explicitly accounts for the evolution of its microstructure during flow. Unlike the most recent models, the focus of this work is on colloidal solutions that exhibit nonlinear rheological behavior, in particular shear thickening. We have modified the constitutive model developed by Stickel et al. [27] by including terms nonlinear in the structure while also greatly reducing the number of free parameters. By varying free parameters, the modified model can predict shear thinning, both continuous and discontinuous shear thickening under shear flow, and extensional thickening under planar extensional flow. We further provide a first insight into the simulation capabilities of the model by conducting simulations in a straight channel and comparing a simple one-dimensional approximation with a more detailed, two-dimensional simulation. The agreement between the simulations, together with the model's ability to capture the nonlinear rheological signatures means that the model is suitable for capturing the response of such materials via simulations in complex geometries.

2. Materials and Methods

SPPC Model

In the Stickel-Phillips-Powell (SPP) [27] model, structure is defined in a suspension in terms of the average distance $l_{mf}(\hat{x})$ that a test particle can move in a direction denoted by the unit vector $\hat{x}$ before colliding with another particle. A dimensionless structure function is then defined inversely proportional to this mean free-path length. Using a functional expansion with spherical harmonics as the basis set written in terms of a Cartesian tensor, the structure may be described as a spherical surface represented by a symmetric, second order tensor, $\mathbf{Y}$. In equilibrium, the structure is in an isotropic state given by:

$$f(\phi) = \left(\frac{1}{\phi} - \frac{1}{\phi_m}\right)^{-1}, \tag{1}$$

where ϕ is the volume fraction of particles and ϕ_m is the maximum packing fraction. Under flow, the structure evolves according to

$$\overset{\circ}{\mathbf{Y}} = \mathbf{Y}(\mathbf{Y}, \dot{\gamma}) \tag{2}$$

where

$$\overset{\circ}{\mathbf{Y}} = \frac{\partial \mathbf{Y}}{\partial t} + \mathbf{v} \cdot \nabla \mathbf{Y} - \mathbf{Y}\omega + \omega\mathbf{Y} \tag{3}$$

is the corotational derivative representing the time derivative in a reference frame rotating with the local vorticity; the corotational derivative has been defended as an appropriate choice for particulate suspensions [35]. The strain rate and vorticity tensors are respectively given by

$$\dot{\gamma} = \frac{1}{2}(\nabla \mathbf{v} + \nabla \mathbf{v}^T) \tag{4}$$

$$\omega = \frac{1}{2}(\nabla \mathbf{v} - \nabla \mathbf{v}^T). \tag{5}$$

General forms of the function on the right side of Equation (2) that satisfy the constraint of frame indifference are well known [43] and form the basis of, for example, the second-order fluid constitutive model. For hard-sphere suspensions of particles at low Reynolds number, a linear dependence on the rate of strain is expected because of the linearity of the Stokes equations. Under these assumptions, the equation governing the structure in the SPP model is given by:

$$\begin{aligned}\overset{\circ}{\mathbf{Y}} &= \dot{\gamma}h(Pe,\overset{*}{\dot{\gamma}})[c_1 f(\phi) + c_2 \mathrm{tr}\mathbf{Y}/3 + c_4(\mathbf{Y} - \mathrm{tr}\mathbf{Y}/3)] \\ &+ c_3 \mathrm{tr}(\mathbf{Y}\dot{\gamma})\mathbf{I} + (c_5 f(\phi) + c_6 \mathrm{tr}\mathbf{Y}/3)\dot{\gamma} + c_7(\mathbf{Y}\dot{\gamma} + \dot{\gamma}\mathbf{Y} - 2/3\mathrm{tr}(\mathbf{Y}\dot{\gamma})\mathbf{I}) \\ &+ \hat{\alpha}_4 \mathbf{Y}^2 + \hat{\alpha}_7(\mathbf{Y}^2\dot{\gamma} + \dot{\gamma}\mathbf{Y}^2). \end{aligned} \tag{6}$$

In the interest of simplicity, terms nonlinear in the structure tensor $\mathbf{Y}$ are also neglected, i.e., $\hat{\alpha}_4 = \hat{\alpha}_7 = 0$. As noted in the original paper [27], keeping these terms permits the existence of multiple steady-state structures for a given rate of deformation that may allow for stick-slip behavior or sudden jamming.

In order to generate a model with less adjustable parameters, while including terms nonlinear in the structure, that will allow for predictions of strong shear thinning and thickening, the following modification of the SPP model, called here the SPPC model, is proposed:

$$\begin{aligned}\overset{\circ}{\mathbf{Y}} &= \dot{\gamma}h(Pe,\overset{*}{\dot{\gamma}})[3f(\phi)\mathbf{I} - \mathbf{Y}] + c_3 \mathrm{tr}(\mathbf{Y}\dot{\gamma})\mathbf{I} + c_7(\mathbf{Y}\dot{\gamma} + \dot{\gamma}\mathbf{Y} - 2/3\mathrm{tr}(\mathbf{Y}\dot{\gamma})\mathbf{I}) \\ &+ c_8 \mathrm{tr}(\mathbf{Y}\dot{\gamma})\mathbf{Y}^2 + c_9(\mathbf{Y}^2\dot{\gamma} + \dot{\gamma}\mathbf{Y}^2), \end{aligned} \tag{7}$$

where $\dot{\gamma} = \sqrt{2\dot{\gamma} : \dot{\gamma}}$ is the magnitude of the strain rate tensor. In order to directly relate back to the SPP model, we have kept the names of the constant coefficients the same. On the right hand side (RHS) of the modified equation, we have eliminated several constants, and the first three terms are similar to the original model formulation. For the first term on the RHS, we have simplified the relaxation of the material back to its isotropic state by eliminating the parameters c_1, c_2, c_4 in the SPP model. The term linear to the strain rate in the SPP model does not play a significant role in the development of nonlinear rheological behavior, thus we have neglected this term ($c_5 = c_6 = 0$). The terms proportional to c_8 and c_9 were included in the original model formulation, but ultimately neglected in subsequent analysis for simplification reasons. The simplified, original SPP model can predict a slight shear thickening behavior [27], however, the linearity in the structure prevents the prediction of highly nonlinear rheological behavior, such as a strong shear thickening observed in many particulate suspensions. Similar nonlinear terms appear in a model constitutive model by Goddard [28], but parameters were not tested showing shear thickening behavior. As we will show, incorporating these nonlinear terms allow for the capability to predict the complex rheological behavior of shear thickening colloidal solutions.

The SPP model accounts for structure rearrangement that occurs in the absence of flow, such as the relaxation of a Brownian suspension to an isotropic state following cessation of flow, and hydrodynamic diffusion, which is determined by the suspension structure and a diffusion coefficient determined by the local rate of deformation. It also accounts for rearrangements in the structure that are driven by the imposed flow. The function h contains information about long-range repulsive forces and Brownian diffusion. Thus far, the dependence of the function h on the Peclet number, $Pe \sim \dot{\gamma}$, which contains information about Brownian motion driving the system towards equilibrium, has been neglected. Including this Pe dependence will allow for simulations to probe the complete transition from Brownian (slow flow) to hydrodynamic (fast flow) dominant, similar to the model of [42]. The function h should be inversely related to Peclet number, thus, as an addition to the original SPP model, we assume

$$h(Pe, \overset{*}{\gamma}) = Pe^{-1} + \overset{*}{\gamma}^{-1/8}. \tag{8}$$

The Peclet number describes the ratio of hydrodynamic to Brownian effects

$$Pe = \frac{6\pi\eta_S a^3 \dot{\gamma}}{kT} = S\dot{\gamma}, \tag{9}$$

and the dimensionless number $\overset{*}{\gamma}$ is the ratio of hydrodynamic to repulsive forces

$$\overset{*}{\gamma} = \frac{6\pi\eta_S a^2 \dot{\gamma}}{F_0} = F\dot{\gamma}, \tag{10}$$

where S and F are material parameters having units of time. Here a is the particle radius, and F_0 is the characteristic magnitude of the repulsive interaction force. The exponent $-1/8$ in Equation (8) is the same choice made in the prior work [27,37], which was chosen empirically to match particle simulations and to limit the effect of $\dot{\gamma} \ll 1$ to a weak singularity in their expression for the stress:

$$\begin{aligned}\mathbf{\Pi} &= \eta_S\dot{\gamma}[(k_1 f(\phi) + k_2 \mathrm{tr}\mathbf{Y}/3)\mathbf{I} + k_4(\mathbf{Y} - \mathrm{tr}\mathbf{Y}/3\mathbf{I}) + k_3 \frac{1}{\overset{*1/8}{\gamma\gamma}}\mathrm{tr}(\mathbf{Y}\dot{\gamma})\mathbf{I} + (k_5 f(\phi) \\ &+ k_6 \mathrm{tr}\mathbf{Y}/3))\frac{1}{\overset{*1/8}{\gamma\gamma}}\dot{\gamma} + k_7 \frac{1}{\overset{*1/8}{\gamma\gamma}}(\mathbf{Y}\dot{\gamma} + \dot{\gamma}\mathbf{Y} - 2/3\mathrm{tr}(\mathbf{Y}\dot{\gamma})\mathbf{I})]. \end{aligned} \tag{11}$$

In order to simplify the model further and make it more amenable to complex flow simulations, one major assumption we have made is the stress in the material is a function solely of the structure, i.e., the stress relaxes instantaneously to the local structure of the suspension:

$$\mathbf{\Pi} = \eta_S \left[k_1 (f(\phi) - \mathrm{tr}\mathbf{Y}/9)\mathbf{I} + k_4(\mathbf{Y} - \mathrm{tr}\mathbf{Y}/3\mathbf{I}) \right]. \tag{12}$$

In this formulation, in order to eliminate to direct dependence of the stress on the shear rate, the parameters k_1, k_4 have units of inverse time. The dependence of the stress on the shear rate comes indirectly through its relationship with the microstructure. The simplification of the stress expression eliminates the singularity in the original SPP model, thus, we could in principle choose any exponent in Equation (8); at this point, we follow the original model.

3. Results

3.1. SAOS

Ultimately, we have reduced the model to six parameters: $c_3, c_7, c_8, c_9, k_1, k_4$. While several of these are adjustable, some can be defined from experiment, e.g., small amplitude oscillatory shear. Let $\gamma = \gamma_0 \sin \omega t$ thus $\dot{\gamma} = \gamma_0 \omega \cos \omega t$. For $\gamma_0 \ll 1$, let $\mathbf{Y} = \mathbf{Y}^0 + \mathbf{Y}^1 \sin \omega t + \mathbf{Y}^2 \cos \omega t$, where $\mathbf{Y}^0 = 3f(\phi)\mathbf{I}$ is the equilibrium structure. Plugging in we find:

$$Y_{12}^1 = (6f(\phi)c_7 + 6f(\phi)^2 c_9)\frac{\gamma_0 S^2 \omega^2}{S^2 \omega^2 + 1} \tag{13}$$

$$Y_{12}^2 = -(6f(\phi)c_7 + 6f(\phi)^2 c_9)\frac{\gamma_0 S\omega}{S^2 \omega^2 + 1} \tag{14}$$

which gives:

$$G' = \Pi_{12}^1 = \eta_S k_4 (6f(\phi)c_7 + 6f(\phi)^2 c_9) \frac{\gamma_0 S^2 \omega^2}{S^2 \omega^2 + 1} \tag{15}$$

$$G'' = \Pi_{12}^2 = \eta_S \left[-k_4 (6f(\phi)c_7 + 6f(\phi)^2 c_9) \frac{\gamma_0 S \omega}{S^2 \omega^2 + 1} + 2\gamma_0 \omega \right]. \tag{16}$$

In Figure 1, we plot an example prediction of the storage, G', and loss, G'', moduli predicted by the modified model. The interesting thing to note is that G'' is always larger than G', which has been observed in experiment [44], indicating the colloidal dispersion depicts a sol-like behavior consisting of distinct non-flocculated units. The same experiment showed that throughout the frequency range the plot of loss modulus is nearly linear, qualitatively similar to the predictions of the SPPC model. In the SPPC model, G' plateaus at high frequencies: as $\omega \to \infty$, $G' \sim 6\eta_S f(\phi) k_4 c_7 S^2 \gamma_0$ and $G'' \sim 2\eta_S \gamma_0 \omega$, i.e., constant and linear at high shear rates, respectively. In the experiment, however, the storage modulus curves upward at high frequency showing that at higher frequencies the material storage capacity increases. While the SPPC model does not predict the behavior of G' and G'' over the full range of frequencies, that it captures many of the dominant features suggests that it could be useful under nonlinear conditions. Because the model predicts some experimental behavior, there are several features we can state about the model parameters. In particular, if we neglect the c_9 term, which we will do throughout much of this paper, it may be possible to extract k_4 and c_7 from experimental SAOS data. Neglecting c_9, the model, in essence, only has three adjustable parameters, now on the same order as the model by [31].

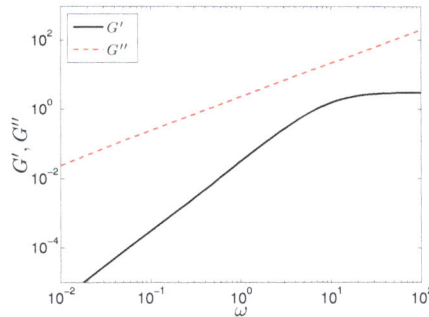

Figure 1. Storage, G', and loss, G'', moduli as a function of frequency, ω. Parameter values are: $S = 10^{-1}$, $c_7 = -0.47, c_9 = 0, \eta_S = 1, f(\phi) = 1.0957, k_4 = -1$.

3.2. Non-Linear Rheology

The primary goal of this paper is to present a model that can capture the nonlinear rheological behavior of shear thickening colloidal dispersions. In the following we consider two flows: simple shear flow and planar extensional flow. We assume linear velocity profiles $\mathbf{v} = (\dot{\gamma}y, 0, 0)$ and $\mathbf{v} = (\dot{\epsilon}x, -\dot{\epsilon}y, 0)$, respectively, and the resulting system of ordinary differential equations describing $\mathbf{Y}$ are solved in time using the Matlab solver ode45 (Version r2016b, MathWorks, Natick, MA, USA). In the following, we investigate the role of individual parameters in the prediction of nonlinear rheological behavior, in particular to show the full range of the model's predictive capabilities. The base set of parameters, $S = 10^{-1}, F = 10, c_3 = -23.5, c_7 = -0.47, c_8 = 10^{-4}, c_9 = 0, \eta_S = 1, f(\phi) = 1.0957, k_1 = -10^3$, $k_4 = -1$, are chosen to align with the SPP model (c_3, c_7), and to predict desired qualitative trends, namely shear and extensional thickening, and negative second normal stress difference.

3.2.1. Simple Shear Flow

By varying the adjustable parameters, the model is capable of predicting the full range of shear rheological behavior, including shear thinning and shear thickening of various degrees, Figure 2. The results for $c_9 = 0$ in Figure 2a–c show a continuous shear thickening (CST) behavior, as observed in experiments, e.g., [45–47]). Additionally, at high shear rates, the model predicts a maximum in the viscosity followed by shear thinning behavior [48]. By varying model parameters, the degree and onset of the shear thickening can be controlled. Varying c_8, the coefficient of the $\mathbf{Y}^2$ term in the SPPC model, the degree of thickening is decreased until eventually predicting shear thinning. Increasing S corresponds to decreasing the contribution of Brownian forces relative to hydrodynamic forces (note that $S \to \infty$ in the original SPP model), and the model predicts that the onset of thickening, a hydrodynamic response, decreases while also decreasing the ratio between the maximum and zero-shear viscosities. Decreasing F corresponds to increasing the repulsive force magnitude relative to hydrodynamic forces, and, as a result, the SPPC model predicts an initial shear thinning prior to the onset of shear thickening for $F \gg 1$. Including $c_9 \neq 0$ can lead to drastically different predictions, namely a jump in the shear viscosity, i.e., discontinuous shear thickening (DST). This additional feature of the model will allow us the ability to predict and analyze the differences between CST and DST under different flow conditions, while providing motivation for future developments of continuum models for particulate suspensions that incorporate a physical mechanism underlying DST.

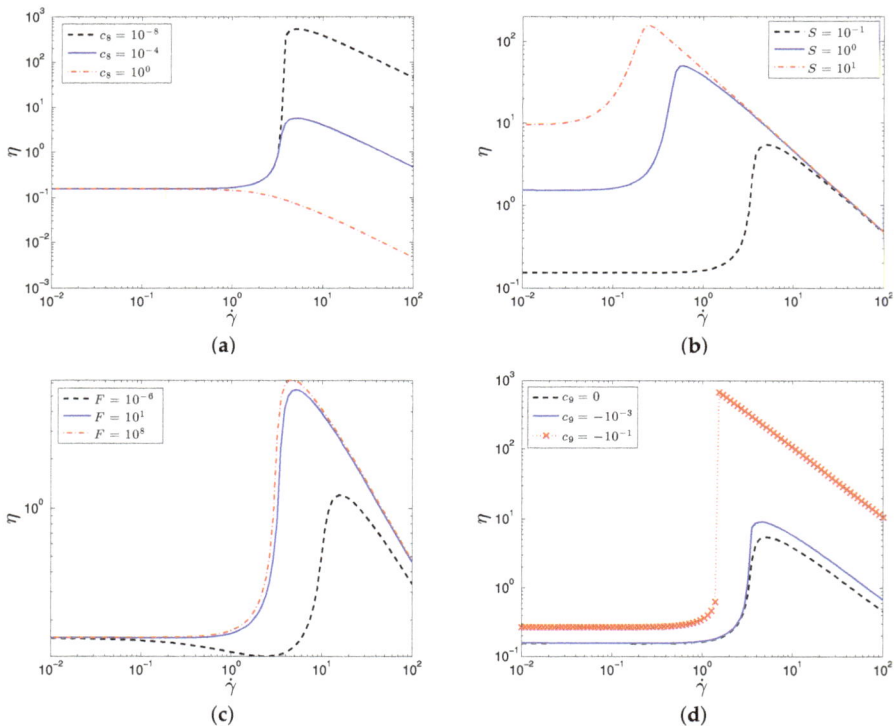

Figure 2. Plot of the particle contribution to the viscosity, $\eta = \Pi_{xy}/\dot{\gamma}$, against shear rate for varying values of (a) c_8, (b) S, (c) F, and (d) c_9. Parameter values are: $S = 10^{-1}, F = 10, c_3 = -23.5, c_7 = -0.47, c_8 = 10^{-4}, c_9 = 0$, $\eta_S = 1, f(\phi) = 1.0957, k_1 = -10^3, k_4 = -1$.

In Figure 3, we plot the first and second normal stress differences, N_1, N_2, for the same sets of parameters. The SPPC models predicts $N_1 = \eta_S k_4 (Y_{11} - Y_{22}) > 0$ and $N_2 = \eta_S k_4 (Y_{22} - Y_{33}) < 0$. The latter prediction is consistent with experiments and particle simulations [45]. The sign of N_1 is still less well understood [45,47]. Recently, it has been shown that a positive value of N_1 is expected for highly dense suspensions, which exhibits a discontinuous shear thickening (DST) behavior, while a negative value of N_1 is expected for CST [13,45,47]. In theories and simulations where some limiting degree of Brownian motion and/or interparticle repulsion has been included, N_1 is still found to be negative [49,50]. The results are in line with the available experiments, except that the two normal stress differences are predicted to be approximately equal. Simulations that incorporate friction indicate that this can reduce $|N_1|$ with respect to $|N_2|$ [50]. More recent simulations have found that these frictional forces for dense suspensions leads to a positive value of N_1 [13,23]. The SPPC model is designed for dense colloidal suspensions, thus a positive value of N_1 is expected; a more complex relationship between stress, structure and strain rate could lead to varying signs of N_1, however, this would lead to more parameters, and it is potentially less suitable for complex simulations.

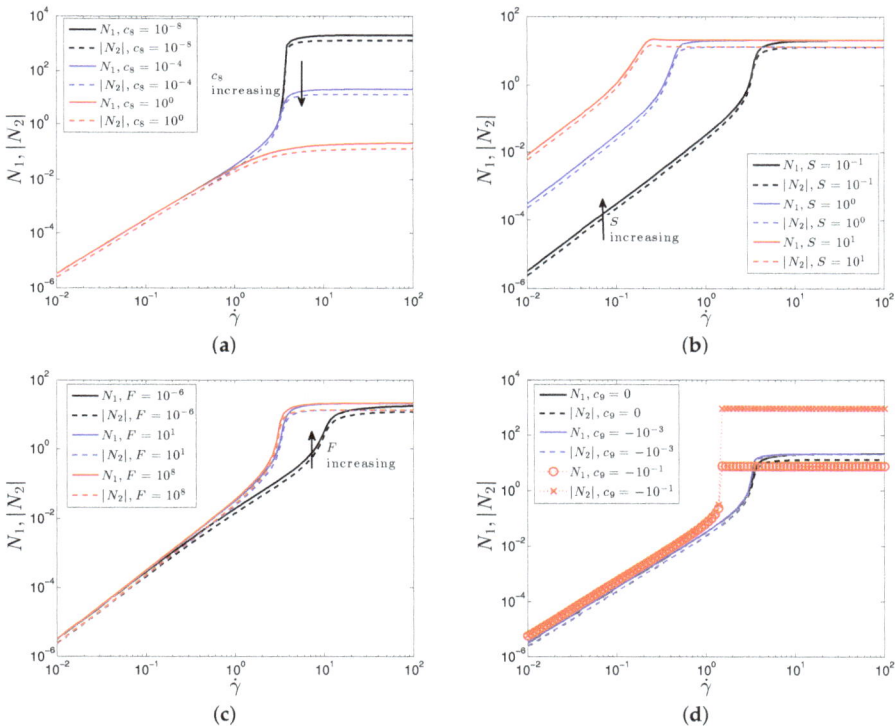

Figure 3. Plot of the first and second normal stress difference, N_1, N_2, against shear rate for varying values of (a) c_8, (b) S, (c) F, and (d) c_9. Parameter values are: $S = 10^{-1}, F = 10, c_3 = -23.5, c_7 = -0.47, c_8 = 10^{-4}, c_9 = 0$, $\eta_S = 1, f(\phi) = 1.0957, k_1 = -10^3, k_4 = -1$.

The macroscopic predictions are determined by the evolution of the microstructure described by the SPPC model. The values of the structure tensor ultimately describe the evolution of a spherical surface determined by the proximity of neighboring particles, in particular, larger values of $\mathbf{Y}$ along the diagonal correspond to a larger degree of clustering of particles. The off-diagonal components correspond to a rotation of the cluster. In order to relate the shear thinning and thickening behavior to the microstructure, in Figure 4 we plot the functions $Z_1 = Y_{11} - Y_{22}$ and $Z_2 = Y_{22} - Y_{33}$, which are the normal structure differences. The negative and positive values of Z_1 and Z_2, respectively, show that Y_{22} is the largest component of the normal structure components, which indicates that the thickening is due the clustering of particles perpendicular to the flow. The increased difference between Z_1 and Z_2 due to the large increase in Z_2 for the DST result, Figure 4d, suggests that the jump in the viscosity is due to a larger degree of clustering of the particles.

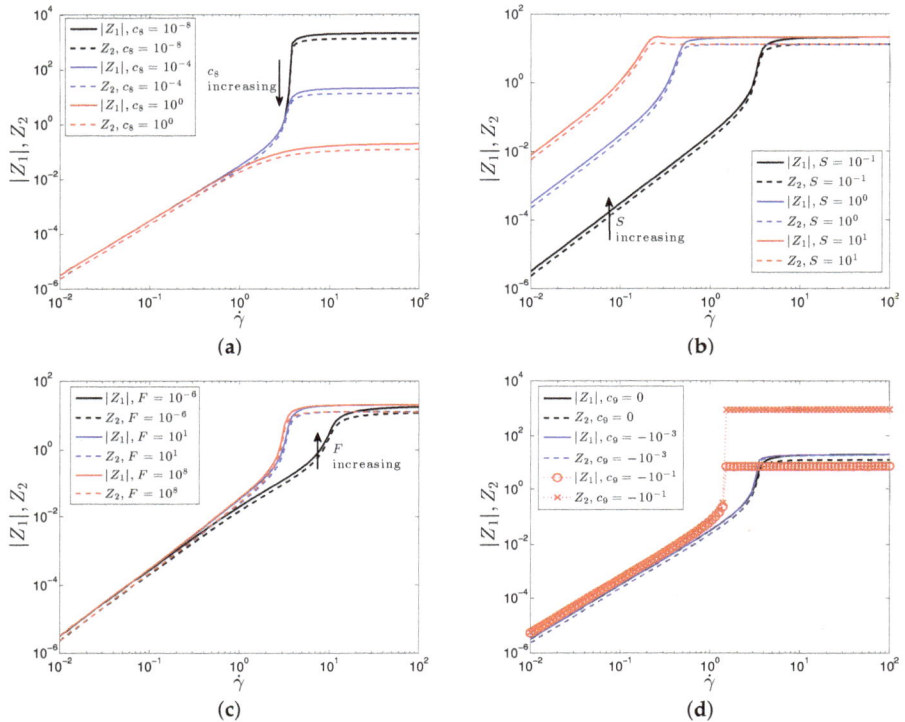

Figure 4. Plot of the first and second normal structure differences, Z_1, Z_2, against shear rate for varying values of (**a**) c_8, (**b**) S, (**c**) F, and (**d**) c_9. Parameter values are: $S = 10^{-1}, F = 10, c_3 = -23.5, c_7 = -0.47, c_8 = 10^{-4}, c_9 = 0, \eta_S = 1, f(\phi) = 1.0957, k_1 = -10^3, k_4 = -1$.

3.2.2. Planar Extensional Flow

In agreement with limited experiments in extensional flow [6,7], the SPPC predicts that shear thickening fluids also exhibit a strong extensional thickening behavior, Figure 5. Furthermore, the extensional thickening occurs at lower rates than those corresponding to shear thickening, as also observed in the same experiments [6,7]. Just as in the case with the shear flow, the extensional thickening is due to the clustering of particles perpendicular to the flow direction, as indicated by a negative value of Z_1 (the magnitude of Z_1 is plotted in Figure 6). The trends predicted by the SPPC model for varying parameters are the same as the above for the shear flow.

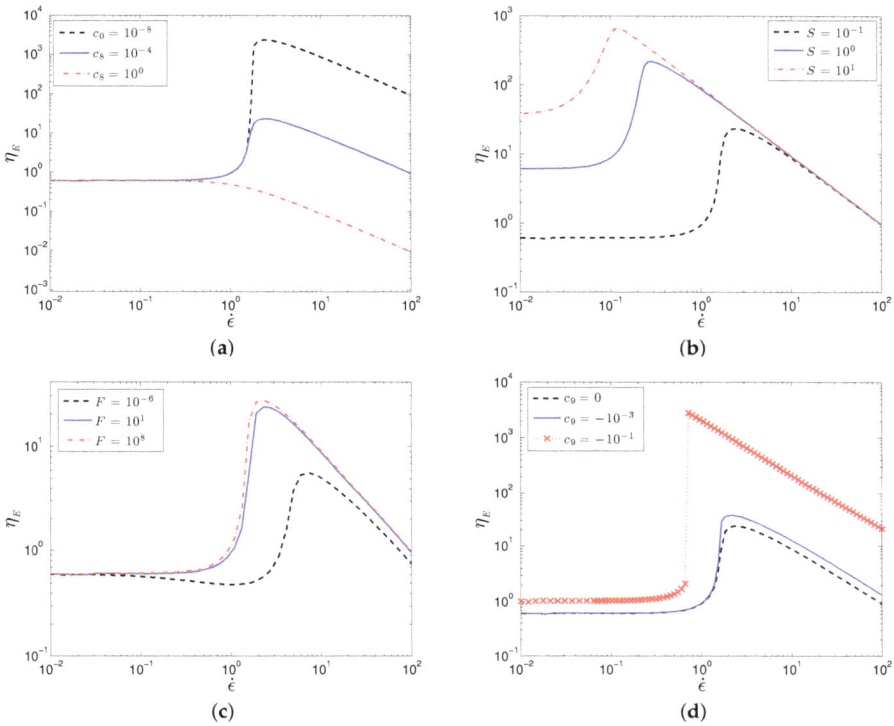

Figure 5. Plot of the particle contribution to the extensional viscosity, $\eta_E = (\Pi_{xx} - \Pi_{yy})/\dot{\epsilon}$, against extension rate for varying values of (**a**) c_8, (**b**) S, (**c**) F, and (**d**) c_9. Parameter values are: $S = 10^{-1}, F = 10, c_3 = -23.5$, $c_7 = -0.47, c_8 = 10^{-4}, c_9 = 0, \eta_S = 1, f(\phi) = 1.0957, k_1 = -10^3, k_4 = -1$.

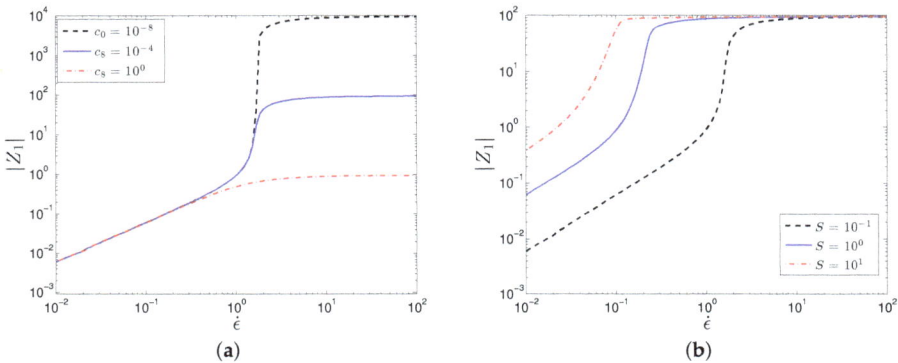

Figure 6. *Cont.*

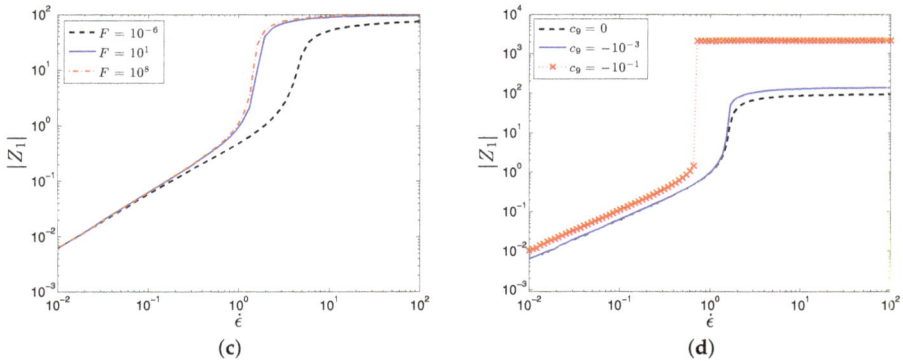

Figure 6. Plot of the first normal structure difference, Z_1, against extension rate for varying values of (**a**) c_8, (**b**) S, (**c**) F, and (**d**) c_9. Parameter values are: $S = 10^{-1}, F = 10, c_3 = -23.5, c_7 = -0.47, c_8 = 10^{-4}, c_9 = 0, \eta_S = 1$, $f(\phi) = 1.0957, k_1 = -10^3, k_4 = -1$.

3.3. Results for Steady Poiseuille Flow

The primary goal of this work is to provide a constitutive model that describes the microstructure of a colloidal solution and its shear thickening behavior, and is suitable for complex flow simulations. In order to first test the model and the numerical solver, in this work we simulate the flow of an STF through a straight channel. Two scenarios are considered: (1) a one-dimensional (1D) assumption, in which we assume the flow moves in the x-direction but only varies in the gradient direction, y, and (2) a full two-dimensional (2D) simulation, for which the profiles across the gap in the fully-developed flow should match with the 1D predictions. There is very limited data on this flow scenario, thus, as an additional benchmark, we check our velocity predictions against one particular set of experimental data [51].

3.3.1. One-Dimensional

We consider a straight channel of length L and rectangular cross-section of width W and height $2H$. We assume the channel is long enough that entrance and exit effects are negligible and we assume that $p_{,x} = -\frac{\Delta p}{L}$, where $\Delta p = p_0 - p_L$ is the pressure drop. The height of the channel is $2H$ where $-H \leq y \leq H$ and we neglect variations in the z-direction. Under these assumptions the velocity, in the positive x-direction, has the form $\mathbf{v} = (u(y), 0, 0)$ and satisfies conservation of mass, $\nabla \cdot \mathbf{v} = 0$.

We assume inertialess flow, and the resulting momentum balance becomes

$$0 = \mathcal{P} + \eta_S u_{,yy} + \Pi_{xy,y}, \tag{17}$$

where $\mathcal{P} = \frac{\Delta p}{L}$. Note that integration of the momentum equation across the channel gives $-\mathcal{P}y = \tau_{xy}$, or at the wall $\mathcal{P}y = \tau_{xy} |_{y=\pm H}$ where $\boldsymbol{\tau}$ is the total stress so $\tau_{xy} = \Pi_{xy} + \eta_S u_{,y}$. Thus the total shear stress is linear and monotone across the channel. The applied pressure is directly related to the wall shear stress $\pm \mathcal{P}H = \tau_{xy} |_{wall}$.

In this 1D channel flow, the equations form a system of coupled, nonlinear partial differential equations to be solved using appropriate initial conditions and boundary conditions. We impose the no slip boundary conditions at the walls, $u |_{y=\pm H} = 0$. The method of lines is used to solve the time-dependent system where first the spatial dimension is discretized, in this case using a second order central finite difference scheme, and the resulting differential-algebraic system of equations is then marched forward in time until steady-state is achieved. The discretized system is marched forward in time using the Matlab solver *ode15s*.

In Figure 7a,b we plot the velocity and viscosity across the gap for a single set of shear thickening parameters. We find the expected triangular velocity profile associated with a viscosity that increases from the center to wall, opposite of what occurs for the typical shear thinning materials. As expected, for a given a pressure drop, the shear thickening reduces the flow rate compared to constant and shear thinning viscosity simulations. This behavior is clearly seen in Figure 8, which plots the computed volumetric flow rate, $\mathcal{Q} = \int_{-H}^{-H} u\,dy$, against the imposed pressure drop. As expected, as the degree of thickening increases, it becomes more difficult for the more viscous fluid to flow, thus reducing the flow throughput.

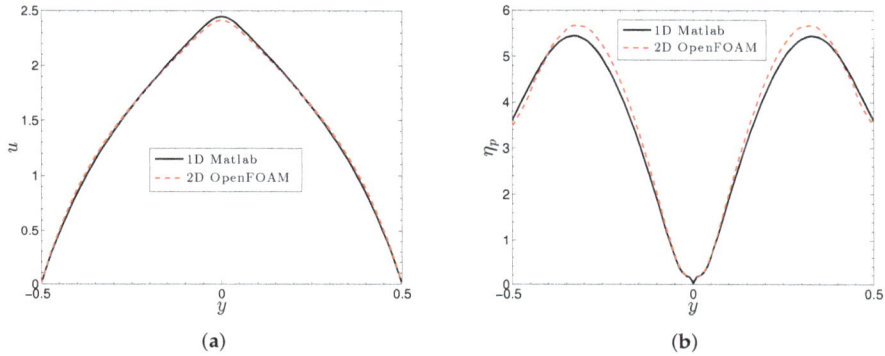

Figure 7. Plots of the (**a**) velocity and (**b**) particle viscosity across the gap from the 1D simulation versus the fully-developed 2D flow. The imposed pressure drop for the 1D simulation is $\mathcal{P} = 10^2$; The inlet velocity for the 2D flow is $U = 1.5133$, which, over a unit gap width, is also the flow rate. Parameter values are: $S = 10^{-1}, F = 10, c_3 = -23.5, c_7 = -0.47, c_8 = 10^{-4}, c_9 = 0, \eta_S = 1, f(\phi) = 1.0957, k_1 = -10^3, k_4 = -1$.

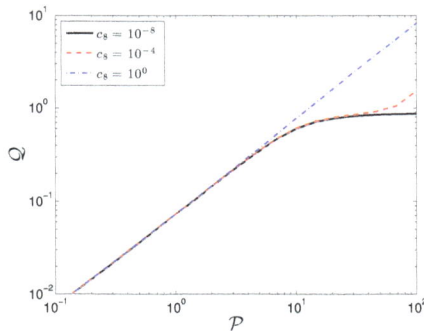

Figure 8. Computed volumetric flow rate, $\mathcal{Q}$, as a function of imposed pressure drop, $\mathcal{P}$. Parameter values are: $S = 10^{-1}, F = 10, c_3 = -23.5, c_7 = -0.47, c_8 = 10^{-4}, c_9 = 0, \eta_S = 1, f(\phi) = 1.0957, k_1 = -10^3, k_4 = -1$.

3.3.2. Two-Dimensional

For two and higher dimensional flows, the governing equations are solved using then numerical solver *viscoelasticFluidFoam* [52], which can be found in the extended version of OpenFOAM® (version 3.0). The *viscoelasticFluidFoam* solver solves the governing equations sequentially in a segregated manner, where the momentum equation is solved first, followed by the pressure-correction equation, and finally the constitutive equations. It uses the Finite Volume Method (FVM) to discretize and solve all governing equations of the particular problem. For those terms involving divergence operators, the volume integral is converted to a surface integral over the control volume surfaces by applying the Gauss' theorem. The convection terms are discretized using the limited linear differencing scheme, which is based on the Sweby scheme [53]. The remaining terms are discretized using the central differencing scheme (CDS). The resulting system of equations is solved by iterative matrix solvers, including the conjugate gradient method for the pressure-correction equation and the biconjugate gradient method for all other equations. The pressure-velocity coupling is ensured using the Pressure-Implicit Split Operator (PISO) algorithm [54]. The equations are solved in time using the built-in "Euler" scheme, which is first-order accurate. Finally, the improved both sides diffusion method [55] is implemented to stabilize the calculations.

At the inlets we impose uniform conditions on the velocity and structure, and a zero gradient condition on the pressure (and vice versa at the outlets). We use a 10:1 length:width ratio, which is sufficiently large to allow for fully-developed flow prior to reaching the outlet. Along the solid walls, the boundary conditions imposed include the no slip boundary condition on the velocity:

$$\mathbf{v} = 0. \tag{18}$$

We use the zero gradient boundary condition on the conformation tensor at the walls, consistent with the original solver benchmark calculations [52]:

$$\mathbf{n} \cdot \nabla \mathbf{Y} = 0. \tag{19}$$

In Figure 9 we show the contour plot of the fully-developed velocity magnitude. Here we see the more triangular shaped velocity profile, expected for shear thickening fluids [51]. Associated with this velocity profile is a viscosity profile in which the viscosity is smallest in the center and increases towards the wall, Figure 10, unlike the typical behavior of constant and shear thinning materials.

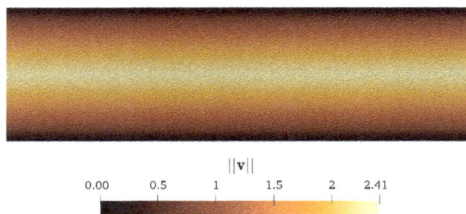

Figure 9. Contour plot of the velocity profile for the 2D channel flow with $U = 1.5133$. Parameter values are: $S = 10^{-1}, F = 10, c_3 = -23.5, c_7 = -0.47, c_8 = 10^{-4}, c_9 = 0, \eta_S = 1, f(\phi) = 1.0957, k_1 = -10^3, k_4 = -1$.

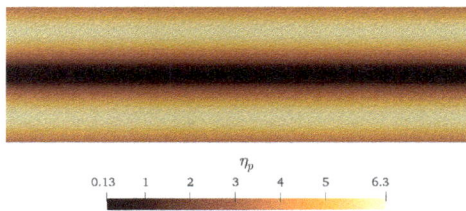

Figure 10. Contour plot of the particle viscosity profile for the 2D channel flow with $U = 1.5133$. Parameter values are: $S = 10^{-1}, F = 10, c_3 = -23.5, c_7 = -0.47, c_8 = 10^{-4}, c_9 = 0, \eta_S = 1, f(\phi) = 1.0957, k_1 = -10^3, k_4 = -1$.

The important question is whether the 1D and 2D predictions coincide. In Figure 7 we plot the velocity and viscosity across the gap (for the 2D simulation we choose a location far enough downstream such that the flow is fully developed). The main differences between the one and two dimensional flows are that the former is pressure-driven whereas the latter is flow-rate-driven, and the fact the used 2D solver requires boundary conditions [52] on the structure, yet none are imposed in the 1D flow. In order to attempt to match the data, we choose a single 1D simulation with $\mathcal{P} = 10^2$.

The computed flow rate is thus $Q = 1.5133$, which is used for the uniform velocity inlet condition in the 2D simulation. The agreement between the 1D and 2D do not perfectly overlap, however they are sufficiently close that we feel confident that the 2D simulations are predicting the expected behavior, thus proving a viable tool for future, more complex studies.

4. Discussion

The model presented in this paper, a modification of the SPP model [27], is a contribution to the ongoing work for the development of constitutive models describing the flow of particulate suspensions. The model is still phenomenological in the sense that the form is dictated by tensor symmetries and invariances, and fit parameters are needed to match the model to experimental data, and little can be said about the physical meaning of these parameters. While the SPPC model is connected to particle physics based upon a spatially-dependent mean free-path length, there is still limited insight that can be derived into the dynamics of the microstructure and the link with the suspension stress. However, the advantage of the provided model is that it has very few free parameters and, due to the inclusion of nonlinear terms in the structure, has the capability of predicting the full range of nonlinear rheological behavior from shear thinning to both continuous and discontinuous shear thickening, and extensional thickening, the effects of which are predicted to be due to the clustering of particles. The model thus includes terms that may need to be considered in future first-principles modeling efforts in order to capture these unique features. Most importantly, having a continuum model that predicts shear and extensional thickening coupled to microstructural dynamics now means that we have the ability to conduct complex flow simulations that will help better understand the dynamics of shear thickening colloids under complex flow conditions [2,5,15–17,56,57].

Author Contributions: Conceptualization, M.C.; Data curation, J.A.G. and M.C.; Formal analysis, J.A.G.; Funding acquisition, M.C.; Investigation, J.A.G. and D.J.R.; Methodology, J.A.G., D.J.R. and M.C.; Project administration, M.C.; Resources, M.C.; Software, J.A.G. and D.J.R.; Supervision, M.C.; Validation, J.A.G., D.J.R. and M.C.; Visualization, J.A.G. and M.C.; Writing—original draft, M.C.; Writing—review & editing, J.A.G. and M.C.

Funding: This research was funded by American Chemical Society Petroleum Research Fund under grant number 56047-UNI9.

References

1. Ding, J.; Li, W.; Shen, S.Z. Research and applications of shear thickening fluids. *Recent Pat. Mater. Sci.* **2011**, *4*, 43–49. [CrossRef]
2. Lee, Y.S.; Wetzel, E.D.; Wagner, N.J. The ballistic impact characteristics of Kevlar® woven fabrics impregnated with a colloidal shear thickening fluid. *J. Mater. Sci.* **2003**, *38*, 2825–2833. [CrossRef]
3. Decker, M.J.; Halbach, C.J.; Nam, C.H.; Wagner, N.J.; Wetzel, E.D. Stab resistance of shear thickening fluid (STF)-treated fabrics. *Compos. Sci. Technol.* **2007**, *67*, 565–578. [CrossRef]
4. Wagner, N.J.; Brady, J.F. Shear thickening in colloidal dispersions. *Phys. Today* **2009**, *62*, 27–32. [CrossRef]
5. Nilsson, M.A.; Kulkarni, R.; Gerberich, L.; Hammond, R.; Singh, R.; Baumhoff, E.; Rothstein, J.P. Effect of fluid rheology on enhanced oil recovery in a microfluidic sandstone device. *J. Non-Newton. Fluid Mech.* **2013**, *202*, 112–119. [CrossRef]
6. Chellamuthu, M.; Arndt, E.M.; Rothstein, J.P. Extensional rheology of shear-thickening nanoparticle suspensions. *Soft Matter* **2009**, *5*, 2117–2124. [CrossRef]
7. White, E.E.B.; Chellamuthu, M.; Rothstein, J.P. Extensional rheology of a shear-thickening cornstarch and water suspension. *Rheol. Acta* **2010**, *49*, 119–129. [CrossRef]
8. Smith, M.I.; Besseling, R.; Cates, M.E.; Bertola, V. Dilatancy in the flow and fracture of stretched colloidal suspensions. *Nat. Commun.* **2010**, *1*, 114. [CrossRef]
9. Roché, M.; Kellay, H.; Stone, H.A. Heterogeneity and the role of normal stresses during the extensional thinning of non-Brownian shear-thickening fluids. *Phys. Rev. Lett.* **2011**, *107*, 134503. [CrossRef]

10. Zimoch, P.J.; McKinley, G.H.; Hosoi, A.E. Capillary breakup of discontinuously rate thickening suspensions. *Phys. Rev. Lett.* **2013**, *111*, 036001. [CrossRef]

11. Majumdar, S.; Peters, I.R.; Han, E.; Jaeger, H.M. Dynamic shear jamming in dense granular suspensions under extension. *Phys. Rev. E* **2017**, *95*, 012603. [CrossRef]

12. Wyart, M.; Cates, M.E. Discontinuous shear thickening without inertia in dense non-Brownian suspensions. *Phys. Rev. Lett.* **2014**, *112*, 098302. [CrossRef]

13. Mari, R.; Seto, R.; Morris, J.F.; Denn, M.M. Shear thickening, frictionless and frictional rheologies in non-Brownian suspensions. *J. Rheol.* **2014**, *58*, 1693–1724. [CrossRef]

14. Seto, R.; Giusteri, G.G.; Martiniello, A. Microstructure and thickening of dense suspensions under extensional and shear flows. *J. Fluid Mech.* **2017**, *825*, R3. [CrossRef]

15. Ouriev, B.; Windhab, E.J. Novel ultrasound based time averaged flow mapping method for die entry visualization in flow of highly concentrated shear-thinning and shear-thickening suspensions. *Meas. Sci. Technol.* **2003**, *14*, 140–147.

16. von Kann, S.; Snoeijer, J.H.; Lohse, D.; van der Meer, D. Nonmonotonic settling of a sphere in a cornstarch suspension. *Phys. Rev. E* **2011**, *84*, 060401. [CrossRef] [PubMed]

17. Hasanzadeh, M.; Mottaghitalab, V. The role of shear-thickening fluids (STFs) in ballistic and stab-resistance improvement of flexible armor. *J. Mater. Eng. Perform.* **2014**, *23*, 1182–1196. [CrossRef]

18. Brady, J.F.; Bossis, G. Stokesian dynamics. *Ann. Rev. Fluid Mech.* **1988**, *20*, 111–157. [CrossRef]

19. Foss, D.R.; Brady, J.F. Structure, diffusion and rheology of Brownian suspensions by Stokesian Dynamics simulation. *J. Fluid Mech.* **2000**, *407*, 167–200. [CrossRef]

20. Picano, F.; Breugem, W.P.; Mitra, D.; Brandt, L. Shear thickening in non-Brownian suspensions: An excluded volume effect. *Phys. Rev. Lett.* **2013**, *111*, 098302. [CrossRef] [PubMed]

21. Seto, R.; Mari, R.; Morris, J.F.; Denn, M.M. Discontinuous Shear Thickening of Frictional Hard-Sphere Suspensions. *Phys. Rev. Lett.* **2013**, *111*, 218301. [CrossRef] [PubMed]

22. Mari, R.; Seto, R.; Morris, J.F.; Denn, M.M. Discontinuous shear thickening in Brownian suspensions by dynamic simulation. *Proc. Nat. Acad. Sci. USA* **2015**, *112*, 15326–15330. [CrossRef] [PubMed]

23. Boromand, A.; Jamali, S.; Grove, B.; Maia, J.M. A generalized frictional and hydrodynamic model of the dynamics and structure of dense colloidal suspensions. *J. Rheol.* **2018**, *62*, 905–918. [CrossRef]

24. Wilson, H.J. 'Shear thickening' in non-shear flows: The effect of microstructure. *J. Fluid Mech.* **2018**, *836*, 1–4. [CrossRef]

25. Phan-Thien, N. Constitutive equation for concentrated suspensions in Newtonian liquids. *J. Rheol.* **1995**, *39*, 679–695. [CrossRef]

26. Phan-Thien, N.; X.-J, F.; Khoo, B.C. A new constitutive model for monodispersed suspensions of spheres at high concentrations. *Rheol. Acta* **1999**, *38*, 297–304. [CrossRef]

27. Stickel, J.J.; Phillips, R.J.; Powell, R.L. A constitutive model for microstructure and total stress in particulate suspensions. *J. Rheol.* **2006**, *50*, 379–413. [CrossRef]

28. Goddard, J. A dissipative anisotropic fluid model for non-colloidal particle dispersions. *J. Fluid Mech.* **2006**, *568*, 1–17. [CrossRef]

29. Morris, J.F. A review of microstructure in concentrated suspensions and its implications for rheology and bulk flow. *Rheol. Acta* **2009**, *48*, 909–923. [CrossRef]

30. Miller, R.M.; Singh, J.P.; Morris, J.F. Suspension flow modeling for general geometries. *Chem. Eng. Sci.* **2009**, *64*, 4597–4610. [CrossRef]

31. Ozenda, O.; Saramito, P.; Chambon, G. A new rate-independent tensorial model for suspensions of noncolloidal rigid particles in Newtonian fluids. *J. Rheol.* **2018**, *62*, 889–903. [CrossRef]

32. Singh, A.; Mari, R.; Denn, M.M.; Morris, J.F. A constitutive model for simple shear of dense frictional suspensions. *J. Rheol.* **2018**, *62*, 457–468. [CrossRef]

33. Gillissen, J.J.J.; Wilson, H.J. Modeling sphere suspension microstructure and stress. *Phys. Rev. E* **2018**, *98*, 033119. [CrossRef]

34. II, G.G.L.; Denn, M.M.; Hur, D.U.; Boger, D.V. The flow of fiber suspensions in complex geometries. *J. Non-Newton. Fluid Mech.* **1988**, *26*, 297–325.

35. Hand, G.L. A theory of anisotropic fluids. *J. Fluid Mech.* **1962**, *13*, 33–46. [CrossRef]

36. Barthés-Biesel, D.; Acrivos, A. The rheology of suspensions and its relation to phenomenological theories for non-Newtonian fluids. *Int. J. Multiph. Flow* **1973**, *1*, 1–24. [CrossRef]

37. Stickel, J.J.; Phillips, R.J.; Powell, R.L. Application of a constitutive model for particulate suspensions: Time-dependent viscometric flows. *J. Rheol.* **2007**, *51*, 1271–1302. [CrossRef]
38. Yapici, K.; Powell, R.L.; Phillips, R.J. Particle migration and suspension structure in steady and oscillatory pipe flow. *Phys. Fluids* **2009**, *21*, 053302. [CrossRef]
39. Chacko, R.N.; Mari, R.; Fielding, S.M.; Cates, M.E. Shear reversal in dense suspensions: The challenge to fabric evolution models from simulation data. *J. Fluid Mech.* **2018**, *847*, 700–734. [CrossRef]
40. Nott, P.R.; Brady, J.F. Pressure-driven flow of suspensions: Simulation and theory. *J. Fluid Mech.* **1994**, *275*, 157–199. [CrossRef]
41. Morris, J.F.; Boulay, F. Curvilinear flows of noncolloidal suspensions: The role of normal stresses. *J. Rheol.* **1999**, *43*, 1213–1237. [CrossRef]
42. Frank, M.; Anderson, D.; Weeks, E.R.; Morris, J.F. Particle migration in pressure-driven flow of a Brownian suspension. *J. Fluid Mech.* **2003**, *493*, 363–378. [CrossRef]
43. Truesdell, C.; Noll, W. The non-linear field theories of mechanics. In *The Non-Linear Field Theories of Mechanics*; Springer: Berlin/Heidelberg, Germany, 2004; pp. 1–579.
44. Passey, P.; Mehta, R.G. Study of Pre-Shearing Protocol and Rheological Parameters of Shear Thickening Fluids Containing Nano Particles. Ph.D. Thesis, Thapar University, Patiala, India, 2016.
45. Mewis, J.; Wagner, N.J. *Colloidal Suspension Rheology*; Cambridge University Press: Cambridge, UK, 2012.
46. Maranzano, B.J.; Wagner, N.J. Flow-small angle neutron scattering measurements of colloidal dispersion microstructure evolution through the shear thickening transition. *J. Chem. Phys.* **2002**, *117*, 10291–10302. [CrossRef]
47. Royer, J.R.; Blair, D.L.; Hudson, S.D. Rheological signature of frictional interactions in shear thickening suspensions. *Phys. Rev. Lett.* **2016**, *116*, 188301. [CrossRef] [PubMed]
48. Kawasaki, T.; Ikeda, A.; Berthier, L. Thinning or thickening? Multiple rheological regimes in dense suspensions of soft particles. *Europhys. Lett.* **2014**, *107*, 28009. [CrossRef]
49. Brady, J.F.; Morris, J.F. Microstructure of strongly sheared suspensions and its impact on rheology and diffusion. *J. Fluid Mech.* **1997**, *348*, 103–139. [CrossRef]
50. Sierou, A.; Brady, J.F. Rheology and microstructure in concentrated noncolloidal suspensions. *J. Rheol.* **2002**, *46*, 1031–1056. [CrossRef]
51. Ouriev, B.; Windhab, E.J. Rheological study of concentrated suspensions in pressure-driven shear flow using a novel in-line ultrasound Doppler method. *Exp. Fluids* **2002**, *32*, 204–211.
52. Favero, J.L.; Secchi, A.R.; Cardozo, N.S.M.; Jasak, H. Viscoelastic flow analysis using the software OpenFOAM and differential constitutive equations. *J. Non-Newton. Fluid Mech.* **2010**, *165*, 1625–1636. [CrossRef]
53. Sweby, P.K. High resolution schemes using flux limiters for hyperbolic conservation laws. *SIAM J. Numer. Anal.* **1984**, *21*, 995–1011. [CrossRef]
54. Issa, R.I. Solution of the implicitly discretised fluid flow equations by operator-splitting. *J. Comput. Phys.* **1986**, *62*, 40–65. [CrossRef]
55. Fernandes, C.; Araujo, M.S.B.; Ferrás, L.L.; Nóbrega, J.M. Improved both sides diffusion (iBSD): A new and straightforward stabilization approach for viscoelastic fluid flows. *J. Non-Newton. Fluid Mech.* **2017**, *249*, 63–78. [CrossRef]
56. Lim, A.S.; Lopatnikov, S.L.; Wagner, N.J.; Gillespie, J.W., Jr. Investigating the transient response of a shear thickening fluid using the split Hopkinson pressure bar technique. *Rheol. Acta* **2010**, *49*, 879–890. [CrossRef]
57. Lomakin, E.V.; Mossakovsky, P.A.; Bragov, A.M.; Lomunov, A.K.; Konstantinov, A.Y.; Kolotnikov, M.E.; Antonov, F.K.; Vakshtein, M.S. Investigation of impact resistance of multilayered woven composite barrier impregnated with the shear thickening fluid. *Arch. Appl. Mech.* **2011**, *81*, 2007–2020. [CrossRef]

MDPI

St. Alban-Anlage 66

4052 Basel

Switzerland

Tel. +41 61 683 77 34

Fax +41 61 302 89 18

www.mdpi.com

Fluids Editorial Office

E-mail: fluids@mdpi.com

www.mdpi.com/journal/fluids

www.ingramcontent.com/pod-product-compliance
Lightning Source LLC
Chambersburg PA
CBHW051841210326
41597CB00033B/5729